"This is a timely, comprehensive contribution to the literature and practice at the nexus of international environmental law and human rights, that boldly addresses critical questions on the sovereignty and stewardship of biodiversity across a broad range of regional perspectives."

Elisa Morgera, *Professor of Global Environmental Law, University of Strathclyde Law School, Glasgow, United Kingdom*

"Environmental jurisprudence over the last two decades has been radically transformed. This epistemic shift is symbolized by the waning of the ideas of ownership and the ascent ideas of stewardship when it comes to lands and waters. The shift has been the result of a growing realization that the dominant discourse of private property has played a key role in the collapse of ecosystems and changing climate. Confronted with the existential question of survival of our species, communities, activists and academics have begun to ask ontological questions regarding the nature of the juridical subject. Specifically, what does it mean to be human and what is our relationship to the natural world. The book you have in your hands is a glorious map of stories, histories and analyses of what is arguably the most critical conversation of the Anthropocene. It consists of riveting essays by some of the best contemporary cartographers of political ecology. It is metacognition at its finest and I urge you to read it and let it transform you."

Sanjay Kabir Bavikatte, *Co-founder of Natural Justice, Lawyers for Communities and the Environment and author of* Stewarding the Earth: Rethinking Property and the Emergence of Biocultural Rights

"In the late 1980s, Darrell Posey and others made the world aware of the inextricable link between biological and cultural diversity. This suggested the possibility of new legal and ethical frameworks, and broad-based actions especially at local level. This exceptional volume builds on Dr Posey's visionary work, showcasing the latest thinking on 'bioculturalism', an issue whose positive resolution all of us has a major stake in."

Dr Graham Dutfield, *Professor of International Governance, University of Leeds, United Kingdom*

BIOCULTURAL RIGHTS, INDIGENOUS PEOPLES AND LOCAL COMMUNITIES

This volume presents a comprehensive overview of biocultural rights, examining how we can promote the role of indigenous peoples and local communities as environmental stewards and how we can ensure that their ways of life are protected.

With Biocultural Community Protocols (BCPs) or Community Protocols (CPs) being increasingly seen as a powerful way of tackling this immense challenge, this book investigates these new instruments and considers the lessons that can be learnt about the situation of indigenous peoples and local communities. It opens with theoretical insights which provide the reader with foundational concepts such as biocultural diversity, biocultural rights and community rule-making. In Part Two, the book moves on to community protocols within the Access Benefit Sharing (ABS) context, while taking a glimpse into the nature and role of community protocols beyond issues of access to genetic resources and traditional knowledge. A thorough review of specific cases drawn from field-based research around the world is presented in this part. Comprehensive chapters also explore the negotiation process and raise stimulating questions about the role of international brokers and organizations and the way they can use BCPs/CPs as disciplinary tools for national and regional planning or to serve powerful institutional interests. Finally, the third part of the book considers whether BCPs/CPs, notably through their emphasis on "stewardship of nature" and "tradition", can be seen as problematic arrangements that constrain indigenous peoples within the Western imagination, without any hope of them reconstructing their identities according to their own visions, or whether they can be seen as political tools and representational strategies used by indigenous peoples in their struggle for greater rights to their land, territories and resources, and for more political space.

This volume will be of great interest to students and scholars of environmental law, indigenous peoples, biodiversity conservation and environmental anthropology. It will also be of great use to professionals and policymakers involved in environmental management and the protection of indigenous rights.

Fabien Girard is an Associate Professor in the Faculty of Law at Université Grenoble Alpes (UGA), France, and also a former Research Fellow, Maison Française d'Oxford (MFO), UK. He is the co-editor of *The Commons, Plant Breeding and Agricultural Research* (2018).

Ingrid Hall is an Associate Professor of anthropology at the University of Montréal, Canada, and an associate member of the Unité mixte de recherché Savoirs, Environnements, Sociétés (UMR SENS), France. She is the co-editor of *Savoirs Locaux en Situation* (2019).

Christine Frison is an FNRS Post-Doctoral Researcher and Lecturer at the Institute for Interdisciplinary Research in Legal Sciences of the UCLouvain, Belgium. Her latest books are *The Commons, Plant Breeding and Agricultural Research* (2018) and *Redesigning the Global Seed Commons* (2018).

Routledge Explorations in Environmental Studies

Ecocritical Geopolitics
Popular Culture and Environmental Discourse
Elena dell'Agnese

Environmental Defenders
Deadly Struggles for Land and Territory
Edited by Mary Menton and Philippe Le Billon

Globalization, Environmental Law and Sustainable Development in the Global South
Challenges for Implementation
Edited by Kirk W. Junker and Paolo Davide Farah

The Political Ecology of Austerity
Crisis, Social Movements, and the Environment
Edited by Rita Calvário, Maria Kaika and Giorgos Velegrakis

Sociology Saves the Planet
An Introduction to Socioecological Thinking and Practice
Thomas Macias

Polluting Textiles
The Problem with Microfibres
Edited by Judith S. Weis, Francesca De Falco and Mariacristina Cocca

Biocultural Rights, Indigenous Peoples and Local Communities
Protecting Culture and the Environment
Edited by Fabien Girard, Ingrid Hall and Christine Frison

For more information about this series, please visit: www.routledge.com/Routledge-Explorations-in-Environmental-Studies/book-series/REES

BIOCULTURAL RIGHTS, INDIGENOUS PEOPLES AND LOCAL COMMUNITIES

Protecting Culture and the Environment

Edited by Fabien Girard, Ingrid Hall and Christine Frison

Routledge
Taylor & Francis Group
LONDON AND NEW YORK

earthscan
from Routledge

Cover image: Getty Images

First published 2022
by Routledge
4 Park Square, Milton Park, Abingdon, Oxon OX14 4RN

and by Routledge
605 Third Avenue, New York, NY 10158

Routledge is an imprint of the Taylor & Francis Group, an informa business

British Library Cataloguing-in-Publication Data
A catalogue record for this book is available from the British Library

Library of Congress Cataloging-in-Publication Data
A catalog record has been requested for this book

ISBN: 978-1-032-00085-5 (hbk)
ISBN: 978-1-032-00081-7 (pbk)
ISBN: 978-1-003-17264-2 (ebk)

DOI: 10.4324/9781003172642

Typeset in Bembo
by codeMantra

CONTENTS

ILLUSTRATIONS

Figures

Tables

Maps

Boxes

NOTES ON CONTRIBUTORS

Reia Anquet is an English Lecturer (PRCE) at Sciences Po Grenoble, where she teaches English for special purposes on Bachelor's and Master's courses and lecture content-based courses on Australian Identity, the Rights of Indigenous Peoples and Environmental Policy and Knowledge. She is a PhD candidate in Grenoble, France, but initially, her research journey began in Tasmania, Australia. Her thesis topic investigates Australian Indigenous Environmental Governance Discourse in Australian common law, legislation and public policy.

Kelly Bannister is a consultant and Co-Director of the POLIS Project on Ecological Governance at the Centre for Global Studies, University of Victoria in British Columbia, Canada. She is an ethnobiologist and biocultural ethicist who specializes in applied intercultural ethics, focusing on ethical and legal issues in research and education involving biodiversity, Indigenous knowledge and cultural heritage. Much of her work has examined the role of governance tools and processes (e.g., ethical codes, community protocols and research agreements) to address power relations, support Indigenous self-determination and facilitate equitable practices in collaborations. Her current work explores new approaches to ethics education, policy and practice through guidance offered by Indigenous wisdom traditions, relational ethics, intercultural communication and zen-based conflict resolution.

Kelly has been involved in ethics policy research, analysis, development and education from local to national and international levels for over 20 years, working with and for Indigenous communities and Indigenous organizations, universities, government and not-profit organizations. She holds BSc and MSc degrees in Microbiology and a PhD in Botany, and she has completed two Post-docs in Applied Ethics. Kelly is a fourth generation Canadian with ancestors who emigrated from Ukraine and England. She lives with her family in unceded Coast

Salish territory on the southwest coast of British Columbia, Canada where she is an avid practitioner of Aikido, the martial art of peace.

Christine Frison carried out legal studies in France and England and obtained a master's degree in public international law at the *Université Libre de Bruxelles*, Belgium. She specializes in international environmental law and worked as a consultant for international institutions (UN-related) and national governments. She also worked as a legal research fellow at the Center for International Sustainable Development Law (CISDL), Montreal (Canada, based at McGill University). She defended her joint-PhD entitled Towards *Redesigning a Global Seed Commons* both at UC Louvain and at KU Leuven (Belgium, under the co-supervision of professors Tom Dedeurwaerdere, Olivier De Schutter and Geertrui Van Overwalle), which took interest in the intellectual property rights and collective action implications of the International Treaty on Plant Genetic Resources for Food and Agriculture, using the Commons as a theoretical framework.

Christine currently holds a postdoctoral fellowship from the Belgian Research Foundation (FNRS, 2020–2023) and is an associate researcher at the "Government & Law Research Group" of the University of Antwerp Law Faculty. She has taught the course "International and European Environmental Law" at the *Université Libre de Bruxelles*, and part of the "Advanced course on International Sustainable Development Law" at the University of Antwerp. She also supervises master and PhD students in four Belgian universities.

She has published several books, including a monograph published by Routledge:

Redesigning the Global Seed Commons. Law and Policy for Agrobiodiversity and Food Security, 1st Edition 2019.

Her latest publication "Bringing access and benefit-sharing into the digital age", *Plant People, Planet*, 2021, makes policy proposals to untie the Gordian knot around digital sequence information in the access and benefit-sharing international regime.

Fabien Girard is an Associate Professor (tenured) at the Faculty of Law, Université Grenoble Alpes (UGA) in France, where he teaches legal philosophy, agrobiodiversity and the law, and comparative law.

Between 2011 and 2013, he was appointed as a Project Manager to the Haut Conseil des Biotechnologies in Paris (France), where he worked on industrial property, plant biotechnology and farmers' seed networks.

While on secondment at the *Maison Française d'Oxford* (MFO – Oxford, UK) from 2014 to 2016, he conducted a research program called "Theory of Law and Legal Anthropology". As part of that responsibility, he organized several conferences and workshops on legal and political philosophy, as well as on the international governance of biodiversity. He has written extensively on the connections between biotechnology, plant innovation and industrial property, as well as on formal and informal seed systems and crop genetic diversity.

His last co-edited book, *The Commons, Plant Breeding and Agricultural Research. Challenges for Food Security and Agrobiodiversity* (with Christine Frison), was published in 2018 with Routledge.

Garrett Graddy-Lovelace Born and raised in a farming family and community in Kentucky, Garrett Graddy-Lovelace is Associate Professor at American University's School of International Service in Washington, DC (Piscataway lands), where she researches and teaches agricultural policy and agrarian politics. Her forthcoming book, *The Power of Seeds & Agricultural Biodiversity Policy* (MIT), explores crop diversity conservation, governance, movements and regeneration. Through community-partnered research, Graddy-Lovelace traces the colonial origins and decolonial potential of farm and seed policies. As coPI of an NSF-SESYNC Pursuit on agrobiodiversity and food security/sovereignty/nutrition, she leads the intercultural shared analysis: "Agrobiodiversity Nourishes Us/*La Agrobiodiversidad Nos Nutre:* Research-Action for Agroecological Transformations" (special feature at *Elementa: Science of the Anthropocene*). She co-leads the community-based public scholarship project: "Disparity to Parity to Solidarity: Balancing the Scales of Agricultural Policy for Justice & Resilience" (disparitytoparity.org) and co-founded the Agroecology Research-Action Collective.

Ingrid Hall Associate Professor, Department of anthropology, University of Montréal, Canada

Ingrid Hall is an anthropologist who specializes in environmental and political anthropology. She has been previously trained as an agronomic engineer. She has been conducting fieldwork in rural communities of Peru since 1998 and she has focused for the last 10 years on the politics of conservation of native potatoes in the Peruvian Andes. She is doing multi-site ethnography, departing from the communities of the Potato Park of Pisac (Peru), taking into account the work of NGOs and of the International Potato Center. She is now focusing on the global governance of biodiversity, especially the CDB. She gives an account of the complexity of the formulation and recognition of biocultural rights by actors rooted at different scales, from the local to the global. She is concerned by the processes of recognition of culture (cultural politics) and the deconstruction of the concept of nature (politics of nature). She coedited the book *Savoirs locaux en en situation. Retour sur une notion plurielle et dynamique* in 2019, which explores how the category of Indigenous knowledge is mobilized in specific contexts by different kinds of actors.

Leslé Jansen is the Chief Executive Officer of Resource Africa (South Africa) and a South African Indigenous lawyer with over 15 years' experience in the areas of environmental and social justice, with special focus on traditional knowledge, local communities and resource rights. Prior to joining Resource Africa, she was part of the management team of Natural Justice where she worked for nine years supporting local communities' resource and related rights across southern Africa. Ms Jansen holds a Master's degree in Indigenous Peoples in International Law from the University of Arizona (USA) and a Master's degree in the Rule of Law for Development from Loyola University (Chicago), completed in Rome, Italy. She is an Indigenous expert member to the African Commission on Human and Peoples' Rights' special mechanism – the Working Group on Indigenous Populations/Communities in Africa.

Pía Marchegiani Latin American School of Social Sciences (FLACSO-Argentina); Environmental Policy Director, FARN (Fundación Ambiente y Recursos Naturales), Argentina.

Pía Marchegiani holds a Law Degree from the University of Buenos Aires (2005) and a Master's in Social Sciences from the Global Studies Programme granted jointly by the University of Freiburg, Germany and the University of Kwazulu-Natal, South Africa (2010). She is currently a PhD candidate at the Latin American Faculty of Social Sciences (FLACSO-Argentina), the Environmental Policy Director at FARN and, since 2011, a lecturer at the Environmental Law Legal Clinic run by FARN and the Faculty of Law of the University of Buenos Aires. She works to support communities asserting their rights in the context of lithium mining.

Louisa Parks Associate Professor of Political Sociology, School of International Studies and Department of Sociology and Social Research, University of Trento, Italy.

Louisa Parks is the author of *Benefit-sharing in Environmental Governance: Local Experiences of a Global Concept,* published in 2020 by Routledge, and a former research fellow of the BeneLex project on benefit-sharing in global environmental law. She has published articles on community protocols, the discursive spaces for Indigenous peoples' and local communities' views in the Convention on Biological Diversity, and their participation in the Convention on Biological Diversity. Her past research has focused on transnational social movements, civil society and the impacts of campaigning in the European Union.

Manohisoa Rakotondrabe is a trained agronomist with a focus on emerging social and economic systems in rural areas. She holds a PhD in natural resource management and development. She is currently a postdoctoral fellow with the ANR Bioculturalis project, based at the Legal Research Center (CRJ) of the University of Grenoble Alpes, where she works on issues around the development of Biocultural Community Protocols for the protection of genetic resources and traditional knowledge in Madagascar. She had previously worked in Madagascar as a consultant for some natural resource conservation projects such as UNDP and GIZ.

Miri (Margaret) Raven is a Scientia Fellow affiliated with the Social Policy Research Centre; the Environment and Society Group; and The George Institute for Global Health at UNSW. Margaret is a member of the Australian Research Council College of Experts, and sits on the Australian Institute for Aboriginal and Torres Strait Islander Studies (AIATSIS) Human Research Ethics Committee. Margaret is also a Chief Investigator with Professor Daniel Robinson on the five-year Australian Research Council (ARC) Discovery Project (DP180100507): Indigenous knowledge futures: protecting and promoting indigenous knowledge (2018–2022). Margaret is a Yamatji-Noongar (and non-Indigenous) woman from Western Australia, with a background in geography, human rights, native title and social policy research. Her research focuses on protocols, Indigenous food and social policy. Her papers cover themes related to protocols, biopiracy and food security.

Daniel Robinson is a Professor in the Environment and Society Group at UNSW. He is also an Academic Lead for the Pacific, for the UNSW Institute for Global Development (IGD). Daniel is currently the Pacific Regional Manager of the Access and Benefit-Sharing Capacity Development Initiative, which has a five-year project implementing the Nagoya Protocol in the Pacific from 2017 to 2021. See http://www.abs-initiative.info/. Daniel is also Chief Investigator with Dr Miri (Margaret) Raven on the five-year Australian Research Council (ARC) Discovery Project (DP180100507): Indigenous knowledge futures: protecting and promoting indigenous knowledge (2018–2022).

Giulia Sajeva In January 2020, Giulia Sajeva begins a Marie Skłodowska-Curie Individual Fellowship on Rights for Ecosystem Services at the Strathclyde Centre for Environmental Law and Governance, University of Strathclyde. Under the supervision of Professor Elisa Morgera, Giulia will aim at developing a theoretical and legal framework – labelled Rights for Ecosystems Services (RES) (echoing the highly debated Payment for Ecosystem Services framework) – to guide policy and legal developments towards reducing the risk of local communities abandoning their sustainable practices due to the lack of effective protection.

Giulia holds a PhD in human rights, wherein she produced a thesis in legal theory on biocultural rights. She collaborated with the NGO Natural Justice: Lawyers for communities and the environment and worked with the Kew Royal Botanic Gardens' Conventions and Policies Section, where she conducted research on the Convention on Biological Diversity and the Convention on the International Trade on Endangered Species, as well as on Ethical Guidelines for research with Indigenous peoples.

She was the Vice-President of the Religion and Conservation Biology Working Group of the Society of Conservation Biology, which is interested in the intersection between religions and environmental protection and is now an honorary member of the ICCA Consortium, for which she recently co-compiled ICCA Consortium Policy Brief no.7.

Giulia has a Master's degree in Conservation Science (Imperial College London), a Master's in Global Rule of Law and Constitutional Democracy (*Università degli Studi di Genova*), a PhD in human rights from the Law Department of the *Università degli Studi di Palermo* and a Bachelor of Art in Political Sciences and International Relations from the *Università degli Studi di Palermo.* She recently published her first book *When Rights Embrace Responsibilities. Biocultural Rights and the Conservation of Environment* (2018).

Rayna Sutherland (she/her) is a Research Coordinator at Resource Africa, a non-profit organization which works with communities in Southern Africa to advocate for people's rights to sustainably use their natural resources, respecting their cultures, worldviews and knowledge systems. Rayna has over four years' experience working with grassroots environmental and social justice organizations and socially just research hubs in Tanzania, South Africa and Canada, across the thematic areas of food justice, gender-based justice, land and resource rights, protected areas policies and Access and Benefit Sharing. Rayna holds an Honours Bachelor of Arts in International Development and Women's and Gender Studies from the University of Toronto and is based in Vancouver, Canada.

ACKNOWLEDGEMENTS AND DEDICATIONS

It is rare to carry out a book project that, as it unfolds, is shaken by the climate and health crises which also form its thematic background. It took a great deal of effort for our authors – sometimes struck by the tragedy of the loss of a loved one, sometimes under strict lockdown regulations and unable to continue their fieldwork, sometimes stunned by a heat dome or blazes tearing through hectares of forestland – to respond to the call we editors made.

We are all the more indebted to them for having maintained their commitment and contributed with great energy and insight into the realization of this project, which began three years ago.

The research project no. ANR-18-CE03-0003 ("Biocultural Community Protocols: Justice, Biodiversity and Law") – BioCulturalis, funded by the French National Research Agency (ANR), supported the realization of this edited volume by funding the fieldwork which provides the background of some of the chapters as well as funding the holding of an Online Panel Discussion (pandemic obliges) in 2021. This two-day gathering was a decisive impetus and an opportunity to bring together – in addition to most of the contributors to this volume – seasoned academics, development professionals and activists: Sanjay Kabir Bavikatte, Miranda Forsyth, Tom Suchanandan, Krystyna Swiderska and Brendan Tobin. The virulence and persistence of the health crisis, coupled with various professional and personal commitments, did not allow them all to contribute to the volume, but their collaboration was extremely valuable, and we wish to thank them most warmly here. Finally, the ANR project also covered the full open access fee of the book.

As is often the case, a work of this magnitude owes its success – at least its completion – entirely to the investment, support and skills of colleagues and friends to whom we also wish to express our sincere gratitude: Fitiavana Ranaivoson and Manohisoa Rakotondrabe who assisted in the gathering of data information in Madagascar were first in line, as well as Benjamin Coudurier and Kim Mazenot who provided helpful research assistance. Mélanie Congretel commented on early versions of some chapters and her astuteness allowed some ideas to progress considerably. In addition, exchanges with members of the Bioculturalis team have been a constant source of renewal, and we must also thank them: Jean Foyer, Geoffroy Filoche, Reia Anquet and Patricia Guzmán Aguilera.

F.G. has a tender thought for its two princesses, Tany & Babette, whose softness, undecipherable (but comforting) babbles and constant encouragement have enabled him to overcome moments of doubt. F.G. would also like to express his gratitude to his two co-editors who have allowed him to confirm that work and friendship are far from – to use a popular French expression – the (improbable) wedding of the carp and the rabbit.

C.F wishes to dedicate this book to all the Indigenous people and local communities, peasants' groups and citizens' movements who work every day for a more equitable and sustainable life on Earth. She thanks her two co-editors for the great adventure this research project has been over the last three years.

I.H. wants to acknowledge the importance of the work done by different persons of goodwill, Indigenous or not, towards the recognition of the rights of Indigenous people and local communities over their natural resources, especially the people mobilized around the Potato Park, farmers, as well as the ANDES NGO. She has a special thought for her two co-editors and the participants of this project: this journey has truly been enriching. Finally, she wants to thank Enrique and Mathilda for their patience during this process, which stretched over a very challenging period.

TABLE OF CASES AND LEGISLATION

Table of Legislation

International Instruments

African Charter on Human and Peoples' Rights ("Banjul Charter") (27 June 1981, OAU Doc. CAB/LEG/67/3 rev. 5, 21 I.L.M. 58 (1982))

ARIPO Swakopmund Protocol on the Protection of Traditional Knowledge and Expressions of Folklore (Diplomatic Conference of ARIPO at Swakopmund (Namibia) on August 9, 2010)

Cartagena Protocol on Biosafety to the Convention on Biological Diversity (Montreal, 29 January 2000, 2226 UNTS 208)

Convention on Biological Diversity (Rio de Janeiro, 5 June 1992, 1760 UNTS 79)

Convention on the Elimination of All Forms of Discrimination against Women (New York, 18 December 1979, 1249 UNTS 13)

Convention on the International Trade in Endangered Species of Wild Fauna and Flora (CITES) (Washington, 03 March 1973, 993 UNTS 243)

Declaration on the Right to Development, Adopted by General Assembly resolution 41/128 of 4 December 1986 (A/RES/41/128)

ILO Convention (No. 169) concerning indigenous and tribal peoples in independent countries (Geneva 27 June 1989, ILO, Official Bulletin, Vol. LXXII (1989) Ser. A, No. 2)

International Convention for the Protection of New Varieties of Plants of December 2, 1961, as Revised at Geneva on November 10, 1972, on October 23, 1978, and on March 19, 1991, (UPOV/PUB/221)

International Convention on the Elimination of All Forms of Racial Discrimination (New York, 7 March 1966, 660 UNTS 195)
International Covenant on Civil and Political Rights (New York, 16 December 1966, 999 UNTS 171)
International Covenant on Economic, Social and Cultural Rights (New York, 16 December 1966, 993 UNTS 3)
International Treaty on Plant Genetic Resources for Food and Agriculture (Rome, 3 November 2001, 2400 UNTS 303)
International Tropical Timber Agreement (Geneva, 27 January 2006, 2797 UNTS 75)
International Undertaking on Plant Genetic Resources, adopted 23 November 1983 (FAO Resolution 8/83).
Nagoya Protocol on Access to Genetic Resources and the Fair and Equitable Sharing of Benefits Arising from their Utilization to the Convention on Biological Diversity (Nagoya, 29 October 2010, UNEP/CBD/COP/DEC/X/1; 3008 UNTS 3).
Paris Agreement (Paris, 12 December 2015, C.N.92.2016. TREATIES-XXVII.7.d of 17 March 2016)
Rio Declaration on Environment and Development, UN Conference on Environment and Development Rio de Janeiro, 3–14 June 1992 (A/CONF.151/26/Rev.l (Vol. l))
UN Convention on the Elimination of All Forms of Discrimination against Women (CEDAW) (New York, 18 December 1979, 1249 UNTS 13)
UN Convention on the Rights of Persons with Disabilities (CRPD) (New York, 13 December 2006, 2515 UNTS 3)
UN Convention to Combat Desertification in those Countries Experiencing Serious Drought and/or Desertification, Particularly in Africa (Paris, 14 October 1994, 1954 UNTS 3)
UN Declaration of the United Nations Conference on the Human Environment (Stockholm Declaration) (1972), A/CONF.48/14/Rev.1
UN Declaration on Permanent Sovereignty over Natural Resources, A/Res/1803 (XVII) (14 December 1962)
UN Declaration on the Right to Development, UN Doc A/RES/41/128 (4 December 1986)
UN Declaration on the Rights of Indigenous Peoples, Resolution adopted by the General Assembly on 13 September 2007 (A/RES/61/295, 2 October 2007)
UN Declaration on the Rights of Peasants and Other People Working in Rural Areas, Resolution adopted by the Human Rights Council on 28 September 2018 (A/HRC/RES/39/12, 8 October 2018)
UN Declaration on the Rights of Persons Belonging to National or Ethnic, Religious and Linguistic Minorities (1992), Adopted by

General Assembly resolution 47/135 of 18 December 1992 (A/RES/47/135, 3 February 1992)
UN Framework Convention on Climate Change (New York, 9 May 1992, 1771 UNTS 107)
UN Johannesburg Declaration on Sustainable Development, UN Doc A/CONF.199/20 (4 September 2002)
UNESCO Convention concerning the Protection of the World Cultural and Natural Heritage (World Heritage Convention) (Paris, 16 November 1972, 1037 UNTS 151)
UNESCO Convention for the Safeguarding of the Intangible Cultural Heritage (Paris, 17 October 2003, 2368 UNTS 3)
Universal Declaration of Human Rights: Adopted and proclaimed by UN General Assembly Resolution 217 A (III) of 10 December 1948 (UN Document A/810, p. 71 (1948))
Vienna Convention on the Law of Treaties (VCLT) (Vienna, 23 May 1969, 1155 UNTS 331)
WTO Trade-Related Aspects of Intellectual Property Rights, Marrakesh Agreement Establishing the World Trade Organization, Annex 1 C (Marrakesh, 15 April 1994, 1869 U.N.T.S. 299)

Domestic Legislation

Argentina

Constitución Nacional [CN], Boletín Oficial [BO], Jan, 3, 1995
Ley General del Ambiente, Ley N.° 25.675, B.O. del 28/11/2002

Australia

Aboriginal Land Rights (Northern Territory) Act 1976
Native Title Act 1993, No. 110, 1993
Environmental Protection and Biodiversity Conservation Act (EPBC) 1999
Environment Protection and Biodiversity Conservation Regulations 2000
Biological Resources (Northern Territory) Act 2006

Benin

Décret N° 2018–405 du 07 septembre 2018, portant directives nationales pour l'accès et le partage des avantages issus de l'utilisation des ressources génétiques et des connaissances traditionnelles associées en République du Bénin
Arrêté interministériel N° 0121/MEHU/MDGLAAT/DC/SGM/DGFRN/ SA du 16/11/2012 fixant les conditions de gestion durable de la forêt sacrée en République du Bénin

Bolivia

Constitución Política del Estado Plurinacional de Bolivia, 7-Febrero-2009

Canada

Constitution Act 1982, enacted as Schedule B to the Canada Act 1982, 1982, c. 11 (U.K.), (entry into force 17 April 1982).
Impact Assessment Act (S.C. 2019, c. 28, s. 1)
House Government Bill, Bill C-15, An Act respecting the United Nations Declaration on the Rights of Indigenous Peoples, 2021

Cook Islands

House of Ariki Act 1966, as amended 1972
Traditional Knowledge Act 2013 (TK Act)

Ecuador

Constitución de la República del Ecuador, Registro Oficial 449 de 20-oct-2008

India

Biological Diversity Rules, 2004

Kenya

Constitution of Kenya, 2010
Environmental Management and Coordination Act No. 8 (1999) (EMCA)
Protection of Traditional Knowledge and Cultural Expressions Act, 2016 (No. 33 of 2016)
Environmental Management and Co-ordination (Conservation of Biological Diversity and Resources, Access to Genetic Resources and Benefit Sharing) Regulations, 2006

Madagascar

Loi 96-025 du 10 septembre 1996 relative à la gestion locale des ressources naturelles renouvelables, (JO 14.10.1996, p. 2377–2385)
Décret N° 2004–299 du 3 mars 2004 fixant l'organisation, le fonctionnement et les attributions du *Fokontany* (JO n°2898 du 12.04.04, p.1594)
Décret N° 2017 –066 du 31/01/2017 portant réglementation de l'accès et du partage des avantages découlant de l'utilisation des ressources génétiques

Mexico

Constitución Política de los Estados Unidos Mexicanos (1917) (Última reforma publicada DOF 28-05-2021)
Ley General del Equilibrio Ecológico y Protección al Ambiente, DOF 28-01-1988
Ley Agraria, DOF 26-02-1992
Ley de Aguas Nacionales, DOF 01-12-1992
Ley General de Vida Silvestre, DOF 03-07-2000
Ley General de Desarrollo Forestal Sustentable, DOF 25-03-2003
Ley General de Bienes Nacionales, DOF 20-05-2004
Ley General de Pesca y Acuacultura Sustentable, DOF 24-07-2007
Ley General de Cultura y Derechos Culturales DOF 19-06-2017
Ley General de Desarrollo Forestal Sustentable, DOF 05-06-2018
Ley Federal para el Fomento y Protección del Maíz Nativo, DOF 13-04-2020
Decreto por el que se reforman diversas disposiciones de la Ley General de Desarrollo Forestal Sustentable, DOF 25-04-21
Constitución Política del Estado Libre y Soberano de Oaxaca, 04-04-1922
Ley de Derechos de los Pueblos y Comunidades Indigenas del Estado de Oaxaca, 21.03.1998 (Última reforma 19-12-2020)

New Zealand

Te Awa Tupua (Whanganui River Claims Settlement) Act, 2017

Panama

Constitución Política de la República de Panamá (1972)
Ley 10 de 7 de Marzo de 1997, Por la cual se crea la Comarca Ngäbe-Buglé y se toman otras Medidas, Gaceta Oficial No. 23242 (11-03-1997)
Ley 41 General de Ambiente de la República de Panamá, De 1 de julio de 1998, Gaceta Oficial N° 23578 (03-07-1998)
Ley 20, de 26 de Junio de 2000, del régimen especial de propiedad intelectual sobre los derechos colectivos de los pueblos indígenas, para la protección y defensa de su identidad cultural y de sus conocimientos tradicionales, y se dictan otras disposiciones, Gaceta Oficial, N° 24083 (27-06-2000)
Ley 11 de 26 de Marzo de 2012, Que establece un régimen especial para la protección de los recursos minerales, hídricos y ambientales en la comarca Ngäbe-Buglé, Gaceta Oficial N° 27001 (26-03-2012)
Ley 17 del 27 de Junio de 2016, Que establece la protección de los conocimientos de la medicina tradicional indígena, Gaceta Oficial Digital N° 28063-B (29-06-2016)

Decreto Ejecutivo N° 12 de 20 de Marzo de 2001, Por la cual se Reglamenta la Ley No. 20 de 26 de junio de 2000, del Régimen Especial de Propiedad Intelectual sobre los Derechos Colectivos de los pueblos indígenas para la Protección y Defensa de su Identidad Cultural y de sus conocimientos tradicionales, y se dictan otras disposiciones, Gaceta Oficial N° 24270 (28-03-2001)

Decreto Ejecutivo N° 19 de 26 de Marzo de 2019, Por el Cual se Reglamenta el Acceso y Control del Uso de los Recursos Biologicos y Geneticos en la Republica de Panama y Se Dictan otras Medidas, Gaceta Oficial Digital N° 28741-A (27-03-2019)

Peru

Ley N° 27262- Ley general de semillas, 13-05-2000

Ley N° 27811- Ley que establece el régimen de protección de los conocimientos colectivos de los pueblos indígenas vinculados a los recursos biológicos, 10-08-2002

Ley N° 28216- Ley de protección al acceso a la diversidad biológica peruana y los conocimientos colectivos de los pueblos indígenas, 01-04-2004

Decreto Supremo N° 003–2009-MINAM, Eleva al rango de Decreto Supremo la Resolución N° 087–2008-MINAM y ratifica la aprobación del Reglamento de acceso a recursos genéticos, efectuada por dicha Resolución, 8-02-2009

Decreto Supremo N°006–2012-AG, Aprueban Reglamento General de la Ley General de Semillas, 01-06-2012

Ordenanza del Gobierno Regional de Cusco No. 048–2008 sobre Acceso a los Recursos Geneticos y a los Conocimientos Tradicionales, 15-01-2009

South Africa

Constitution of the Republic of South Africa, 1996

Hottentots Code of 1809

Apprentice Act of 1812

Masters and Servants Act of 1856

Race Classification and Population Registration Act of 1950 (Act No. 30 of 1950)

Group Areas Act of 1950 (Act No. 41 of 1950)

Restitution of Land Rights Act 22 of 1994 (Act No. 22 of 1994)

National Environmental Management: Biodiversity Act 10 of 2004 (Act No. 10 of 2004)

Bioprospecting, Access and Benefit-Sharing Regulations, 2008 (No. R. 138 of 2008)
Traditional and Khoi-San Leadership Act 3 of 2019 (Act No. 3 of 2019)

Vanuatu

Environmental Protection and Conservation Act [CAP 283]
Protection of Traditional Knowledge and Expressions of Culture Act No. 21 of 2019, (2019)
Traditional Knowledge Act (2020)

Table of Cases

International

African Commission/Court on Human and Peoples' Rights

African Commission on Human and Peoples' Rights v Kenya, App 006/2012, (006/2012) [2017] AfCHPR 28; (26 May 2017)
Centre for Minority Rights Development (Kenya) and Minority Rights Group International on behalf of Endorois Welfare Council v Kenya, Comm No 276/03 (Merits) (4 February 2010), [2009] AHRLR 75

Inter-American Court of Human Rights

Case of Saramaka People v Suriname (Preliminary Objections, Merits, Reparations, and Costs), IACtHR Series C No 172 (28 November 2007)
Case of the Mayagna (Sumo) Awas Tingni Community v. Nicaragua (Merits, Reparations, and Costs), IACtHR Series C No. 79 (31 August 2001)
Case of the Moiwana Community v Suriname (Preliminary Objections and Merits), Judgment, IACtHR Series C No 124 (15 June 2005)

UN Human Rights Committee

Ángela Poma Poma v Peru, Comm No. 1457/2006, UN Doc CCPR/C/95/D/1457/2006 (2009)
Apirana Mahuika v New Zealand, Comm No 547/1993, UN Doc CCPR/C/70/D/547/1993 (27 October 2000)
Ilmari Länsman et al. v Finland, Comm No. 511/1992, UN Doc. CCPR/C/52/D/511/1992 (1994)

National

Argentina

Corte Suprema de Justicia de la Nación, *Comunidad Aborigen de Santuario Tres Pozos y otros c/ Provincia de Jujuy y otros s/ amparo*, Expte. C.1196.XLVI, 18/12/2012

Australia

John Bulun Bulun & Anor v R & T Textiles Pty Ltd [1998] IndigLawB 87; (1998) 4(16) Indigenous Law Bulletin 24

Canada

Delgamuukw v. British Columbia, 1997 CanLII 302 (SCC), [1997] 3 SCR 1010

Colombia

Colombia. Corte Constitucional. (2016). Sentencia T-622/16 Principio de precaución ambiental y su aplicación para proteger el derecho a la salud de las personas – Caso de comunidades étnicas que habitan la cuenca del río Atrato y manifiestan afectaciones a la salud como consecuencia de las actividades mineras ilegales. Magistrado Ponente: Jorge Iván Palacio Palacio. Bogotá: La Corte

United States of America

Diamond v. Chakrabarty (447 U.S. 303, 1980)
Ex parte Hibberd, 227 U.S.P.Q. 443, 1985

ACRONYMS AND ABBREVIATIONS

ABS	Access and benefit-sharing
ABSCH	Access and Benefit-Sharing Clearing-House
ACHPR	African Commission on Human and Peoples' Rights
AfCHPR	African Court on Human and Peoples' Rights
AIATSIS	Australian Institutes for Aboriginal and Torres Strait Islander Studies
ANDES	Asociación para la naturaleza y el desarrollo sostenible (Peru)
ANR	Agence Nationale de Recherche (France)
ARC	Australian Research Council
AREI	Aboriginal Research Ethics Initiative
Artt	Articles
BC	Biocultural
BCP	Biocultural Community Protocol
CBD	Convention on Biological Diversity
CBMIS	Community-based Monitoring and Information Systems
CBNRM	Community-based natural resource management
CERD	Committee on the Elimination of Racial Discrimination
CESCR	Committee on Economic, Social and Cultural Rights
CGRFA	Commission on Genetic Resources for Food and Agriculture (FAO)
CIHR	Canadian Institutes for Health Research
CITES	Convention on the International Trade in Endangered Species
COP	Conference of the parties
CP	Community Protocol
CRIAA	SA-DC Centre for Research, Information, Action in Africa - Development & Consulting Namibia

CRSH	Social Sciences and Humanities Research Council (Canada)
DNA	Desoxyribonucleic Acid
DOI	Digital Object Identifier
DSI	Digital Sequencing Information
eDNA	environmental DNA sequencing
EIA	Environmental Impact Assessment
EU	European Union
FAIR	Findable, Accessible, Interoperable, Reusable
FAO	Food and Agriculture Organization (UN)
FCCC	UN Framework Convention on Climate Change
FOFIFA	Foibem-pirenena momba ny Fikarohana ampiharina amin'ny Fampandrosoana ny eny Ambanivohitra – Centre National de la Recherche Appliquée au Développement Rural (Madagascar)
FPCI	Fundación para la Promoción del Conocimiento Indígena (Peru)
FPIC	Free, Prior and Informed Consent
FRQSC	Fonds de recherche du Québec – société et culture (Canada – Québec)
GATT	General Agreement on Tariffs and Trade (WTO)
GEF	Global Environment Facility
GLIS	Global Information System (ITPGRFA)
GMO	Genetically Modified Organism
HRComm	Human Rights Committee
HTS	high-throughput DNA sequencing
IACtHR	Inter-American Court on Human Rights
ICCAs	Territories and areas conserved by indigenous peoples and local communities
ICOMOS	International Council on Monuments and Sites
IGC	Inter-governmental Committee on Intellectual Property, Genetic Resources, Traditional Knowledge and Folklore (WIPO)
IIED	International Institute for Environment and Development
IIFB	International Indigenous Forum on Biodiversity
ILK	Indigenous and Local Knowledge
ILO	International Labour Organization
INAI	Instituto Nacional de Asuntos Indígenas (Argentina)
INDECOPI	Instituto Nacional de Defensa de la Competencia y de la Proteción de la Propiedad Intelectual (Peru)
INIA	Instituto Nacional de Innovación Agraria (Peru)
INRAB	National Agricultural Research Institute of Benin
IP	Intellectual Property

IPBES	Intergovernmental Science-Policy Platform on Biodiversity and Ecosystem Services
IPCC	Intergovernmental Panel on Climate Change
IPLCs	Indigenous Peoples and Local Communities
IPR	Intellectual Property Right
IRD	Institut de recherche pour le développement (France)
ISE	International Society of Ethnobiology
ISSD	International Institute for Sustainable Development
ITPGRFA	International Treaty on Plant Genetic Resources for Food and Agriculture
IU	International Undertaking on Plant Genetic Resources (FAO)
IUCN	International Union for Conservation of Nature
LINKS	Local and Indigenous Knowledge Systems (UNESCO)
MAB	Man and the Biosphere (Programme – UNESCO)
MAEP	Ministry of Agriculture, Livestock and Fisheries (Madagascar)
MAT	Mutually Agreed Terms
MEDD	Ministry of the Environment and Sustainable Development (Madagascar)
MLS	Multilateral System of access and benefit-sharing (Plant Treaty)
MRC	Medical Research Council of Canada
MTA	Material Transfer Agreement
Nagoya Protocol	Nagoya Protocol on Access to Genetic Resources and the Fair and Equitable Sharing of Benefits Arising from their Utilization to the CBD
NAHO	National Aboriginal Health Organization
NFP	National Focal Point
NGO	Non-Governmental Organization
NGS	Next Generation Genomic Sequencing
NHMRC	National Health and Medical Research Council (Australia)
NKSC	National Khoi and San Council (South Africa)
NPO	Non-Profit Organization
NSERC	Natural Sciences and Engineering Research Council (Canada)
OCAP	Ownership, Control, Access, Possession
PGR	Plant Genetic Resource
PGRFA	Plant Genetic Resources for Food and Agriculture
PIC	Prior Informed Consent
PRE-TACAR	Panel on Research Ethics-Technical Advisory Committee on Aboriginal Research (Canada)

R&D	Research & Development
RCAP	Canada & Royal Commission on Aboriginal Peoples
REDD+	Reducing Emissions from Deforestation and forest Degradation, Plus the sustainable management of forests, and the conservation and enhancement of forest carbon stocks
RES	Resolution
SARC	South African Rooibos Council
SBSTTA	Subsidiary Body on Scientific, Technical and Technological Advice
SCBD	Secretariat of the Convention on Biological Diversity
SDGs	Sustainable Development Goals
SMTA	Standard Material Transfer Agreement (ITPGRFA)
SPDA	Sociedad Peruana de Derecho Ambiental (Peru)
SSHRC	Social Sciences and Humanities Research Council of Canada
TCPS	Tri-Council Policy Statement: Ethical Conduct for Research Involving Humans (Canada)
TCPS2	Tri-Council Policy Statement: Ethical Conduct for Research Involving Humans 2 (Canada)
TEK	Traditional Ecological Knowledge
TK	Traditional Knowledge
TKECA	Traditional Knowledge and Expressions of Culture Authority (Vanuatu)
TRC	Truth and Reconciliation Commission of Canada
TRIPS	Trade related Aspects of Intellectual Property Rights (Agreement, WTO)
TRRs	Traditional Resources Rights
UN	United Nations
UNCED	United Nations Conference on Environment and Development
UNCTAD	United Nations Conference on Trade and Development
UNDP	United Nations Development Programme
UNDRIP	United Nations Declaration on the Rights of Indigenous People
UNDROP	United Nations Declaration on the Rights of Peasants and Other People Working in Rural Areas
UNEP	United Nations Environment Programme
UNESCO	United Nations Educational, Scientific and Cultural Organization
UNFCCC	UN Framework Convention on Climate Change
UNTS	United Nations Treaty Series
UPOV	International Union for the Protection of New Varieties of Plant
USPQ	United States Patents Quarterly
VKS	Vanuatu Cultural Centre

VOI	Vondron'Olona Ifotony ("Basic Community") (Madagascar)
WG8j	Ad Hoc Open-ended Working Group on Article 8(j) and Related Provisions of the Convention on Biological Diversity
WIPO	World Intellectual Property Organization
WTO	World Trade Organization
WWF	World Wildlife Fund

1

COMMUNITY PROTOCOLS AND BIOCULTURAL RIGHTS

Unravelling the Biocultural Nexus in ABS

Fabien Girard, Ingrid Hall and Christine Frison[1]

To introduce this collection of chapters, we explore the "biological diversity/cultural diversity" nexus against the backdrop of biocultural community protocols (BCPs) and biocultural rights. BCPs and biocultural rights signal the development of biocultural approaches in biodiversity conservation. However, by no means, they can be confined to the sphere of conservation. They speak to the heart of sovereignty and the politics of identity, as much as they have a bearing on land claims and touch upon issues that we may venture to place under the heading of political ontology. The coming together of BCPs and biocultural rights, the success of which is in large part credited to the lawyer and activist Kabir Bavikatte (Bavikatte, 2014), shows that we are treading new ground. In the Mo'otz Kuxtal Voluntary Guidelines, (Biocultural) Community Protocols (CPs) are broadly defined as a

> [...] term that covers a broad array of expressions, articulations, rules and practices generated by communities to set out how they expect other stakeholders to engage with them. They may reference customary as well as national or international laws to affirm their rights to be approached according to a certain set of standards. Articulating information, relevant factors, and details of customary laws and traditional authorities helps other stakeholders to better understand the community's values and customary laws. Community protocols provide communities an opportunity to focus on their development aspirations vis-a-vis their rights and to articulate for themselves and for users their understanding of their bio-cultural heritage and therefore on what basis they will engage with a variety of stakeholders. By considering the interconnections of their land rights, current socio-economic situation, environmental concerns, customary laws and

DOI: 10.4324/9781003172642-1

> traditional knowledge, communities are better placed to determine for themselves how to negotiate with a variety of actors.[2]

While these guidelines address "prior and informed consent", "free, prior and informed consent", or the "approval and involvement" of Indigenous peoples and local communities (IPLCs) in order to access their knowledge, innovations, and practices, the expansive acceptance of CPs shows the extent to which theoretical work on biocultural rights has percolated into the framing of BCPs/CPs. Bavikatte conceived of biocultural rights as a "bundle" encompassing (i) the right to land, territory, and natural resources; (ii) the right to self-determination, principally understood here in its "internal" dimension, i.e. the right of communities to autonomy and self-administration; and (iii) cultural rights. Additionally, "stewardship" (or "guardianship"), which Bavikatte saw as the cornerstone of biocultural jurisprudence (Bavikatte & Bennett, 2015), is now enshrined in another document emanating from the Convention on Biological Diversity[3] (CBD), the Tkarihwaié:ri Code of Ethical Conduct.[4] It is also reflected in the first decision to build on biocultural rights and BCPs, the *Atrato River Case* from the Constitutional Court of Colombia (Macpherson et al., 2020).[5]

The remaining part of this introductory chapter investigates the popularisation of BCPs within the Access and Benefit-Sharing (ABS) context, linking it to the rise of biocultural jurisprudence and against the short history of the interlinkages between cultural diversity and biological diversity in conservation. It begins with the context, marked by the high profile which has been progressively gained by "traditional" communities at the end of the last century. It situates this progression within an institutional and political context markedly concerned with North-South imbalances in access to genetic resources and the shifting ground in conservation which saw a dramatic reappraisal of the role of IPLCs and peasants/farmers in the management of biodiversity. This series of shifts, to which biocultural approaches gave decisive impetus, have gone so far as to open fresh, sweeping debates on self-determination and sovereignty over resources for IPLCs.

The chapter then moves on to introduce the scope of the book. To begin with, the first section traces the history of BCPs/CPs from the emergence of the so-called "cultural protocols" to the enshrinement of CPs in the Nagoya Protocol. BCPs/CPs are then successively assessed as legal and political tools, against the backdrop of biocultural jurisprudence, while a final note investigates the contentious concept of "stewardship" in conservation. The last section briefly outlines the content of the remaining chapters in the volume.

While this book is not devoid of Indigenous voices, it does not claim to speak on behalf of Indigenous peoples, local communities, or peasants. We do hope, at the very least, that it speaks to the depth of knowledge and perspectives shared with us by Indigenous peoples, local communities, and peasants over the course of our research.

The Context: "Traditional" Communities and Biodiversity Conservation

Since the late 1980s, the key role played by IPLCs in the sustainable management of complex ecological systems has increasingly been recognised. As early as 1987, the Brundtland Report took the view that IPLCs (referred to as "indigenous or tribal peoples") are "the repositories of vast accumulations of traditional knowledge [TK] and experience that links humanity with its ancient origins", and warned that "their disappearance is a loss for the larger society, which could learn a great deal from their traditional skills in sustainably managing very complex ecological systems" (Brundtland and the World Commission on Environment and Development 1987). At about the same time, the Declaration of Belém (1988), very much imbued with Posey's tenacious work on debunking misconceptions about Indigenous peoples, emphatically declared that "[...] native peoples have been stewards of 99 percent of the world's genetic resources" and forcefully stressed the "link between cultural and biological diversity".[6]

In the following years, as concerns on environmental deterioration and the erosion of biodiversity were beginning to reach the broader public, and alternative, much more participative, people-centred, and place-based approaches were gaining momentum (Altieri et al., 1987; Freeman, 1989; Oldfield & Alcorn, 1987; Toledo, 1990), "attention has expanded to include a wide range of local communities, including forest peoples, farmers, fishers, herders, pastoralists, diversely manifested around the world" (IPBES, 2019). The increasing awareness to local communities is probably best captured by the CBD, Article 8(j) of which requires each contracting party to

> respect, preserve and maintain knowledge, innovations and practices of indigenous and local communities embodying traditional lifestyles relevant for the conservation and sustainable use of biological diversity and promote their wider application with the approval and involvement of the holders of such knowledge, innovations and practices.[7]

According to estimates, there are around 5,000 groups of Indigenous peoples in the world (Hall & Patrinos, 2012), amounting to 476 million people (ILO, 2021, p. 10). Some have never been in contact with other human societies, others are isolated (sometimes voluntarily after disastrous contacts), and yet others have had brief contact. Other much larger groups can be found across the globe, including in peri-urban areas and urban centres. Local communities represent an even larger population, and are as diverse and widely distributed around the world. When considered together, IPLCs[8] represent about 1 billion people and it is estimated that they hold, either under customary tenure or a community-based regime (formally recognised in domestic law), between half and two-thirds of the world's lands (Wily, 2011).[9]

In recent years, much effort has been undertaken to document and appraise the role of IPLCs in shaping the ecologies and resource of vast regions of the world, for instance, in the management of forests, soil fertility, grasslands, mountains, watersheds, and coastal areas (IPBES, 2019). Furthermore, there is now ample evidence that farmer-managed seed systems have been instrumental in building viable and diverse crop types over millennia and that they continue to provide more than 70% of the seeds used around the world today (McGuire & Sperling, 2016).

In addition, after centuries of disregard or contemptuous treatment, Indigenous and Local Knowledge (ILK), also referred to as Traditional Ecological Knowledge (TEK) or simply TK, has recently stirred up a great deal of interest from conservation biologists, ecologists, "sustainability" scientists, and, of course, geographers, anthropologists, sociologists, and political scientists. TEK refers to

> [...] the worldviews, knowledge, practices, and innovations embedded in the relationship between people and nature, as expressed in local knowledge about the natural world, techniques and technologies of resource management, as well as in local institutions governing social relations and relationship to nature.
>
> *(IPBES, 2019, p. 37)*

TEK is placed-based, is embodied in social structures, and has a holistic dimension. In contrast to the way it was previously commonly pictured as backward, static, and self-contained, it is now commonly described as open and hybrid, dynamically evolving, as IPLCs appropriate new forms of knowledge and interact with animals, plants, and land (Berkes, 2012, p. 7; Berkes & Berkes, 2009, p. 7).

These significant changes in perspective have contributed decisively towards IPLCs being recognised as major actors in the struggle against climate change (IPCC, 2015, pp. 758, 765–766) and the erosion of biodiversity (Secretariat of the Convention on Biological Diversity, 2020, p. 115). The IPBES' most recent Global Assessment Report on Biodiversity and Ecosystem Services underlines, in singularly forceful fashion, that:

> While local in action, IPLC management of nature and biodiversity provides contributions to the larger society, in rural and urban areas alike, including the provisioning of food, fibers, material, and medicine to local and to export markets, and the management of agrobiodiversity of major regional and global crops. In many regions IPLC lands contribute to the conservation of watersheds that supply large regional populations.
>
> *(IPBES, 2019, p. 42)*

Traditionally owned or occupied IPLCs' landscapes are home to much of the world's biodiversity (Brondizio & Tourneau, 2016); and evidence is mounting that biodiversity is declining at much lower rate in Indigenous lands than elsewhere (Garnett et al., 2018). Recent studies also stress the prominent role that small-scale farmers play as innovators in society. They have ensured and continue to ensure the spatial and social development and distribution of genetic, morphological, and varietal diversity – diversity that is increasingly being recognised as critical for farming productivity and climate change adaptation (Coomes et al., 2015).

IPLCs, farmers, rangers, pastoralists, and foresters, who are the focus of this book, are very diverse in the way they produce food and products, manage landscapes, safeguard agrobiodiversity, build knowledge about food and medicines, and pass this and associated intangible heritage onto future generations. Nevertheless, they share many common concerns and have often been subjected to similar centuries-long ordeals: poverty, discrimination and violence, displacement and land confiscation, limited access rights to land and resources, lack of or limited access to culturally appropriate healthcare services and education, and a lack of or limited access to basic services such as potable water, energy, and sanitation. They are among those who are at a greater risk from environmental harm (Knox, 2018, para. 41). These are but a few of the problems that IPLCs face, to which may be added more recent encroachments upon their customary rights, together with legal and physical conflicts with mining companies, oil corporations, logging companies, and the agri-food industry.

Biodiversity and Access and Benefit-Sharing

None of the issues cited above fall outside the scope of the study. However, this book is primarily concerned with biodiversity in the context of ABS.

The CBD, which came into force in 1993, reaffirmed "that States have sovereign rights over their own biological resources" (on this issue, see Mgbeoji, 2003), thereby steering the international community away from the idea that genetic resources are a "common heritage of mankind" – an international legal concept that was previously fashionable and which was conveniently, not to say strategically, recast by industrialised countries to satisfy their extractive agenda (ibid.).

The first objective of the CBD is to conserve biodiversity and, through Article 8(j), the instrument recognises the role of IPLCs in this conservation. Yet, this new ABS regime has laid down a new framework that problematically centred around the sustainable use of resources (Rosendal, 1991, p. 28) and the market value of the components of biodiversity. As Daniel Robinson recalls, the CBD is a hard-fought compromise negotiated by biodiverse and mega-diverse countries, i.e. the "Global South" (Robinson, 2014, p. 3). In the midst of the controversial Uruguay round of negotiations held by the General Agreement on Tariffs and Trade (GATT) that were to result in an unprecedented extension of intellectual property rights (IPRs),[10] the aim of the CBD was to ensure that access to genetic materials from plants, animals, and microbes with potential value by researchers and industry (mainly) from the "Global North" was sought with the permission of the provider country.[11] An additional and central goal was that access permission could be granted subject to a benefit-sharing agreement made under mutually agreed terms (MATs) between the user (e.g. researcher, pharmaceutical industry, and plant breeding industry) and the provider country (i.e. inter-state benefit-sharing) and then between the State and a community (i.e. intra-state benefit-sharing) (Morgera et al., 2014, p. 25).[12] The backbone of the new regime was dubbed by scholars as the "Grand Bargain", i.e. a grand (and oversimplified) narrative striving to harmoniously articulate efforts at "bioprospecting" or "biodiscovery" (collection and screening activities for R&D) and conservation. As one of the champions of "bioprospection" put it, "[s]ystematic screening, by developed and developing countries working together could pay off for both while aiding conservation efforts" (Eisner, 1989, p. 31).

The underlying assumption is transparent: the CBD links the conservation of biodiversity with the market value of its components – the "biological resources" – which are amenable to the protection of IPRs (Boisvert & Vivien, 2012, p. 1166). The reasoning is that biodiversity-rich countries (the Global South), now able to capture part of the benefits arising out of IPRs on "biodiscovery" or to benefit directly from new ("environmentally sound") technologies through "technology-for-nature swaps",[13] are deemed better equipped and, above all, incentivised to tackle the erosion of biodiversity (Sedjo, 1992).[14]

This initial focus was significantly altered by the nascent international human rights-based approach of the United Nations Conference on Environment and Development (UNEP) negotiations defending Indigenous peoples (Halewood, 1999, pp. 955, 965). For IPLC advocates, the impending strengthening of IPRs[15] made it more pressing to reinforce communities' control over the use of their knowledge and innovations and to secure some form of benefit-sharing. Hence, there were strong pushes from several quarters for intellectual property-style (*sui generis*) rights for IPLCs and small-scale farmers,[16] but not without serious objections as to the feasibility of IPRs on genetic resources and TK held by IPLCs. Questions were raised about the appropriateness of relying on instruments underwritten by certain assumptions about what is "nature" (Hamilton, 2008), "creation", or "cultures" (Coombe, 1998, p. 247), about the role of market economy in conservation (McAfee, 1999, p. 144), and about communities' "interest in commercializing their knowledge" (Brush, 2001, p. 521). More far-reaching proposals were also brought to the forefront, as some IPLC proponents argued that land rights and the right to self-determination[17] were the only way forward.[18]

None of these concerns are obvious in the provisions of the CBD. Admittedly, Article 10(c) of the CBD (like Article 8(j)) voluntarily retained an open-textured nature (Halewood, 1999, p. 978; Posey, 2004, p. 163), leading to a twofold (heavily qualified) (Glowka et al., 1996, p. 62) obligation being put on contracting parties, namely to "[p]rotect *and* encourage customary use of biological resources [...]" (emphasis added). In reality, however, there was an almost exclusive focus throughout the negotiations on "encouraging", i.e. on making IPLCs and farmers "participate"[19] in conservation activities as defined and financed by the Global North. The few inroads made into "protection" are primarily concerned with *incentivising* traditional resource management systems,[20] rather than with shielding against the unauthorised commercial use of biological resources and associated TK.[21] As there is a growing recognition that IPLCs "have a vital role in environmental management and development because of their knowledge and traditional practices" (Rio Declaration, Principle 22), the main challenge and matter of concern is to harness this potential towards conservation and rural development through bilateral negotiations and "Coasean contracts" (Sedjo, 1992, pp. 207–208) between bioprospectors and communities.

At first sight, IPLCs sit rather uncomfortably within this new framework which is also geared towards treating genetic resources and TK held by IPLCs "as a commodity that will be traded by [them] in exchange for monetary and non-monetary benefits [...]" (Bavikatte et al., 2010, p. 294). In particular, as Bavikatte et al. remarked,

> the acknowledgement of market-alienable aspects of TK such as ILCs' [indigenous and local community's] ownership of their TK tends to mask the existence of inalienable aspects of TK, such as IPLCs' rights to their traditional lands and to practise their cultures which are in effect the wellspring of their TK
>
> *(ibid., 296; Nemogá, 2019, p. 262).*[22]

The observation echoes criticisms that were levelled against proposals to endow IPLCs with "tribal rights" (Greaves, 1996) or "community intellectual rights" (Egzibher, 1996), in other words new kinds of *sui generis* rights supposedly better suited to collective creations and innovation, but which remained largely modelled upon the Western intellectual property tradition. Certainly, these new ideas were laudable – notably in the appeal to the "embedding concept" of tradition (Strathern, 1996, p. 22) – and they are regularly summoned up in debates on "biopiracy" (on which see Mgbeoji, 2006, p. 13; Hamilton, 2008) every time tensions between activists and bioprospectors resurface. They nevertheless raised serious concerns and limitations. First, insofar as farmers' varieties (landraces) are concerned, intellectual property-style rights may have a disruptive effect on the flow and exchange of valuable genetic materials and a propensity to create tensions between communities (Correa, 2016; Srinivasan, 2016).[23] Second, as the concept of ownership over seeds, plants, and TK is alien to some cultures (Tsosie, 2007), it was felt that their inception might adversely affect the cosmologies of certain communities (Anderson, 2015, pp. 769, 771, 777; Posey, 1995), and they might actually unwillingly serve to expand the religion of property.[24]

On the face of it, as the CBD began to be implemented, neither ABS mechanisms nor *sui generis* intellectual property laws appeared to offer robust and uncontroversial tools for IPLCs to secure their rights over their own resources and knowledge. Even access laws, suffused with ideas about *incentivising* IPLCs' "participation" in biodiversity conservation and *assigning* economic value to resources and TK, were met with severe reservations as they shared some of the shortcomings that affect *sui generis* IPRs.

At the same time, one element of paramount importance is worth stressing: the Preparatory Committee and Working Groups tasked with the preparation of the Rio Earth Summit, as well as the Ad Hoc Working Group and then the Intergovernmental Negotiating Committee (INC) for a Convention on Biological Diversity (see McGraw, 2017), had been working in a distinctive political and intellectual environment asserting new bonds between "two strands of international law [that] were being developed in relative isolation from one another for quite some time" (Halewood, 1999, p. 965), namely human rights law[25] and environmental law. Besides, as Halewood showed, based on interviews and a review of archival materials, some delegates to the Ad Hoc Working Group (primarily from Canada and Sweden) pushed for the inclusion of knowledge and innovations of IPLCs in the CBD without having "clear ideas about the kinds of national laws that would eventually be required to realize the objectives of including the term in the agreement in the first place" (Halewood, 1999, n. 104). This sort of "Trojan-horsing" of the CBD, whose primary aim was to ensure "[...] that indigenous and local peoples would be included in future national and international processes wherein such mechanisms would be defined" (ibid.), also contributed towards bringing hermeneutic openness to the instrument, in particular when read, as Posey astutely suggested, in conjunction with the Rio Declaration and the Agenda 21 (UNCED, 1992; see Posey, 2004, p. 163).

By and large, it can be argued that, taken in their broad intellectual and political contexts and read together with soft law instruments also adopted at the Rio Conference, Articles 8(j) and 10(c) represented a landmark which widened the number of epistemic communities, players/stakeholders, constituencies allowed and called upon to debate what Donna Haraway refers to as "nature-in-the-making" (Harvey & Haraway, 2016), while Stengers (2005) and Latour (2004) propose the term "cosmopolitical". This has led to the establishment of hard-won new exchange relationships through forms of resistance to cultural homogenisation, unfolding a "cosmopolitanism from below" as Appadurai puts it (2013), and opening up new discursive spaces for addressing complex ontological issues and initiating translation processes and boundary work (Fisher, 1988; Löfmarck & Lidskog, 2017; Mollinga, 2010) across diverse epistemic boundaries.

The Politics of Biocultural Diversity

At a time when contracting parties are gradually implementing the CBD and now the Nagoya Protocol,[26] "genetic resources" and "traditional knowledge" are arguably still seen and treated by some as "resources" to be tapped into for the development of new technologies amenable to the protection of IPRs (see Bonneuil, 2019). However, the Earth Summit signalled a watershed moment, heralding the concept of "biocultural diversity". This is a complex history – which is still in the making – but three movements seem to have prompted its emergence and have moulded and refined its meaning over time.

The first movement is most directly related to nature conservation and shakes up the deeply ingrained dogma of conservation science, the so-called "fortress conservation" model that had presided over the fate of IPLCs living in or near protected areas for almost three centuries (Colchester, 2003; Doolittle, 2007). As the model met with mounting criticism (Freeman, 1989), new bottom-up and community-based approaches gained traction from the early 1980s onwards (Buergin, 2013, p. 5 and the literature referred to therein).[27] The notion that relationships between humans, their social organisation, and their natural environment cannot be universally understood along the nature-culture divide, irrelevant for numerous populations around the world, underpins and drives new approaches to conservation. The so-called "ontological turn" in philosophy and anthropology (Castro, 1992; Descola, 1996 [1986]; Descola & Sahlins, 2014; Kohn, 2015; MacCormack & Strathern, 1978) gave ground to the academic recognition of non-Western ontologies (Blaser, 2009b, 2013) and evidence of how those populations shape the worlds they know, see, and interact with, and especially the way they consider elements of the so-called "natural world" beyond the subject-object divide. These reflections have undoubtedly contributed to the idea that IPLCs are, due to their non-naturalist ontologies, repositories of vast bodies of knowledge that may be relevant for biodiversity conservation (Buergin, 2013, p. 4).

This was already contemplated in the World Heritage Convention (1972)[28] which is certainly one of the first international instruments to recognise that conservation of nature goes hand in hand with the protection of cultural identities, thereby confirming the "relationship between indigenous peoples and the environment" (Ulloa, 2005, p. 137; also see Kari & Rössler, 2017). Also of note is the major step taken through the 1980 World Conservation Strategy, in which the IUCN, WWF, and UNEP stressed the importance of "traditional knowledge", even though there is no mention of Indigenous peoples in the text, only of "rural communities" (IUCN et al., 1980, Chapter 14, para. 10).

In the 1970s and 1980s, the coming together of the notions of capital, progress, and technology ushered in a period of certainty known as "development". The second movement arose against the backdrop of the prescriptions produced by major institutions such as the World Bank and Western universities, which were meant to actively transform "traditional" into "modern" societies through major investments and new technologies (Escobar, 1995). The concept of development has engendered essential critical theories – liberal, Marxist, post-structuralist, post-development – which are of particular interest within the context of the construction of environmental issues as a global problem (Buergin, 2013, p. 7). Their differences apart, these movements share a common base: the "intrinsic values of culture and cultural diversity as well as opportunities this diversity provides regarding sustainable or alternative ways of development [received] broader attention, frequently addressed in terms of 'traditional ecological knowledge' or local knowledge systems" (Buergin, 2013, p. 8). Linking the "development crisis" to the "escalation of environmental problems" and the "struggle for natural resources" (Banerjee, 2003, p. 151) opened the way for the concept of "sustainable development" in the late 1980s. The Brundtland Report (Brundtland & World Commission on Environment and Development, 1987, para. 74), which constitutes a milestone in this history, thus signalled, albeit in an ambiguous fashion, the shift in focus to the local level, the link with cultural diversity, while promoting a new model of "participatory" and "inclusive" development (Buergin, 2013, p. 8; Peet & Watts, 1996). Admittedly, the report is still pervaded with the idea – very much in tune with the concept of "progress" – of a time arrow, with IPLCs sitting on the first point – distant and fading away (the "ancient origins" of humanity[29]) and, of course, the reference to "productivity" (Brundtland & World Commission on Environment and Development, 1987, para. 76) which is yet more proof, if proof was needed, of the ambiguity of the concept of "sustainable development". However, the text encapsulates a series of conceptual shifts that were underway: the awareness of the important of "traditional" ways of life, "traditional rights to land", "own institutions to regulate rights and obligations" which have been translated into "harmony with nature" and an "environmental awareness" and, therefore, the acceptance of the pressing need to recognise and sustain "traditional rights", which "must go hand in hand with measures to protect the local institutions that enforce responsibility in resource use" (ibid., para. 75).

The last movement is inseparable from the international recognition of Indigenous peoples. As Anaya recounted, a new generation of Indigenous women and men, well-versed in the dominant legal system, "began drawing increased attention to demands for their continued survival as distinct communities with historically based cultures, political institutions, and entitlements to land" (Anaya, 2004, p. 46). This counter-narrative, which articulates new identities based on distinct cultures and entrenched in the land, was disseminated by a vast campaign that took place during major international conferences and was relayed by academics and NGOs from the 1970s onwards (ibid.). The adoption of ILO Convention No. 169 of 1989[30] represents "international law's most concrete manifestation of the growing responsiveness to indigenous peoples' demands" (Anaya, 2004, p. 47), followed, about 20 years later, by the 2007 UN Declaration on the Rights of Indigenous Peoples.[31] Importantly, the connection between cultural diversity and biodiversity conservation underpinned the work of ILO, as Buergin remarks:

> [c]ultural diversity and environmental conservation were crucial issues in the arguments about 'indigenous peoples' and their rights to lands, local resources, self-determination, and particular identities from the beginning. A particular relationship to the places they inhabit, often related to historical continuity, is at the core of their claims to lands and territories and

> discussed in the context of particular conceptualizations of and relations to 'nature' different from 'modern' environmental relations.
>
> *(Buergin, 2013, p. 11)*[32]

As mentioned above, the Declaration of Belém (1988)[33] represented a further milestone by stressing the "inextricable link" between cultural and biological diversities, calling for the recognition and protection of "cultural and linguistic identity", and articulating the idea that Indigenous peoples are "stewards" of biodiversity. Agreements and documents adopted at the Earth Summit in 1992 abound in similar references to "traditional lifestyles" dependent on biological resources[34] and, conversely, on "traditional" knowledge, innovations, and practices of Indigenous and local communities "relevant for the conservation and sustainable use of biological diversity".[35] This echoes Principle 22 of the Rio Declaration on Environment and Development (1992), which states that "Indigenous people and their communities and other local communities have a vital role in environmental management and development because of their knowledge and traditional practices". Additionally, Agenda 21, the action plan adopted at the Rio Conference, stresses that "Indigenous people and their communities have an historical relationship with their lands [...]" and "[t]hey have developed over many generations a holistic traditional scientific knowledge of their lands, natural resources and environment" (Chapter 26.1) that needs to be recognised (Chapter 26.3 (iii)) (UNCED, 1992).

In sum, at the beginning of the 1990s, the main features that would go on to feed into the "biocultural diversity" concept began to be fleshed out. Biological diversity was conceived of as being linked to culture, in particular that of IPLCs which have developed a holistic (Nelson, 2009) knowledge of their land, resources, and environment. This specific knowledge is conceived of as being particularly relevant for the conservation of biodiversity and begins to support the idea that IPLCs are "stewards" of their environment. In 1996, as it became possible to pair maps of biodiversity with maps of cultural (and particularly linguistic) diversity (Maffi, 2001b; UNESCO et al., 2003), the concept of biocultural diversity received the final impetus[36] that enabled it to gain a foothold in the scientific vocabulary (Cámara-Leret & Bascompte, 2021). The 1996 Berkley Conference on "Endangered Languages, Endangered Knowledge, Endangered Environments",[37] organised by Luisa Maffi (the proceedings only appeared in 2001) (Maffi, 2001a), is an important academic landmark in this respect. Posey accelerated the dissemination of the concept by entrusting Luisa Maffi with the chapter on "Linguistic Diversity" (Maffi, 1999) published in his *Cultural and Spiritual Values of Biodiversity* (Posey, 1999), a stand-alone volume, but which was part of the influential Global Biodiversity Assessment undertaken by UNEP in 1995 (Heywood et al., 1995). As Figure 1.1 (Annex) illustrates, references to biocultural diversity – or connections between cultural heritage and natural heritage/biodiversity – have multiplied since the 1990s in binding international instruments, soft law, and declarations from major international and regional conferences, sometimes endorsed by the CBD and UNESCO. The CBD[38] and IPBES[39] have recently started integrating the concept in their respective work, and dedicated international programmes strive to bridge the gaps between cultural and natural heritage.[40] Another step in this direction lies in the "Nature-Culture Alliance" to be launched at the forthcoming CBD COP 15, which will extend – and probably take the place of – the joint SCBD-UNESCO nature-culture programme.[41]

Some also expressed their discontent with the way the issue was raised. For example, Brosius and Hitchner regretted at some point that the "biocultural perspective" and the "biocultural diversity" concept be "entirely the product of the crisis narrative", thereby preventing their designers and advocates from acknowledging the "dynamic, creative possibilities that can emerge from human agency and processes of hybridity" (Brosius & Hitchner, 2010, p. 143). They therefore argued for examining the potentialities and implications – in terms of practices, policy, and governance – of what they saw as a new "trans-disciplinary field" engaged in conservation (Brosius & Hitchner, 2010, p. 142). By and large, their call was heeded by scholars (Bridgewater & Rotherham, 2019). With hindsight, there is little doubt that the "biocultural axiom" (Nietschmann, 1992) opened a political space (Mulrennan & Bussières, 2020, p. 293; Nemogá, 2019, p. 260) for IPLCs by showing that *biodiversity*, beyond its "concrete biophysical elements" (Escobar, 1998, p. 53), is a "discursive invention of recent origin" (ibid.). At the very least, this "biocultural axiom" has nurtured the notion that biodiversity forms part of complex territorialised networks, living systems, inextricably linking worldviews and praxis, expressing attachments to the earth and more-than-humans (Tsing, 2013); and that, therefore, alternative approaches to biodiversity conservation ought to be sustained, bounding up considerations of human rights, equity, ethics, and ontologies (see Mulrennan & Bussières, 2020, p. 299).

To return to the three avenues that were explored during the CBD negotiations to tackle the issue of TK's "protection" – *sui generis* intellectual property laws vested in communities, access laws, and "recognition (and elevation to levels of national and international recognition)" of communities' own traditions and customs (Halewood et al., 2006, p. 185; also see Halewood, 1999) – there is little doubt that the last two have undergone significant changes as engagements with biocultural diversity have intensified. In the field of access law, with the "rediscovery" of the "work" (Hayden, 2007) of IPLCs that has been too long belittled or denied (Younging, 2010), and the "reappearance" of humans within a supposedly "inanimate" or "wild" nature (Posey, 2001, pp. 384–385), new debates have emerged with the Nagoya Protocol[42] at the forefront. Under the new legally binding agreement, all parties now have to take measures with the aim of ensuring that TK and genetic resources are accessed with the prior and informed consent (PIC) or approval and involvement of IPLCs, and that MATs have been established (Nagoya Protocol, Articles 6 & 7). States are also obliged to take measures for the purpose of ensuring that benefits arising from the use of these TK[43] and genetic resources are shared in a fair and equitable way with the IPLCs concerned (Morgera et al., 2014). In recent years, academic research has particularly focused on three main and interwoven issues with a view to securing IPLCs' engagements through contracts with stakeholders on access to genetic resources and TK. First, making sure that bioprospecting contracts do not arbitrarily "cut collectives", in other words create some sort of *ad hoc* collectives or "political sociality" aimed at distinguishing between "deserving" and "underserving" "contributors" to the production of a new technology, and exclusively made of those relatively few groups who are, in fact, best able to articulate their demands (Hayden, 2007). Second, guaranteeing that bioprospecting contracts *do* involve IPLCs and that they *do* address the asymmetrical relations and inequitable bargaining power that characterise ABS negotiations (Robinson, 2014, p. 12). Third, securing the "market-inalienability" (e.g. that which touches on language, sacred sites, and ancestral ties) of certain aspects of seeds, plants, and TK (Bavikatte et al., 2010, p. 298; Gilbert, 2018, p. 83).

Furthermore, biocultural approaches have naturally rekindled proposals made at the start of the UNCED and CBD negotiations (Halewood, 1999, p. 955) aimed more radically at steering the debate away from considerations about contracts on, and trade in, TK, and to refocus it on what should be deemed fundamental prerequisites to protect the ecological values and traditional lifestyles that sustain the conservation of biological diversity. Clear evidence of this is the work of the Ad Hoc Open-ended Working Group on Access and Benefit-Sharing – which prepared the ground for negotiations on the Nagoya Protocol – replete as it is with calls to take new ground in the field of the protection of TK, and framing the discussion in terms of ontological conflicts, inalienability, divergences between legal traditions, and cultural values.[44]

ABS within the Biocultural Nexus: From Participation to Sovereignty and Self-Determination

This broad-strokes account manages to capture the pivotal role played by "participation" on a new intellectual canvas which is dominated by the concept of "sustainable development". Interlinkages between "participation" and "development" were already under way in the Declaration on the Right to Development of 1986, Article 2 of which states that "the human person is the central subject of development and should be the active participant and beneficiary of the right to development".[45] This articulation – and in particular the connexion between participation and benefits – has paved the way for a rights-based approach to development and, above all, further provided the decisive impetus for moving beyond participation and to embrace more radically the issue of consent, sovereignty, and self-determination (Gilbert, 2018, pp. 65–66).

Under ILO Convention No. 169, references to participation are still encumbered by a great deal of ambiguity. Along the lines of the 1980 World Conservation Strategy (IUCN et al., 1980, Chapter 14, para. 10), it is believed that, in addition to the benefits of relying on local ecological processes, people-centred conservation is cost-effective and more readily accepted than command-and-control conservation projects. Furthermore, as local populations in the South are still seen as a major source of resource depletion (Peet & Watts, 1996, p. 7), participation coupled with benefits cannot but align with the development blueprint laid out by experts from the late 1980s. This can be drawn from Article 15(1) which states that the rights of the "peoples concerned to the natural resources pertaining to their lands" "include the right of these peoples to participate in the use, management and conservation of these resources". At the same time, what is well entrenched is the relatively new idea that Indigenous peoples have the right to seek development on their own terms (Posey, 2001, p. 388). This implies, at the very least, giving them the right to be consulted whenever legislative or administrative measures or plans and programmes for national or regional development "may affect them directly" (Articles 6.1(a) and 7.1).

As the links between TK and conservation deepened, the rights-based approach to development gave rise to procedural rights which were deemed crucial to achieving "sustainable development" (Gilbert, 2018, p. 65). It also pushed for the recognition of substantive rights based on the premise that good management of land and resources implies a level of "control" thereof.[46] Against this background, the progressive shift in international instruments and regional jurisprudence from *consultation* to *consent* (free prior and informed consent – FPIC – or prior and informed consent – PIC) – and then the advent of the principle of benefit-sharing – expresses difference of kind rather than of degree.[47] The first inroads into PIC and benefit-sharing were visible in Articles 8(j) and 10(c) of the CBD, and the emergence of the right to FPIC/PIC was confirmed in the UNDRIP. The major breakthrough can be credited to the Nagoya Protocol, the stipulations of which, however qualified they remain when it comes to IPLCs' rights over genetic resources (Morgera et al., 2014, pp. 122–125), strongly support the view that, for the "custodians of biodiversity"[48] to carry on stewarding the seeds and plants and cultivating their knowledge, they must be able to "control" and therefore to decide (collectively) whether, when, with whom, and in consideration of what they want to share elements of their heritage (Posey, 2001, pp. 388–389).

In other words, as some scholars have suggested, PIC and benefit-sharing in the Nagoya Protocol are

> implicitly underpinned by a substantive environmental right of indigenous and local communities to their genetic resources. It embodies an obligation owed directly to them, deriving from established international human rights, in their collective dimension, to indigenous peoples' self-determination, ownership and cultural identity.
>
> *(Morgera et al., 2014, p. 118; also see ibid., pp. 42, 113)*

By extension, PIC and benefit-sharing reverberate through, and reinforce, the vast array of rights which are deemed essential to maintain this stewardship role and which cannot be uncoupled from rights over genetic resources and TK, namely the right to land and territory, the right to self-determination, and the right to cultural identity. Admittedly, some of these rights, as now enshrined in the UNDRIP,[49] have been considerably reinforced by human rights treaty bodies and, most of all, by regional human rights courts over the past decades (Charters, 2018; Errico, 2018; Gilbert, 2018; Saul, 2016). But progress achieved by the Inter-American Court of Human Rights notwithstanding (Gilbert, 2018, pp. 72–73, 76–79), provisions on PIC and benefit-sharing remain fragile in international human rights law. For instance, the right to property and natural resources continues to be built on the assumption – traced back to the doctrine of eminent domain – that "[...] the state has underlying title to the land and the natural resources it contains" (Gilbert, 2018, p. 42). From this, it follows that, albeit subject to compensation (as against benefit-sharing),[50] states often retain both the "[...] overall power [...] to expropriate any private property in land without the owner's consent, even if such a right is legally and constitutionally protected" (Gilbert, 2018, p. 37) and the right to dissociate the regime of property in land and property over natural resources (Almeida, 2017). PIC and benefit-sharing obligations enshrined in the Nagoya Protocol certainly go further by imposing PIC and benefit-sharing obligations in respect of land, resources, and cultural heritage even in the absence of any restriction or deprivation and sustain the dynamic interpretation that PIC/FPIC comes very close to a veto right.[51] PIC and benefit-sharing also give new support to the right to self-determination over natural resources (Gilbert, 2018, pp. 26–28). Obviously, all these developments are of considerable interest to groups that do not have the status of Indigenous peoples (Morgera et al., 2014, p. 120),[52] i.e. local communities[53] whose rights could be extended based on the inseparable link between territory, genetic resources, TK and cultural identity, and their role in the preservation of biodiversity,[54] a reading that is now reinforced by the centrality of (communal) land in the UN Declaration on the Rights of Peasants and Other People Working in Rural Areas (Errico & Claeys, 2020; Le Teno et al., Forthcoming).[55]

Scope of the Book: Key Themes in (Biocultural) Community Protocols and Biocultural Rights

Genesis: From Cultural Protocols to Biocultural Community Protocols

The preceding pages provide the context within which BCPs or CPs first appeared in scholarly work and in international law. Their advent is inseparable from the debates around ABS law on how to have "a more level-playing field among the parties" (Morgera et al., 2014, p. 222) in negotiating PIC and MATs, in a context of growing disagreement over practices denounced as "biopiracy"

(Mgbeoji, 2006; Robinson, 2011), blatant violations of customary law, or seen as irremediably tainted by power asymmetries. These three themes recurred several times throughout the *travaux préparatoires* to the Nagoya Protocol (see Table 1.1, Annex).

BCPs/CPs emphasise a community's customary rules and laws and its cultural heritage, while at the same time making visible and explicit the local norms to be followed for the negotiations. Hence, they have been heralded as powerful tools for achieving substantive equity (e.g. a just distribution of benefits taking IPLCs' perspectives into account) and procedural fairness (e.g. through a process avoiding misunderstanding, allowing enough time and money) in respect of access to genetic resources and associated TK (Raven, 2006, p. 14). Still in the context of the CBD and the Nagoya Protocol, BCPs/CPs have been showcased as useful instruments to harness the potential of IPLCs in biodiversity conservation.

Following heated debates and strong advocacy from the African Group (via Namibia's representation) and active non-profit organisations, CPs were eventually transcribed into the protocol which was signed on 29 October 2010 in Nagoya (see Art. 12, for example). The concept of BCP/CP itself was introduced into international negotiations by the active role of Natural Justice, an NGO based in South Africa which operates globally for the conservation and sustainable use of biodiversity through the self-determination of IPLCs (Bavikatte & Jonas, 2009; Bavikatte & Robinson, 2011).

Although BCPs/CPs are new tools, they are commonly presented as things that have existed since time immemorial. Indeed, for their promoters, BCPs/CPs are nothing but emanations of these customary rules and procedures through which a community usually regulates conducts and interactions both within and outside the community (Shrumm & Jonas, 2012). Some have indeed been in place for centuries, but it remains difficult to isolate them from their wider socio-cultural context, embedded as they are in different fields, including religion, personhood, kinship, medicine, agriculture, and so on. They might find an expression in different media such as songs, dances, carvings, drawings, and oral traditions, despite the fact that *even so translated* and *mediatised*, they may remain elusive and ungraspable for those who do not belong to these communities. In sum, drafting a protocol always implies a complex process of translation, all the more so, given that BCPs/CPs are immersed in global environmental politics and run a greater risk of uncontrolled equivocations (Blaser, 2009a; Viveiros De Castro, 2004).

In fact, as the last point indicates, the novelty and distinctiveness of BCPs/CPs lie in the fact that IPLCs are increasingly engaging with external stakeholders such as government agencies, international organisations, researchers, NGOs, and private companies. Engagements with IPLCs, though at times adequate and in line with communities' procedures/protocols, priorities, and visions, nevertheless quite often prove to be driven by outsiders' goals and defined exclusively according their own terms. As a result, "communities often have to act defensively in response to imposed plans or threats" (Shrumm & Jonas, 2012, p. 13).

One way of limiting the risk of IPLCs' local rules and procedures from being subverted or ignored, and thus their tangible and intangible heritage from being misappropriated, has been suggested, namely to articulate these communities' protocols in a form that can be acknowledged by external actors (Anderson, 2015, p. 776). There has been talk and initiatives have been taken about new protection mechanisms, especially for cultural heritage, in Canada and Australasia for decades. Among the vast array of non-legal (*contra*, Bowrey, 2006) – or better still, "primarily ethical in nature" (Gray, 2004) – and community-level mechanisms of protection for intangible culture heritage that have been considered and proposed by scholars, "Community Research Protocols" or "(Indigenous) cultural protocols"[56] have garnered considerable attention over the years (even though the open-textured nature of the expression that covers a "diverse spectrum of instruments related to principles, processes, and rules of conduct" (Bannister, 2009, pp. 285–286) is sometimes regretted). These locally developed instruments were a response to the increasing use of the intangible cultural heritage of First Nations/Aboriginal peoples in research, technology, and mainstream culture (Anderson & Younging, 2010; Janke, 2005; Nicholas & Bannister, 2004; Raven, 2010; Riphagen & Stolte, 2016). They are without doubt the main source of inspiration for BCPs within the context of biodiversity-related contracts.

Simultaneously, while IPLCs sought to consolidate their position through environmental law, initiatives were multiplying around "Indigenous peoples' declarations and statements on equitable research relationships", "Community research agreements", "Community protocols", and "Community codes of conduct", with a view to setting the foundation for new relationships between IPLCs on the one hand, and researchers and private companies on the other, in the field of biodiversity (see, in particular, Laird, 2002). Among these initiatives, those around "community protocols" – soon to be known as BCPs – gained in popularity due to the considerable publicity given to the Intercommunity Benefit Sharing Agreement in the Parque de la Papa, in Pisac (in the Cuzco region of Peru)[57] with the support of the ANDES association (ANDES et al., 2012)[58] and the International Institute for Environment and Development (IIED),[59] before being promoted internationally by Kabir Sanjay Bavikatte and Harry Jonas, two international lawyers and the founders of the NGO Natural Justice.

Although the thread of history seems easy to follow, care should be taken not to overstate the intellectual continuities. As will be seen, the Potato Park biocultural protocol is entrenched in a specific theoretical framework and is but one part of a larger system built around the notion of biocultural heritage (Argumedo, 2008; Argumedo & Pimbert, 2008; Graddy, 2013; Hall, this volume, Chapter 3).

BCPs as Legal Tools

While community protocols are broadly defined in the Mo'otz Kuxtal Voluntary Guidelines (2016),[60] they are tightly articulated around PIC and benefit-sharing in relation to TK associated with genetic resources under the Nagoya Protocol. For legal scholars, if anything, BCPs/CPs are tools – encompassing relevant views, laws, and procedures – which must be weaved into local PIC procedures and duly acknowledged by domestic legislation along state PIC mechanisms. The common view is that states are obligated to support the implementation of, and ensure compliance with, community protocols, "giving them legal effect in national legal system with a view to ensuring compliance by users and collaboration with user countries in that endeavour" (Morgera et al., 2014, p. 356). In other words, *de lege ferenda* if not *de lege lata*,[61] BCPs ought to be an integral part of domestic ABS legislation and compliance with provisions of community protocols ought to be made mandatory (Nagoya Protocol, Art. 6(2) and 7(2)).

This effect may already be seen in some BCPs related to TEK in Peru[62] and it will probably be reflected in the recent move in Malagasy legislation, which has broadly recognised the legal status of BCPs both for genetic resources and for associated TK. Within the framework of the Darwin Initiative project on mutually supportive implementation of the Nagoya Protocol and the ITPGRFA, Bioversity International, together with international organisations and partners such as Natural Justice, worked from 2015 to 2018 to develop ABS laws and agreements that contribute to pro-poor rural development and offset the cost of conserving genetic resources in Madagascar and Benin. In Madagascar, the main result is the enactment of Decree No. 2017–066 of 31 January 2017 on the regulation of ABS arising out of the use of genetic resources. Importantly, the Decree states that,

> in cases where the subject matters of the application are resources located on lands ("*terres*") managed or occupied by private individuals, the applicant shall need to obtain PIC of any legal or natural person with power to access the land and collect the resources herein. This person may be the private owner, local custodians ("*gestionnaires*") of natural resources, or the holders of TK associated with genetic resources.
>
> *(Art. 12)*

"Local custodians of natural resources" are defined as "groups of inhabitants who legally and/or traditionally manage the resources for which access is requested and whose way of life is relevant for the conservation and sustainable use of biodiversity". The text adds that for the local custodians of natural resources and holders of associated TK, consent shall be given in the form of a written contract. This contract is established according to customary rules, traditional values, and practices as locally prescribed and cannot be contrary to statutes and regulations in force (Art. 14). Finally, a paragraph specifies that in cases where traditional values and practices are already documented in an instrument implemented by the communities, that instrument must be consulted and embedded in the contract (Art. 14). This is a clear reference to BCPs, which is unsurprising, as Natural Justice, as part of the Darwin Initiative project, has backed the efforts of Bioversity International in facilitating the development of BCPs in Madagascar.[63]

In Benin, new national guidelines for ABS arising from the use of genetic resources and associated TK were endorsed by Decree No. 2018–405 of 7 September 2018 (on national guidelines for ABS arising from the use of genetic resources and associated TK in the Republic of Benin). It states that "bio cultural community protocols" are defined as

> tools that establish a set of basic principles for the participation of Local Communities (LCs) in the ABS process and describes how to access or use traditional knowledge and genetic resources held by local communities. These protocols set out procedures that assert customary rights and emphasize the obligation of reciprocity, by involving all the parties concerned.

Article 8 of the national guidelines also provides that "[p]ositive cultural rules of local communities or Bio cultural Community Protocols shall be respected".

In these instances, BCPs/CPs clearly stand as legally binding instruments aiming to set (unambiguous) terms and conditions to governments and the private, research, and non-profit sectors willing to engage with IPLCs on their local resources and knowledge.[64]

Relying on first-hand ethnographic materials and best practices, findings and feedback from practitioners and scholars who have gained significant experience and expertise in the development and implementation of BCPs/CPs the world over, the book investigates these pioneering pieces of legislation and considers the lessons that can be learnt and implications that can be drawn about the situation of IPLCs and the future of the ABS framework.

The book also aims to look at BCPs/CPs beyond these few instances in which they are legally recognised as part of procedures for local PIC. In all these cases, the status of BCPs/CPs is unclear, as they appear to hover on the edge of ABS legislation, if not floating above formal legal systems. How then are BCPs/CPs to be read and understood within these specific contexts? Should they be approached as non-legal instruments, incentives, or ethical tools in the same way as CPs, community research protocols, and ethical codes are seen in Canada and Australasia? If so, what are their functions and aims, their underlying philosophy, and theoretical underpinnings? What is the rationale behind their creation, implementation, and content? What are the benefits expected by their advocates and proponents, together with their shortcomings and perceived risks?

BCPs and Biocultural Jurisprudence

Such questions are all the more legitimate, given that, even within the relatively narrow framework of the CBD and the Nagoya Protocol (with a primary focus on PIC and benefit-sharing) and without even considering those BCPs that have developed in relation to protected areas and land tenure (FAO, 2016; FAO et al., 2017, p. 3; Pritchard et al., 2013, p. 65), extractive industry (Makagon et al., 2016),[65] or REDD+ projects (Tyrrell & Alcorn, 2011, para. 4.7), BCPs/CPs are often broad in scope and ambition. Thus, it is not uncommon that they outline (at times with a great deal of detail) the community's core ecological, cultural, and spiritual values. They are also the vehicle through which a community strives to self-define and/or reaffirm its rights over a land or territory, although this is in no way a straightforward and frictionless process (Blaser, 2010; Ellison et al., 2009; Escobar, 1997). Even in an ABS context, BCPs/CPs can hardly be confined to a proactive or a defensive tool whereby a community either lays down in advance what rules and procedures should apply for negotiations with a researcher or a private company on access to genetic resources and TK or expounds all necessary steps to be followed internally to make decisions on access and the distribution of benefits within the community. Experience indicates, and the Mo'otz Kuxtal Voluntary Guidelines hint at this,[66] that most BCPs/CPs have a protean dimension and go beyond legal relations on PIC and benefit-sharing to encompass questions about how social actors interact, while simultaneously emphasising the community's spiritual, cultural, and reciprocal relationships with nature, or the community's role in the preservation of the environment; in other words, their "stewardship" of biodiversity (Bavikatte, 2014).

In light of the intellectual and political contexts reviewed above, the multifaceted nature of BCPs/CPs should come as no surprise. They are premised on the idea, which owes much to Darrell Posey (1999, p. 7), of a deep interconnectedness between ecological and social systems, the idea that biodiversity is part of the diversity of life, with multi-layered manifestations – biological, cultural, linguistic – which "are interrelated (and likely coevolved) within a complex socio-ecological adaptive system" (Maffi & Woodley, 2010, pp. 5–6). In sum, the "biocultural axiom" interlinks biological diversity and IPLCs' ways of life.[67] Read against this paradigmatic shift that holds biodiversity conservation and certain lifestyles together, BCPs/CPs can certainly not be reduced to mere technical and "legalistic" instruments. If anything, BCPs aim at sustaining IPLCs – to borrow language unmistakably associated with the "biological turn" – in their traditional role as traditional guardians and custodians of ecosystems.

This was, at least, the firm belief of Sanjay Kabir Bavikatte and Harry Jonas, the two founders of Natural Justice, the South African NGO which has been at the forefront of the development of BCPs for the last decade or so. There is no question that these two international lawyers have been decisive in mainstreaming the thought that IPLCs play a key role in the maintenance of world biodiversity, due to their ways of life (deemed "traditional"), world visions, and deep relationships with their lands and environment. Drawing from Posey's inspirational proposal to protect and nurture Indigenous peoples' identity and socioeconomic development alongside environmental conservation (Posey & Dutfield, 1996, p. 95), through the bestowal upon them of a "bundle of rights" (the so-called "traditional resource rights" – Posey, 2004, p. 163; Posey

& Dutfield, 1996), the two lawyers and their collaborators went on to reason that something along the same lines must be wrought to preserve and foster the stewardship role of IPLCs. Seeing the "ethic of stewardship" as the keystone of this role (Bavikatte, 2014; Bavikatte & Bennett, 2015), they further argued for providing IPLCs with a comprehensive bundle of rights, referred to as "biocultural rights" (Bavikatte, 2014; Bavikatte & Robinson, 2011; Sajeva, 2018), aimed at protecting their "traditional" lifestyles" which they saw as the bedrock of their stewardship of nature.[68] Importantly, as mentioned earlier, the basket of biocultural rights consists of three categories of rights to which is added a duty, placing them in an uneasy position within the human rights tradition (Sajeva, 2015): (i) rights to land, territory, and natural resources, i.e. the right to access and use traditional lands and territories, and access to and use rights over biotic and abiotic resources present in the land; (ii) rights to self-determination, i.e. self-governance, itself comprising two strands: (a) the aptitude of each people and community to regulate their internal matters through the use of their traditional legal institutions and rules (Anaya, 2004); (b) protection from the imposition of decisions taken by external actors and regarding matters that can influence the community (e.g. exploitation of resources by extractive industry); (iii) cultural rights, including the rights and conditions necessary to safeguard the integrity of the values, worldviews, institutions, practices, and knowledge of IPLCs. These rights are accompanied by a duty of sustainability, stemming from the stewardship ethic at the core of biocultural rights (Sajeva, 2015). The political pathway already being tread by the PIC principle can be discerned here: from a procedural guarantee to an engagement with foundational issues of sovereignty. Here, much like concentric circles expanding from their central procedural core, CPs/BCPs appear to expand the scope of their politico-legal claim from PIC to cultural identity, right to lands, territories and resources, and normative and institutional autonomy (Gilbert, 2016, p. 69; Morgera et al., 2014, p. 37).

The fact that biocultural jurisprudence and the "bundle of rights" approach underwrite past and current reflections on BCPs/CPs is clearly visible throughout the case studies included in the book. As several examples presented below illustrate, current protocols in the context of ABS frequently outline communities' core ecological, cultural, and spiritual values, as well as their stewardship of nature, and generally do more than simply lay down the rules and procedures to be followed in negotiations with researchers or private companies on access to genetic resources and TK.

What, though, are the implications of BCPs/CPs as an outgrowth of biocultural jurisprudence? Should they be understood as a challenge to "the fragmentary nature of state law" (Bavikatte, 2014, p. 234), or a claim to a non-Westphalian sovereignty (Lenzerini, 2006, pp. 22–23) based "on special dependency on and attachment to the land",[69] biocultural heritage (Shaheed, 2015, para. 35), the right to science and culture (Shaheed, 2012, para. 65), stewardship of biodiversity, right to food[70] and food sovereignty (Edelman et al., 2014),[71] and the right to seed?

The importance of the question cannot be overstated. The answer is key to a clear understanding of the recent overgrowth in bottom-up, biocultural-based, and NGO-driven initiatives helping IPLCs and farmers/peasants to regain political and legal spaces to control their territory, protect their cultural heritage, and continue to use, trade in, and access traditional crop varieties, semi-domesticated varieties, and wild relatives. A few examples illustrate this trend: the "territories and areas conserved by indigenous peoples and local communities" (ICCAs) developed by the ICCA Consortium (Kothari et al., 2012; Stevens, 2014, p. 71)[72]; the Globally Important Agricultural Heritage Systems of the FAO (Harrop, 2009; Santilli, 2016, pp. 289–294); the recent "other effective area-based conservation measures" discussed at the CBD (Jonas et al., 2014)[73]; the labelling system for biocultural heritage-based products (Swiderska et al., 2016); and the TK and Biocultural (BC) Labels to address issues of Indigenous data sovereignty (Anderson & Hudson, 2020; Liggins et al., 2021). These initiatives, some of which have gained high-profile attention and even support in international fora,[74] have in common some distinctive features which are worth stressing: an emphasis on culturally important practices relevant to the maintenance of genetic resources and TK; the importance of social organisation and self-governance through community-level decision-making processes; attachment to land and territories grounding an "ethic of stewardship" (Bavikatte & Bennett, 2015; Mulrennan & Bussières, 2020), itself defining a set of rights and duties between community members and towards non-humans. In sum, all these initiatives share the basic premise that "[...] the conservation of Nature is a result of a holistic way of life" (Bavikatte, 2014, p. 233; also see Reyes-García et al., 2021).

At the same time, as the reference to the Potato Park suggests, we should be wary of over-reliance on too narrow a genealogy of BCPs/CPs. For instance, the work by Darrell Posey, Graham Dutfield, and Alejandro Argumedo on "biocultural heritage" has left a deep mark on current debates on biodiversity conservation and continues to influence international actors, such as the IIED[75] which was involved in the Potato Park project and which continues to develop BCPs worldwide, but as part of a broader strategy revolving around the concept of "Indigenous Biocultural Territories" (Argumedo & Pimbert, 2008). This begs the question of the different influences that may shape the development of BCPs/CPs and how to make sense of and do justice to them.

As a recent opinion of the Mexican Committee on the Environment, Sustainability, Climate Change and Natural Resources (*Comisión de Medio Ambiente, Sustentabilidad, Cambio Climático y Recursos Naturales*), including a draft decree on the protection of biocultural heritage in Mexico, testifies, "biocultural heritage" is understood as the "legacy made up of the environment, the culture and the territory in a reciprocal relationship".[76] The idea of "heritage" certainly carries a holistic way of life (ANDES et al., 2012) and sustains a *praxis* which is grounded in "ethical doings" (Puig de la Bellacasa, 2017, p. 151) rather than following a predefined set of ethical principles – insofar as it ushers in a form of "situated ethics" (ibid. 150). In the Potato Park, the centrality of "*Sumaq kawsay*" (also spelled: "*Sumaq cawsay*") or "well-living" ("*buen vivir*" in Spanish) which stems from a balanced relationship between the three *ayllus* (*runa* or humans, *sallqas* or flora and fauna, and *awquis* or high spirits) (Argumedo & Hall, n.d.; Argumedo & Wong, 2011)[77] that make up the community stresses this "situated" ethical and holistic perspective. But what is to be seen as the mainstay of "biocultural heritage" approaches is the territory as the "base" or "foundation" (Gudeman, 2001, p. 27) which is to be constantly maintained and nurtured as it sustains the identity (Nemogá et al., 2018, p. 24) and must be bequest to future generations so as to ensure the continuity of the community. It echoes the biocultural dimension ascribed to Mexican Indigenous peoples' territories (Martínez Coria & Haro Encinas, 2015), particularly visible in the BCP of Capulálpam de Méndez, Oaxaca (Mexico) and which has a firm foothold in the concept of "communality" ("*comunalidad*") that underlines a human's belonging to the land (Martinez Luna, 2010) and an individual's embeddedness in the community (Polo & Danielson, 2013).[78]

How are these distinct theoretical foundations translated into practice? What are the main differences that can be observed in the development process[79] and in terms of the roles and functions of BCPs/CPs?

BCPs and Stewardship: "The Noble Savage" Redux

Finally, while the focus on the "ethic of stewardship" as the cornerstone of biocultural jurisprudence has been hailed as a quantum leap for IPLCs, there are concerns that this may only be a facade of consideration, masking what are in fact only new forms of colonial representation, such as the "ecologically noble savage" (Ramos, 1994; Raymond, 2007). It should not be forgotten that BCPs/CPs and biocultural jurisprudence are part of a series of shifts that began in the late 1970s, whereby Indigenous peoples "made efforts to represent, distinguish and promote their own environmental practices and development views at the national and international level" (Ulloa, 2005, p. 8), i.e. using Western language and figures. And these Indigenous peoples' environmental views "began also to be related to global environmentalism, which prompted the construction of *ecological natives*" (ibid.). This idea of the ecological native, of which the "stewardship of nature" is a mere offshoot, certainly exposes IPLCs to the risks of being subordinated to Western visions of global environmental development, or worse still, to see their rights being conditional upon demonstrating their alignment with adequate conservation practices (Reimerson, 2013, p. 993; Sajeva, 2021).[80] In particular, consideration should be given as to whether pervasive representations such as "steward of biodiversity", "ecological native", or "natural conservationist" are not used, in domestic policies and international fora, as a strategic means to shift environmental policies from the management of non-human nature to the management of people, and also to replace political negotiations with managerial interventions (Brosius, 1999; Coombe, 2011, p. 83).

This is an issue which deserves full consideration, inasmuch as:

> [...] the concepts of biodiversity and conservation are not indigenous and, indeed, are alien to Indigenous peoples. This does not mean they do not respect and foster living things, but rather that nature is an extension of society. Thus, biodiversity is not an object to be conserved. It is an integral part of human existence, in which utilization is part of the celebration of life.
>
> *(Posey, 1999, p. 7)*

As Posey further clarified, "[t]he problem then is not one of whether indigenous and traditional peoples are or are not 'natural conservationists',[81] but rather who (and how) are we to judge them?" (ibid.; also see Tauli-Corpuz, 2016, para. 15).

Despite all the negative connotations and implications of these representations, there is evidence that IPLCs have been able to use them "as means to gain recognition within national and transnational environmental discourses" and to "transform nonindigenous peoples' ideas of their identities not only within the nation-state, but also in transnational arenas" (Ulloa, 2005, p. 215; also see Coombe, 2011, pp. 89–90). This agency of IPLCs has also to be considered while exploring BCPs/CP.

Overview of the Book

All these issues and more are addressed in this book, which is the first attempt at a comprehensive study of BCPs/CPs mainly within the context of ABS legislation, with a large geographic reach, including case studies from Africa, Latin America, North America, and the Pacific.

The book is organised in three parts. The first part opens with conceptual insights aimed at providing the reader with foundational concepts such as biocultural diversity, biocultural rights, and community rule-making, a grasp of which is a prerequisite for progressing through the second and third parts of the book. In this first part, Kelly Bannister explores the principle of consent within Canadian ethics policy initiatives as an illustrative example of how the relational emphasis of biocultural ethics can complement biocultural rights-based approaches by encouraging not only justice of rights but also relational justice through right relationships. Ingrid Hall then contributes two chapters. The first delves into the Peruvian Potato Park, home to the concept of *Sumaq kawsay*, translated as "well-being", examining how a non-governmental organisation can play the role of an ontological diplomat in the CBD forum. In her second chapter, she analyses how IPLCs challenge the nature-culture dualism within the CBD through ethnographic observations during three CBD meetings in 2019: the First Global Thematic Dialogue for Indigenous Peoples and Local Communities, the 11th Meeting of the Working Group on Article 8j, and the 23rd Meeting of the

Subsidiary Body on Scientific, Technical and Technological Advice. In Chapter 5, Garrett Graddy-Lovelace addresses the contested sovereignties at work and at odds in the colonial settler and neoliberalised context of the United States/Turtle Island. Both the Atateken North American Regional Declaration on Biocultural Diversity and Recommended Actions and the TK and BC Labels initiative are mobilised to expand the emancipatory possibilities for official recognition of Indigenous knowledge as well as data and seed sovereignty, offering key insights for those working to move from the US obstructionism to Turtle Island-led decolonisation and indigenisation within the realm of agrobiodiversity governance and beyond. Giulia Sajeva concludes this first part in Chapter 6, where she explores the pros and cons of claiming biocultural rights for IPLCs as separate groups. While Indigenous peoples hold Indigenous rights, local communities are still fighting to receive protection under international law. This difference places them in dissimilar positions regarding the decision as to whether to frame their claims as biocultural rights claims or not.

The second part of the book moves on to community protocols within the ABS context, while taking a glimpse into the nature and role of community protocols beyond issues of access to genetic resources and TK. In so doing, it strives to investigate, through a thorough review of specific cases drawn from field-based research and/or work with communities around the world, the content and functions of BCPs/CPs. Comprehensive chapters also explore the negotiation process and raise stimulating questions about the role of international brokers and the way multilateral lending agencies, donor institutions, and conservation organisations can use BCPs/CPs as disciplinary tools for national and regional planning or to serve powerful institutional interests. In this second part, Pía Marchegiani and Louisa Parks explore the case of the 2015 Kachi Yupi (Tracks in the Salt) community protocol produced by Kolla and Atacama communities in the Salinas Grandes and Laguna de Guayatayoc areas of northwest Argentina. The case suggests the need for favourable political contextual conditions for community protocols to lead to formal legal change, but also underlines that their potential for underpinning collective action goes beyond this, i.e. by expanding the action repertoires of local communities and strengthening them as collective actors. Chapter 8 outlines a participatory action research project involving communities in Vanuatu and Cook Islands. Miri (Margaret) Raven and Daniel Robinson explore biopiracy and technological innovations as drivers for the use of protocols to protect Indigenous knowledge of genetic resources. They identify challenges in developing and implementing community protocols and caution against essentialising Indigenous peoples, their knowledge, and cultural practices. In Chapter 9, Leslé Jansen and Rayna Sutherland sketch the development, within a colonial and apartheid context, of the Rooibos Biocultural protocol, a legal affirmation of San and Khoikhoi Community's rights that was not recognised in South African ABS legislation. They show that a gap exists between what the BCP affirms and what their lived reality is, which is due to the collective history of trauma, dispossession, pain, disruption, and a legacy of oppression, rather than of their being illegitimate. In Chapter 10, Fabien Girard and Manohisoa Rakotondrabe continue with a Malagasy case. Their research focused on the BCP of the farming communities of Analavory (Madagascar). Their analysis draws on the concept of "boundary work" to describe how the different actors involved negotiate the boundaries between different social worlds; and why this fell short of producing the expected result. They illustrate significant misunderstandings about the exact scope of the protocol and mismatches between vernacular representations of plants and TK and definitions of plant genetic resources as enshrined in the international regime on ABS.

Finally, the third part of the book addresses the subject introduced in the preceding part and endeavours to open up new avenues. In particular, it considers whether BCPs/CPs, notably through their emphasis on "stewardship of nature" and "tradition", can be seen as political tools and representational strategies used by IPLCs in their struggle for greater rights to their land, territories, and resources, and for greater political space, or whether they are problematic arrangements that constrain IPLCs within the Western imagination, without any hope of them taking their fate into their own hands and reconstructing their identities according to their own visions. In this final part, Reia Anquet and Fabien Girard thoroughly analyse seven BCPs to deconstruct the optimistic vision of BCPs as a support tool for IPLC's role in biodiversity conservation and maintenance, protecting their ways of life from being affected by the unfettered extension of market economies. Against the backdrop of political ontology, the chapter questions whether BPCs, with their emphasis on customary law and community rule-making, embedded with non-Western ontologies and their anchorage of the ethics of stewardship, should be read as attempts at challenging Western worldviews and dominant forms of legal modernity. This chapter argues that maintaining IPLCs' identities within the ABS framework would imply a complete overhaul of Western/European legal constructs and their naturalistic underpinnings.

Finally, the book ends with a concluding chapter by Fabien Girard, Christine Frison, and Ingrid Hall, which encapsulates the main lessons from this collective work following the three parts of the book. To open up reflection on future avenues, it proposes a typology identifying four possible functions of BCPs/CPs: as purely technical instruments; as a call for the recognition of local procedures in ABS; as political claims; and as political-ontological claims. This categorisation highlights the multifaceted nature of BCPs/CPs, ranging from technical instruments to tools that the authors venture to place under the heading of political ontology. Above all, the chapter stresses the heuristic virtue of the typology that enables the identification of impending areas of work to give IPLCs the place to which they lay claim in conservation and development policies.

Annex

TABLE 1.1 BCPs and cross-cutting themes in the Nagoya Protocol negotiations (Fabien Girard, original material for the book)

Cross-cutting theme in the Nagoya Protocol negotiations	*Meeting*	*Example*
Biopiracy/ misappropriation	CBD, COP, Decision VIII/4, Annex International Regime on Access and Benefit-Sharing, UNEP/CBD/COP/8/31, 15 June 2006	p. 135
	"Paris Draft" – Report of the Seventh Meeting of the Ad Hoc Open-Ended Working Group on Access and Benefit-Sharing, UNEP/CBD/WG-ABS/7/8, 5 May 2009, Annex	p. 56
	"Montreal I Draft" – Report of the Eighth Meeting of the Ad Hoc Open-ended Working Group on Access and Benefit-Sharing, UNEP/CBD/WG-ABS/8/8, 20 November 2009	Para. 32
	Addendum. Submission by the African Group, UNEP/CBD/WG-ABS/8/3/Add.2, 30 October 2009	p. 7
Respect due to customary law and traditional values	Report of the International Indigenous and Local Community Consultation on Access and Benefit Sharing and the Development of an International Regime, UNEP/CBD/WG-ABS/5/INF/9, 19 September 2007	Para. 29
	CBD, COP, Decision IX/12, Annex II, UNEP/CBD/COP/9/29, 9 October 2008	p. 120
	Compilation of submissions by Parties, governments, international organisations, Indigenous and local communities and relevant stakeholders in respect of the main components of the international regime on Access and Benefit-Sharing listed in Decision IX/12, annex I, UNEP/CBD/WG-ABS/7/INF/1, 2 February 2009	Submission from Namibia on Behalf of the African Group (p. 72)
	Study on Compliance in Relation to the Customary Law of Indigenous and Local Communities, National Law, Across Jurisdictions, and International Law, UNEP/CBD/WG-ABS/7/INF/5, 6 March 2009	p. 19

(Continued)

Cross-cutting theme in the Nagoya Protocol negotiations	*Meeting*	*Example*
	"Paris Draft" Report of the Seventh Meeting of the Ad Hoc Open-Ended Working Group on Access and Benefit-Sharing, UNEP/CBD/WG-ABS/7/8, 5 May 2009, Annex	p. 32
	Report of the meeting of the group of technical and legal experts on traditional knowledge associated with genetic resources in the context of the international regime on Access and Benefit-Sharing, UNEP/CBD/WG-ABS/8/2, 15 July 2009	Para. 60. Para. 34
	Addendum. Submission by The African Group, UNEP/CBD/WG-ABS/8/3/Add.2 30 October 2009	p. 4
	"**Montreal I Draft**" – Report of the Eighth Meeting of the Ad Hoc Open-ended Working Group on Access and Benefit-Sharing, UNEP/CBD/WG-ABS/8/8, 20 November 2009, Annex I	p. 32, pp. 65–66
	Report of the First Part of the Ninth Meeting of the Ad Hoc Open-Ended Working Group on Access and Benefit-Sharing Contents, UNEP/CBD/WG-ABS/9/3, 26 April 2010	Paras 27, 35, 143
	"**Cali Draft**" – "Revised Draft Protocol on Access to Genetic Resources and the Fair and Equitable Sharing of Benefits Arising from their Utilization to the Convention on Biological Diversity", Report of the First Part of the Ninth Meeting of the Ad Hoc Open-Ended Working Group on Access and Benefit-Sharing Contents, UNEP/CBD/WG-ABS/9/3, 26 April 2010, Annex I	Preambular clause 11, Art. 9
	Report of the Second Part of the Ninth Meeting of the Ad Hoc Open-ended Working Group on Access and Benefit-Sharing, UNEP/CBD/COP/10/5/Add.4, 28 July 2010	Para. 73

Cross-cutting theme in the Nagoya Protocol negotiations	*Meeting*	*Example*
	"Montreal II Draft" – Meeting of the Interregional Negotiating Group, UNEP/CBD/WG-ABS/9/ING/1, 21 September 2010, Annex	Art. 9
Balanced negotiations	CBD, COP, Decision IX/12, Annex II, UNEP/CBD/COP/9/29, 9 October 2008	p. 117
	Collation of Operative Text Submitted by Parties, Governments, International Organisations, Indigenous and Local Communities and Relevant Stakeholders in Respect of the Main Components of the International Regime on Access and Benefit-Sharing Listed on Decision IX/12, Annex I, UNEP/CBD/WG-ABS/7/4 28 January 2009	Namibia on behalf of the African Group (p. 32)
	Compilation of submissions by Parties, governments, international organisations, Indigenous and local communities and relevant stakeholders in respect of the main components of the international regime on Access and Benefit-Sharing listed in Decision IX/12, annex I, UNEP/CBD/WG-ABS/7/INF/1, 2 February 2009	p. 71 pp. 71–72
	Report of the meeting of the group of technical and legal experts on traditional knowledge associated with genetic resources in the context of the international regime on Access and Benefit-Sharing, UNEP/CBD/WG-ABS/8/2, 15 July 2009	Para. 43.
	Addendum. Submission by The African Group, UNEP/CBD/WG-ABS/8/3/Add.2, 30 October 2009	p. 8
	"**Cali Draft**" – "Revised Draft Protocol on Access to Genetic Resources and the Fair and Equitable Sharing of Benefits Arising from their Utilization to the Convention on Biological Diversity", Report of the First Part of the Ninth Meeting of the Ad Hoc Open-Ended Working Group on Access and Benefit-Sharing Contents, UNEP/CBD/WG-ABS/9/3, 26 April 2010, Annex I	Preambular clause 11, Art. 9

	TREATIES	BINDING PRACTICES (POST-TREATY RULES)	NGO/ CIVIL SOCIETY DECLARATIONS/ STATEMENTS	SOFT LAW
1990 1995	5 June 1992 **Convention on Biological Diversity:** Preamble. Artt. 8(j) & 10(c)	14 December 1992 **Consecration of the concept of cultural landscape:** World Heritage Committee, WHC-92/CONF.002/12, CONF 002 XIII.1-3: Revision of the Operational Guidelines		10-21 February 1992 **Caracas Action Plan 1992 (Caracas, VE, 10-21 February 1992, 1992-02-10):** Recommendation 6: "Human communities, especially those living in and around protected areas, often have important and long-standing relationships with those areas" 1994 **Revision of the IUCN protected area management categories. Revision of category V Creation of category VI** IUCN Commission on National Parks and Protected Areas with the assistance of the World Conservation Monitoring Centre, Guidelines for Applying Protected Areas Management Categories, 1994
1996 2000		10-18 May 1999 **Ramsar Convention - Guidelines for establishing and strengthening local communities' and indigenous people's participation in the management of wetlands:** Resolution VII.8, 7th Meeting of the Conference of the Contracting Parties to the Convention on Wetlands (Ramsar, Iran, 1971) held at San José, Costa Rica, 10-18 May 1999		
2001 2005	17 October 2003 **Convention on the Safeguarding of Intangible Cultural Heritage, Paris, 17 October 2003:** The preamble recognises "the deepseated interdependence between the intangible cultural heritage and the tangible cultural and natural heritage". Also see, Art. 2(1)	9-20 and 27 February 2004 **Akwé: Kon Guidelines:** "Sacred sites, and areas or places of other cultural significance may have important functions with respect to the conservation and sustainable use of biological diversity and, by extension, the maintenance of the natural resources upon which such communities rely for their well-being" (COP CBD Decision VII/16, F., Annex, UNEP/CBD/COP/DEC/VII/16)		
2006 2010	29 October 2010 **Nagoya Protocol:** Recognition of "custodians of biodiversity" (Preamble) Recognition of "the interrelationship between genetic resources and traditional knowledge, their inseparable nature for indigenous and local communities" (Preamble) Recognition of "community protocols" (Artt. 12.1, 11.3(a) and 21(i))	18-29 October 2010 **The Tkarihwaié:ri Code of Ethical Conduct to Ensure Respect for the Cultural and Intellectual Heritage of Indigenous and Local Communities:** Definition of "Traditional Guardianship/Custodianship" (COP CDB, Decision X/42, UNEP/CBD/COP/DEC/X/4, para. 20; also see paras 2, 5, 17, 18)	8-10 June 2010 **The 2010 Declaration on Bio-cultural Diversity:** "Recognizing also that biological and cultural diversity are intrinsically and inextricably linked and together hold the key to sustainable development and are critical for the achievement of the Millennium Development Goals" (COP CBD, UNEP/CBD/COP/10/INF/3, 2 September 2010) Also see, COP CDB, Decision X/20,UNEP/CBD/COP/DEC/X/20, para. 16	13 September 2007 **United Nations Declaration on the Rights of Indigenous Peoples:** Preamble: "Recognizing that respect for indigenous knowledge, cultures and traditional practices contributes to sustainable and equitable development and proper management of the environment"

FIGURE 1.1 Timeline of the diffusion of the concept of biocultural diversity in law and policy (from 1990 onwards)

NB: Post-treaty rules have legal validity whenever bodies adopting them have been delegated some normative powers by States that are parties to constitutive treaties. Soft law refers to

TREATIES	BINDING PRACTICES (POST-TREATY RULES)	NGO/ CIVIL SOCIETY DECLARATIONS/ STATEMENTS	SOFT LAW
		25 May 2012 **Montpellier Statement: Statement in Support of Indigenous, Traditional and Local Peoples in Policy and Decision-making for Addressing the Global Environmental Challenges,** 13th Congress of the International Society of Ethnobiology, Montpellier, May 2012	
		November 2014 **ICOMOS, Florence Declaration on Heritage and Landscape as Human Values (2014):** Declaration of principles and recommendations adopted to highlight the cultural value of heritage and landscapes with a view to promoting democratic and peaceful societies (para. 2.2.b)	**9-14 December 2013** **Definition of "Nature" & biocultural diversity (IPBES):** Decision IPBES-2/4, Conceptual framework for the Intergovernmental Science-Policy Platform on Biodiversity and Ecosystem Services, IPBES/2/17 Also see, IPBES, Annex II to decision IPBES-4/1 (pollinators), IPBES/4/19, 29 March2016
12 December 2015 **Paris Agreement:** Preamble: "noting the importance of ensuring the integrity of all ecosystems, including oceans, and the protection of biodiversity, recognized by some cultures as Mother Earth, and noting the importance for some of the concept of 'climate justice', when taking action to address climate change"	**13-17 December 2017** **Definition of biocultural diversity (CDB):** WG Art. 8(j), 10/2.Glossary of relevant key terms and concepts within the context of Article 8(j) and related provisions, CBD/WG8J/REC/10/2. Also see, COP CBD, Decision 14/13, CBD/COP/DEC/14/13, 30 November 2018	**27-29 October 2016** **Ishikawa Declaration on Biocultural Diversity:** First Asian Conference on Biocultural Diversity, 27-29 October 2016, UNESCO-SCBD Joint Programme on the Links between Biological and Cultural Diversity, United Nations UNESCO, CBD Secretariat, United Nations University, and the Ishikawa Prefectural Government	
		22-24 November 2018 **Sharm El-Sheikh Declaration on Nature and Culture:** CBD/COP/14/INF/46, 25 November 2018. UNESCO-SCBD Programme, The nature and culture summit. Biological and cultural diversity for humanity and biodiversity resilience: Achieving the sustainable development goals and the Paris Agreement, Sharm El-Sheikh, Egypt, 22-24 November 2018	**17 December 2018** **United Nations Declaration on the Rights of Peasants and Other People Working in Rural Areas, Resolution 73/165:** Preamble: "Recognizing the special relationship and interaction between peasants and other people working in rural areas and the land, water and nature to which they are attached and on which they depend for their livelihood"
	30 June-10 July 2019 **Revision of the Operational Guidelines for the Implementation of the World Heritage Convention. Revision of the "integrity" criteria:** "Biological diversity and cultural diversity can be closely linked and interdependent and human activities, including those of traditional societies, local communities and indigenous peoples, often occur in natural areas" UNESCO, World Heritage Committee, Decision COM 11A,WHC/19/43.COM/ 18, 23 July 2019 8	**May 2019** **The North American Regional Declaration on Biocultural Diversity:** North American dialogue on biological and cultural diversity, 5-8 May, 2019, Montreal, Canada; Center for Indigenous Conservation and Development (Joint program "Linking Biological and Cultural Diversity" between UNESCO and the SCBD)	

instruments which, as such, are not legally binding, but whose content may be binding by virtue, for instance, of their customary status. Statements or declarations adopted by civil society or NGOs are not legally binding, but their political force may be considerable and prompt changes in law.

Notes

1 The authors would like to thank Kim Mazenot and Benjamin Coudurier for their research assistance. *All URLs retrieved on 1 September 2021.

2 COP CBD, Decision XIII/18, CBD/COP/13/25 (17 December 2016). The Mo'otz Kuxtal Guidelines indicate that they "do not apply" to traditional knowledge associated with genetic resources under the Nagoya Protocol on Access to Genetic Resources and Benefit-Sharing (see Morgera, 2018).

3 Convention on Biological Diversity, adopted in Rio de Janeiro, Brazil, on 5 June 1992 and entered into force on 29 December 1993 (1760 UNTS 79). 196 contracting parties.

4 COP CBD, Decision X/42. The Tkarihwaié:ri Code of Ethical Conduct to Ensure Respect for the Cultural and Intellectual Heritage of Indigenous and Local Communities, UNEP/CBD/COP/10/27 (20 January 2011), section 20.

5 Corte Constitucional de Colombia, *Sentencia de revisión de tutela* T-622/16 (2016).

6 https://www.ethnobiology.net/wp-content/uploads/Decl-Belem-Eng-from-Posey.pdf

7 Convention on Biological Diversity, Rio de Janeiro, 5 June 1992. The second of the three Rio Convention, the UN Convention to Combat Desertification in those Countries Experiencing Serious Drought and/or Desertification, Particularly in Africa, Paris, 14 October 1994, also place a strong emphasis on the need to

> protect, integrate, enhance and validate traditional and local knowledge, know-how and practices ensuring, subject to their respective national legislation and/or policies, that the owners of that knowledge will directly benefit on an equitable basis and on mutually agreed terms from any commercial utilization of it or from any technological development derived from that knowledge.
>
> *(Art. 17(c))*

On the tension between "protecting" and "promoting" (in particular through documentation and inventories for the purpose of environmental objectives), see Savaresi, 2018, p. 39.

8 By convention, throughout the introduction, we use the designation instigated with the CBD, bearing in mind that Indigenous peoples and local communities (the latter being very diverse under the law – think of "peasants", "farmers", "forest-dependent communities" (International Tropical Timber Agreement, 2006, Art. 1, r) – also referred to as "ethnic groups living in forests" (CERD, UN Doc. CERD/C/THA/CO/1-3)) have expressed different demands, pursued divergent agendas, and secured different rights under international and domestic laws (Suagee, 1999; and Sajeva, this book). At the same time, it cannot be denied that (i) there is a trend towards extending Indigenous peoples' protection to non-Indigenous communities (see below); (ii) small-scale communities that "do not fit the strict test of indigeneity" (Bessa, 2015, p. 332), nevertheless, tend to be defined according to criteria that resemble those retained for Indigenous peoples (UNEP/CBD/WG8J/7/8/Add.1, Annex "Advice and Recommendations Arising From the Expert Group Meeting of Local-Community Representatives"; also see Knox (2017, paras 52–58, 2018, para. 48)); (iii) firm foothold of Indigenous rights into the "indigenist rhetoric" has been strongly objected to and an alternative foundation has been sought. "Stewardship" rhetoric may offer such an alternative, with potentially less damaging imagery. Beyond this, and as Giulia Sajeva puts it, "[b]iocultural rights rhetoric can sound more politically neutral and hence be more widely accepted [...]" (Sajeva, 2018, p. 126; also see Uddin et al., 2018, p. 6). As Sajeva nevertheless stresses in Chapter 6 of this book, such move could prove hazardous.

9 Legal recognition of ownership is limited to just 10%. It is well established that the "gap between customary rights and legal title is largest in sub-Saharan Africa" (Oxfam et al., 2016, p. 27).

10 The main outcome is the Agreement on Trade-Related Aspects of Intellectual Property Rights (TRIPS), Annex 1C of the Marrakesh Agreement Establishing the World Trade Organization, signed in Marrakesh, Morocco on 15 April 1994.

11 In the wake of the *Diamond v. Chakrabarty* (447 U.S. 303, 1980) case, a first patent with claims covering maize seed, plants, and tissue culture was confirmed by the Patent Office's Board of Appeals in 1985 (*Ex parte Hibberd*, 227 U.S.P.Q. 443, 1985). Throughout the CBD negotiations, the diffusion of patent on life forms, biotechnologies, and the GATT are seen as worrying trends (UNEP/Bio.Div.3/6 20 June 1990, paras 4 and 5), feared even by pro-IPRs economists (Barton & Christensen, 1988, p. 341).

12 ABS relies on a domestic procedure for requiring and granting "prior and informed consent" (PIC), in addition to the requirement to establish "mutually agreed terms" (MAT) for the purpose of benefit-sharing (Morgera et al., 2014, p. 15).

13 UNEP/Bio.Div.3/6 20 June 1990, para. 9; drawn from UNEP/Bio.Div.3/Inf.4, para. 40.

14 UNEP, Governing Council, Decision 15/34 of 25 May 1989, (A/44/25), p. 161.

15 See footnotes 10 and 11.

16 See the Bellagio Declaration from the 1993 Rockefeller Conference "Cultural Agency/Cultural Authority: Politics and Poetics of Intellectual Property in the Post-Colonial Era": https://case.edu/affil/sce/BellagioDec.html.

17 A/CONF.151/PC/100/Add.21, para. 93h; A/CONF.151/PC/104, para. 59.

18 See, for instance, post-Rio Conference, the emblematic Mataatua Declaration on Cultural and Intellectual Property Rights of Indigenous Peoples, June 1993 (International Conference on the Cultural and Intellectual Property Rights of Indigenous Peoples (1st 1993: Whakatane, New Zealand), E/CN.4/Sub.2/AC.4/1993/CRP.5.

19 UNEP/Bio.Div.3/3, 12 June 1990, para. 9, ii; UNEP/Bio. Div. 3/12, 13 August 1990, Annex I, para. 8; UNEP/Bio.Div/WG.2/1/4), 28 November 1990, p. 39; A/CONF.151/PC/100/Add.13, paras 41, 138, 141, 143, 144. "Participation" was the new buzzword for international financial institutions (World Bank, 1989, p. 37) and intergovernmental organisations (see Peet & Watts, 1996, p. 25). See, for instance, United Nations Convention to Combat Desertification in those Countries Experiencing Serious Drought and/or Desertification, Particularly in Africa, Paris, 14 October 1994.

20 UNEP/Bio.Div.3/6 20 June 1990, paras 1 and 2; also see Barton & Christensen (1988); Wood (1988).

21 COP CBD, Decision III/18, UNEP/CBD/COP/3/38; Decision IV/10, UNEP/CBD/COP/4/27 ("A. Incentive measures: consideration of measures for the implementation of Article 11"); Decision V/5, UNEP/CBD/COP/5/23, para. 5; Decision V/15, UNEP/CBD/COP/5/23; Decision VI/15, UNEP/CBD/COP/6/20; Decision VII/12, UNEP/CBD/COP/7/21, Annex. Addis Ababa Principles and Guidelines for the Sustainable Use of Biodiversity, Practical principle 12.

22 By the same token, see the submission to WIPO by a group of Brazilian shamans:

> As traditional indigenous peoples who inhabit diverse ecosystems, we possess knowledge on the sustainable management and use of this biological diversity. The knowledge is collective and is not a commodity that may be commercialized as any good in the market. Our knowledge on biodiversity is not separate from our identities, our laws, our institutions, our system of values and our cosmological view as indigenous peoples [...].
>
> *(WIPO, 2001, para. 2)*

23 In the area of crop genetic resources and landraces held by farmers, see Brush (1993).

24 In spite of these reservations, the number of *sui generis* regimes for the intellectual property protection of TK (and also traditional cultural expressions) has swelled in recent years, as the following compilation from WIPO testifies: https://www.wipo.int/export/sites/www/tk/en/resources/pdf/compilation_sui_generis_regimes.pdf.

25 For Indigenous peoples, a first breakthrough was the ILO Convention Concerning Indigenous and Tribal Peoples in Independent Countries (ILO No. 169), (1989), Art. 23, paras 1, 13, 15, & 27. Also see, in 1985, the joint work of UNESCO and WIPO on the protection of expressions of folklore against illicit exploitation and other prejudicial actions (UNESCO & WIPO, 1985). In parallel, the painstaking work towards the 2007 UN Declaration on the Rights of Indigenous Peoples (Resolution A/RES/61/295) was initiated at the 34th session of the Sub-Commission on Prevention of Discrimination and Protection of Minorities where, by resolution 2 (XXXIV) of 8 September 1981, it recommended to the Economic and Social Council that it had decided to authorise the Sub-Commission to establish a Working Group on Indigenous Populations. This was endorsed by the Commission on Human Rights in Resolution 1982/19 of 10 March 1982 and authorised by the Economic and Social Council, Resolution 1982/34 of 7 May 1982.

26 Nagoya Protocol on Access to Genetic Resources and the Fair and Equitable Sharing of Benefits Arising from their Utilization to the Convention on Biological Diversity, signed in Nagoya, Japan, on 29 October 2010 (UNEP/CBD/COP/DEC/X/1; 3008 UNTS 3), entered into force on 12 October 2014.

27 This has inspired an expansive literature on natural areas, protected sites, and Indigenous peoples. For recent examples, see Liljeblad & Verschuuren (2019); Stevens (2014).

28 UNESCO Convention concerning the Protection of the World Cultural and Natural Heritage (World Heritage Convention) (1972), adopted the same year as the 1972 Stockholm Declaration (A/CONF.48/14/Rev.1). For a critical review, see Blake (2015, p. 115) and Larsen & Wijesuriya (2017). Despites these great strides forward, significant challenges remain in the implementation of these new approaches by IUCN member organisations (see the recent and comprehensive appraisal by Tauli-Corpuz, 2016, paras 42–49). Over the past decades, conflicts have increased on the impact of protected areas with World Heritage statute on IPLCs. See in the context of the inscription of Lake Bogoria National Reserve in Kenya on the World Heritage List: Resolution on the Protection of Indigenous Peoples' Rights in the Context of the World Heritage Convention and the Designation of Lake Bogoria as a World Heritage site, ACHPR/Res. 197, 5 November 2011; IUCN, Implementation of the United Nations Declaration on the Rights of Indigenous Peoples in the context of the UNESCO World Heritage Convention, WCC-2012-Res-047-EN. Also see Special Rapporteur of the Human Rights Council on the rights of Indigenous peoples Tauli-Corpuz (2016, paras 60–64) and Shaheed (2011, para. 80(c)).

29 "These communities are the repositories of vast accumulations of traditional knowledge and experience that links humanity with its ancient origins" (Brundtland & World Commission on Environment and Development, 1987, p. 74).

30 See note 25.

31 See note 25.

32 See ILO Convention No. 169 of 1989, Preamble, recital 6, where the "distinctive contributions" of Indigenous peoples "to the cultural diversity and social" and "ecological harmony of humankind" are acknowledged. Art. 13 further stresses the "special importance for the cultures and spiritual values of the peoples concerned of their relationship with the lands or territories" (see Boer, 2020).

33 See note 6.

34 CBD, Preamble.

35 CBD, Art. 8(j).

36 Prior to 1996, mentions are scarce; see Baer (1989) and Moran (1993). In a 1988 chapter, Posey spoke of Indigenous peoples' "biocultural environments" (Posey, 1998; Posey & Plenderleith, 2004, p. 130).

37 https://ucjeps.berkeley.edu/Endangered_Lang_Conf/Endangered_Lang.html

38 CBD/WG8J/REC/10/2 16 December 2017, p. 6; CBD/COP/DEC/14/13, 30 November 2018: "*Biocultural diversity* is considered as biological diversity and cultural diversity and the links between them" (original emphasis).

39 Decision IPBES-2/4, Conceptual framework for the Intergovernmental Science-Policy Platform on Biodiversity and Ecosystem Services, IPBES/2/17. Also see IPBES, Annex II to decision IPBES-4/1 (pollinators), IPBES/4/19, 29 March 2016, p. 39:

> A number of cultural practices based on indigenous and local knowledge contribute to supporting an abundance and diversity of pollinators and maintaining valued 'biocultural diversity' (for the purposes of this assessment, biological and cultural diversity and the links between them are referred to as 'biocultural diversity').

40 As early as 1971, UNESCO launched the Man and the Biosphere (MAB) Programme, an interdisciplinary programme of research and training aimed at developing the basis for sustainable use and conservation of biological diversity, and at improving relationships between people and their environment (Reed & Price, 2020; Heinze & German MAB National Committee, 2005). Another flagship UNESCO programme, LINKS ("Local and Indigenous Knowledge Systems"), was established in 2002. Cross-cutting and interdisciplinary in nature, the LINKS programme has three main prongs: the active and equitable role of IPLCs in resource management; local and Indigenous knowledge transmission across generations; and the inclusion of local and Indigenous ecological knowledge in biodiversity conservation and management, climate change assessment and adaptation (see, e.g., Nakashima et al., 2018). Strong institutional links exist between the programme and the CBD, the UN Framework Convention on Climate Change (UNFCCC) (1992), the Intergovernmental Panel on Climate Change (IPCC), and the IPBES. In 2010, the Secretariat of the CBD (SCBD), as a focal point for biodiversity, and UNESCO, as a focal point for cultural diversity, initiated a SCBD-UNESCO joint programme on the occasion of the International Conference on Biological and Cultural Diversity: Diversity for Development-Development for Diversity (8–10 June 2010, Montreal, Canada). The conference resulted in the 2010 Declaration on Bio-cultural Diversity and the draft of the joint programme stressing the need to "enhance synergies between interlinked provisions of conventions and programmes dealing with biological and cultural diversity at relevant scales" (Report of the International Conference on Biological and Cultural Diversity for Development, UNEP/CBD/COP/10/INF/3, 2 September 2010, Annex I). The first meeting for the implementation of the programme was held in Florence (Italy) in April 2014 and produced the UNESCO-SCBD Florence Declaration (see Figure 1.1, Annex) (Agnoletti et al., 2016). Two further events in relation to the Joint Programme were the 1st Asian Conference on Biocultural Diversity (Nanao City, Ishikawa Prefecture, Japan, 27–29 October 2016) (UNEP/CBD/COP/13/INF/28, 10 November 2016, Annex I) and the First North American Dialogue on Biocultural Diversity (Montreal, Canada, 5–8 May 2019) (CBD/WG8J/11/INF/6, 12 October 2019) (see Figure 1.1, Annex). The Joint Programme has also links with the ICOMOS-IUCN joint initiative called "A culture/nature journey", formally launched at the World Conservation Congress in Honolulu in September 2016 (McIntyre-Tamwoy, 2018; Pencek, 2017).

41 See para. 2 of the 2018 Sharm El-Sheikh Declaration on Nature and Culture (CBD/COP/14/INF/46), which "[urges] the establishment of a multi-partner International Alliance on Nature and Culture, as a platform for international cooperation on links between biological and cultural diversity to achieve the global vision of humanity 'Living in Harmony with Nature' by 2050".

42 See note 26.

43 On the narrow definition of TK under the Nagoya Protocol, as against the definition enshrined in the CBD and that which is discussed at WIPO (WIPO/GRTKF/IC/40/18, 19 June 2019; WIPO/GRTKF/IC/41/INF/7, 30 December 2020; WIPO/GRTKF/IC/41/5, 30 December 2020), see Morgera et al. (2014, p. 30). Worth considering are also the Operational Directives for the Implementation of the Convention for the Safeguarding of the Intangible Cultural Heritage, adopted

by the General Assembly of the States Parties to the Convention at its second session (UNESCO Headquarters, Paris, 16 to 19 June 2008), 7.GA (2018), Art. 189 ("Knowledge concerning nature and the universe").

44 See UNEP/CBD/WG-ABS/7/INF/5, 6 March 2009, p. 5; and UNEP/CBD/WG-ABS/7/INF/1, 2 February 2009, p. 70.

45 Declaration on the Right to Development, Adopted by General Assembly resolution 41/128 of 4 December 1986 (A/RES/41/128).

46 See also the Declaration on the Right to Development (A/RES/41/128), Article 1(2).

47 This explains why states such as Canada waged a battle for the inclusion of "approval and involvement" or participation rather than PIC in the Nagoya Protocol (iisd Reporting Services, 2009, p. 9).

48 Nagoya Protocol, Preamble; Resolution 2 Plan of Implementation of the World Summit on Sustainable Development, UN Doc A/CONF.199/20 (World Summit on Sustainable Development & United Nations, 2002, para 54, (h)).

49 On the rights to lands, territories, and resources: UNDRIP, Artt. 10, 25, 26, 27 & 8(2)(b). On the rights pertaining to cultural heritage, TK, traditional cultural expressions, as well as intellectual property thereupon: UNDRIP, Art. 31 (genetic resources, seeds, medicines, knowledge of the properties of fauna and flora, oral traditions specifically fall within the purview of the text) (Stoll, 2018, pp. 310–315, 322–325; Xanthaki, 2014, pp. 224–227). Regarding cultural rights: UNDRIP, Artt. 11(1), 12, 13(1), 15 & 34 (Charters, 2018; Gibson, 2011) (Expert Mechanism advice No. 8 (2015): Promotion and protection of the rights of Indigenous peoples with respect to their cultural heritage, A/HRC/30/53, Annex). The meaning of intellectual property must be understood within the context of human right law (right to science and culture), i.e. "the human right to benefit from the protection of the moral and material interests resulting from one's scientific, literary and artistic productions" (Shaheed, 2015, para. 32). It fundamentally aims at safeguarding the "personal link between authors and their creations and between peoples, communities or other groups and their cultural heritage, as well as their basic material interests, which are necessary to enable authors to enjoy an adequate standard of living" (ibid.). For IPLCs, it is therefore closely tied to self-determination and the right to maintain and develop their culture (ibid., para. 37; Anderson, 2015, pp. 772, 775). States should "ensure availability of legal measures and remedies to ensure the control by indigenous peoples and local communities over their biocultural heritage" (Shaheed, 2015, paras 32, 144).

50 In fact, under Article 28(1) of the UNDRIP, "just, fair and equitable compensation" is inherently linked to the right to redress. This is a reparation measure which is totally disconnected to the idea of consent, normative and institutional autonomy that accompanies PIC and benefit-sharing (Morgera et al., 2014, p. 119). See also the *Case of Saramaka People v Suriname* (Preliminary Objections, Merits, Reparations, and Costs), IACtHR Series C No 172 (28 November 2007). See also Gilbert (2018, p. 80).

51 Under the UNDRIP, the FPIC regime varies depending on the nature of legislative or administrative measures or development projects likely to take place in the lands or territories of Indigenous peoples or to affect them. For measures aimed at removing Indigenous peoples from their lands or territories or for the storage or disposal of hazardous materials in the land or territories, Articles 10 and 29(2), respectively, state that "no relocation shall take place" and "no storage or disposal" "shall take place" without "their free, prior and informed consent". In contrast, Articles 19 and 32(2) both provide that "States shall consult and cooperate in good faith [...] in order to obtain their free and informed consent". It remains, as Barelli stressed, based on the negotiating history and the "spirit and normative context of the Declaration", that

> [...] Articles 19 and 32 must be approached with a certain degree of flexibility. Thus, while FPIC should not be read as conferring an overreaching right to veto, it "may mean that, on occasion, Indigenous peoples should be able to say 'no' to

proposed measures or projects, or that, similarly, States will have to provide alternative solutions which would mitigate the negative effects of the proposed plans.
(Barelli, 2018, pp. 253–254)

See also the *Case of the Saramaka People v. Suriname* (Inter-Am. C.H.R. No. 172, Ser. C (2007)), para. 134. In the individual communications of the HRComm: *Ilmari Länsman et al. v Finland*, Comm No. 511/1992, UN Doc. CCPR/C/52/D/511/1992 (1994); *Ángela Poma v Peru,* Comm No. 1457/2006, UN Doc CCPR/C/95/D/1457/2006 (2009). In the reports of the UN Special Rapporteur on the Rights of Indigenous Peoples: Report of the Special Rapporteur on the situation of human rights and fundamental freedoms of Indigenous people, James Anaya, A/HRC/12/34 (15 July 2009), para. 47; Report of the Special Rapporteur on the rights of Indigenous peoples, James Anaya, A/HRC/24/41 (1 July 2013), paras 31–36. In the reports of the Expert Mechanism on the Rights of Indigenous Peoples: Final report of the study on Indigenous peoples and the right to participate in decision-making, A/HRC/18/42 (17 August 2011), paras 22 and 23.

52 On the progressive extension of the right to FPIC to non-Indigenous communities, see The African Commission on Human and Peoples' Rights, Resolution on a Human Rights-Based Approach to Natural Resources Governance - ACHPR/Res.224(LI)2012; CESCR, China, E/C.12/CHN/CO/2 (13 June 2014), para. 31; CESCR, Mauritania, E/C.12/MRT/CO/1 (10 December 2012), para. 8; CESCR, Togo, E/C.12/TGO/CO/1 (3 June 2013), paras 26 (PFIC) and 27 (benefit-sharing).

53 As said (see note 8), criteria used to define Indigenous peoples and local communities tend to overlap (also see African Model Legislation for the Protection of the Rights of Local Communities, Farmers and Breeders and for the Regulation of Access to Biological Resources, endorsed by the 68th Ordinary Session of the Council of Ministers in 1996, Organization of African Unity, https://www.wipo.int/edocs/lexdocs/laws/en/oau/oau001en.pdf, Article 1). Over the last two decades, these similarities have justified, on a case-by-case basis, the extension of protection afforded to Indigenous peoples to local communities whenever there is a "distinct social, cultural and economic group with a special relationship with its ancestral territory", *Case of the Saramaka People v Suriname* (Inter-Am. C.H.R. No. 172, Ser. C (2007)), paras 80–84. Progress in domestic legislation has also been significant, in particular in Ecuador, Brazil, Peru and Colombia (Bessa, 2015; Huarcaya, 2018; Kania, 2016; Tocancipá Falla & Ramírez Castrillón, 2018). See also Knox (2018, Framework principle 15).

54 Laura Westra has proposed building up a single regime for "land-based communities", with the "unifying concept for all these disparate groups" being "their land/culture connection" (Westra, 2013, p. 20). Far-reaching proposals, suggesting the bestowal of collective rights to land upon any groups showing a specific relationship with a territory and natural resources, have also been made by Olivier de Schutter (De Schutter, 2010; De Schutter & Rajagopal, 2019; Morgera et al., 2014, pp. 40–41).

55 Of paramount importance is, indeed, the "right to land" (Article 17), to be read in conjunction with the "right to adequate food and the fundamental right to be free from hunger" (Article 15 – see also the allusion to the controversial concept of "food sovereignty" in the preambular part), the "right to seeds" (Article 19) (Haugen, 2020), as well as "the right to enjoy their own culture and to pursue freely their cultural development, without interference or any form of discrimination". They also have "the right to maintain, express, control, protect and develop their traditional and local knowledge, such as ways of life, methods of production or technology, or customs and tradition" (Article 26.1). The list of right-holders is expansive: not only "peasants" within the meaning of Article 1.1, but also "any person engaged in artisanal or small-scale agriculture, crop planting, livestock raising, pastoralism, fishing, forestry, hunting or gathering, and handicrafts related to agriculture or a related occupation in a rural area" (Article 1.2). It further applies to IPLCs working

on the land, transhumant, nomadic and semi-nomadic communities, and the landless, engaged in the above-mentioned activities (Article 1.3). Finally, it also applies to hired workers, including all migrant workers (Article 1.4).

56 Legal academic and novelist, Stephen Gray, arguably played a pivotal role in discussions on cultural protocols (Gray, 2004). He has subsequently investigated bioprospecting in relation to Indigenous biological resources (see Rimmer, 2015, p. 8).

57 *Acuerdo intercomunal para el reparto y distribución justa y equitativa de los beneficios derivados de la utilización por parte de terceros del patrimonio biocultural colectivo de las comunidades campesinas del Parque de la Papa*, Pisac, Cusco, 6 April 2011 (consultations began in 2007).

58 On the influence of the "Potato Park" on the negotiations of the Nagoya Protocol, see Study on Compliance in Relation to the Customary Law of Indigenous and Local Communities, National Law, Across Jurisdictions, and International Law, UNEP/CBD/WG-ABS/7/INF/5, 6 March 2009, pp. 8–9.

59 See https://biocultural.iied.org/community-biocultural-protocols.

60 See note 2.

61 De *lege ferenda* means "the law as it should be" and de *lege lata* "the law as it exists".

62 Peru has ratified the ITPGRFA and the Nagoya Protocol, and Law No. 27811/2002 (introducing a Protection Regime for the Collective Knowledge of Indigenous Peoples derived from Biological Resources) (2002) creates space for community PIC when access involves associated TK. Specifically, the text provides for the authorisation granted by an Indigenous organisation that is representative of the Indigenous peoples who hold collective knowledge, in accordance with the rules (customs and practices) that are recognised by them (Art. 2). Following the same law, Indigenous peoples or rural communities may submit an application to the *Instituto Nacional de Defensa de la Competencia y de la Proteción de la Propiedad Intelectual* (INDECOPI) in order to have certain collective knowledge related to genetic resources registered in the National Public Register or the National Confidential Register. They can also use local registers of collective knowledge (Nemogá-Soto, 2013). To date, INDECOPI has recognised 6,585 registers (https://www.gob.pe/institucion/indecopi/noticias/320687-el-indecopi-otorgo-96-registros-de-conocimientos-colectivos-sobre-el-uso-de-la-biodiversidad-a-la-comunidad-nativa-alto-mayo). Furthermore, National Decree No. 003-2009-, MINAM (2009), Art. 5, recognises and protects the rights of Indigenous peoples to make decisions concerning their innovations, practices, and knowledge associated with genetic resources. In addition, Law No. 28216 (2004) on access to biological diversity and associated collective knowledge of Indigenous peoples established a National Commission against Biopiracy. Finally, the regional norm on access to genetic resources and associated traditional knowledge in the territories of the farmers and native communities of Cusco should be noted, which essentially establishes an institutional mechanism to prevent and address cases of biopiracy in the region and promotes the conservation of and research into genetic resources (Peru, Regional Ordinance - Ordenanza del Gobierno Regional de Cusco, No. 048-2008).

63 BCPs are also expressly referred to in the draft inter-ministerial order No. .../2019 laying down implementing rules for Decree No. 2017-066 of 31 January 2017, Art. 22 and Annex III.

64 In Kenya, the Protection of Traditional Knowledge and Cultural Expressions Act, 2016 (No. 33 of 2016) provides for the protection of cultural expressions and TK, to the exclusion of genetic resources which remain within the ambit of "relevant legislations relating to genetic resources" (notably the Environmental Management and Coordination Act No. 8 (1999)) (Mwangi, 2019; Nwauche, 2017; Nzomo, 2015). In Panama, a robust framework for the protection of Indigenous peoples was laid down by Executive Decree No. 19 of 26 March 2019 Regulating Access to and Control of the Use of Biological and Genetic Resources in the Republic of Panama (Artt. 16

& 18), to be read in conjunction with Law No. 7 of 27 June 2016 Establishing the Protection of Knowledge of Indigenous Traditional Medicine (Artt. 13, 20, 22, 23, 26) (also see De Obaldia, 2005; Romero, 2005; Lixinski, 2013, p. 123).

65 For a concrete example, see the Protocolo Comunitario Biocultural para el Territorio del Consejo Comunitario Mayor del Alto San Juan Asocasan, Tado, Chocó, Colombia, 2012 (English version: https://europa.eu/capacity4dev/file/12655/download?token=iKhInQBg); and, CESCR, Ecuador, E/C.12/ECU/CO/3 (13 December 2012), para. 9 which, in the original Spanish version, refers to "*los protocolos de consultas comunitarias*" (protocols of community consultations) that should be respected in the processes of consultation (mining and hydrocarbon resource exploration).

66 See note 2.

67 See above, p. 20.

68 In reality, there are reasons to believe that Posey had arrived at the same conclusion in the early 2000s, linking IPLCs' fight for their "survival" to the conservation of the world's biological diversity (see Posey, 2001, p. 385).

69 Declaration on the Rights of Peasants and Other People Working in Rural Areas, Art. 1.1.

70 Interestingly, Mexico recently passed the *Ley Federal para el Fomento y Protección del Maíz Nativo* (Federal Law for the Promotion and Protection of Native Maize), DOF 13-04-2020. It declares the activities of production, commercialisation, and consumption of native and constantly diversified maize as a cultural manifestation in accordance with Art. 3 of the General Law on Culture and Cultural Rights (DOF 19-06-2017) (Art. 1, para. I). It further provides that

> [t]he protection of Native Maize and Maize in Constant Diversification is recognised as an obligation of the State to guarantee the human right to nutritious, sufficient and quality food, as established in the third paragraph of Article 4 of the Political Constitution of the United Mexican States.— The State shall guarantee and promote, through all competent authorities, that all people have effective access to informed consumption of Native and Constantly Diversified Maize, as well as derived products thereof, under GMO-free conditions.
>
> *(Art. 4)*

71 On biocultural approaches to food sovereignty, see Pimbert (2018); Sarrazin & Scott (2020).

72 Now endorsed by the IUCN and the CBD: Recommendation V.26 (IUCN & The World Conservation Union, 2005); Resolution 3.049 (IUCN & World Conservation Congress, 2005); (IUCN & World Commission on Protected Areas (WCPA), 2019); UNEP/CBD/COP/DEC/XI/24; CBD/COP/DEC/XIII/2, para. 7; CBD/COP/DEC/14/8 Annexe II.

73 CBD/COP/DEC/14/8, para. 2; (IUCN & World Commission on Protected Areas (WCPA), 2019).

74 See above, notes (38–39).

75 https://biocultural.iied.org/about-biocultural-heritage.

76 Dictámenes para declaratoria de publicidad de la *Comisión de Medio Ambiente, Sustentabilidad, Cambio Climático y Recursos Naturales, con Proyecto de decreto por el que se reforman y adicionan diversas disposiciones de la Ley General del Equilibrio Ecológico y la Protección al Ambiente, en materia de protección del patrimonio biocultural* (*Gaceta Parlementaria*, Año XXIV, Número 5664-IX – 2 December 2020). In the proposed Art. 3, para. XXIX of the act, "Biocultural Heritage" (*Patrimonio Biocultural*) is defined as the "Legacy made up of the environment, culture and territory in a reciprocal relationship that propitiates a landscape in equilibrium, whose attributes are heterogeneity, diversity, connectivity, stability and resilience".

77 See also Chapter 3 in this book.

78 See, this volume, Chapter 11.
79 For instance, ANDES and IIED, which facilitated the development of the Potato Park, attached great importance to the methodology of the development process and deployed a framework for research action, while considering ways to "decolonise" methodologies (Smith, 2012).
80 Also see UNESCO Convention for the Safeguarding of the Intangible Cultural Heritage (2003), art. 2.1:

> For the purposes of this Convention, consideration will be given solely to such intangible cultural heritage as is compatible with existing international human rights instruments, as well as with the requirements of mutual respect among communities, groups and individuals, and of sustainable development.

In the report of the independent expert in the field of cultural rights, Farida Shaheed (2011, para. 74) clearly refers to the limits set on cultural heritage protection by international human rights law. And yet, the aforementioned article also opens the door to environmentally conditioned cultural rights (Savaresi, 2018, pp. 41–42).
81 This is increasingly acknowledged in the field of on-farm conservation: along a continuum ranging from "intentionality by default" to "conscious intentionality" (Almekinders et al., 2019, p. 122), small-scale farmers "do not typically choose agrobiodiversity for its own sake but rather because it fits with underlying farming rationales or trait preferences" (ibid.). Intentionality is more visible at the natural landscape level where subsistence farmers, driven by their needs, operate as "multi-use strategists" and seem to "play the game of subsistence through the manipulation of ecological components and processes (including forest succession, life cycles, and movement of materials)" (Toledo, 1990, pp. 55–56, 2001, p. 460). As Toledo stressed, "[t]he acclaimed and somewhat enigmatic ecological rationality of the peasant and traditional producer is just a subsistence strategy developed within a non-commodity system of production" (Toledo, 1990, p. 57).

References

Agnoletti, M., Emanueli, F., Agnoletti, M., & Emanueli, F. (Eds.). (2016). Biocultural Diversity and Landscape in Europe: Framing the Issue. In *Biocultural Diversity in Europe* (pp. 1–18). Springer. https://doi.org/10.1007/978-3-319-26315-1

Almeida, F. (2017). *Legislative Pathways for Securing Community-Based Property Rights*. Rights and Resources Initiative.

Almekinders, C. J. M., Stone, G. D., Baranski, M., Carney, J. A., Hanspach, J., Krishna, V. V., Ramirez-Villegas, J., Etten, J. van, & Zimmerer, K. S. (2019). Socioecological Interactions Amid Global Change. In K. S. Zimmerer & S. de Haan (Eds.), *Agrobiodiversity Integrating Knowledge for a Sustainable Future* (pp. 117–143). MIT Press. https://research.wur.nl/en/publications/socioecological-interactions-amid-global-change

Altieri, M. A., Anderson, M. K., & Merrick, L. C. (1987). Peasant Agriculture and the Conservation of Crop and Wild Plant Resources. *Conservation Biology, 1*(1), 49–58.

Anaya, S. J. (2004). *Indigenous Peoples in International Law* (2nd ed.). Oxford University Press.

Anderson, J. E. (2015). Indigenous Knowledge and Intellectual Property Rights. In J. D. Wright (Ed.), *International Encyclopedia of the Social & Behavioral Sciences* (2nd ed., pp. 769–778). Elsevier. https://doi.org/10.1016/B978-0-08-097086-8.64078-3

Anderson, J., & Hudson, M. (2020). The Biocultural Labels Initiative: Supporting Indigenous Rights in Data Derived from Genetic Resources. *Biodiversity Information Science and Standards, 4*, e59230. https://doi.org/10.3897/biss.4.59230

Anderson, J., & Younging, G. (2010). *Discussion Paper on Protocols. Professional Development Meeting on Aboriginal Arts*. https://works.bepress.com/jane_anderson/11/

ANDES, the Potato Park Communities, & IIED. (2012). *Community Biocultural Protocols: Building Mechanisms for Access and Benefit Sharing Among the Communities of the Potato Park Based on Quechua Customary Norms*. International Institute for Environment and Development (IIED). https://pubs.iied.org/G03168/

Appadurai, A. (Ed.). (2013). *The Future as Cultural Fact: Essays on the Global Condition*. Verso Books.

Argumedo, A. (2008). The Potato Park, Peru: Conserving Agrobiodiversity in an Andean Indigenous Biocultural Heritage Area. In T. Amend, J. Brown, A. Kothari, A. Phillips, & S. Stolton (Eds.), *Protected Landscapes and Agrobiodiversity Values* (Vol. 1, pp. 45–58). Kaspareg Verlag, IUCN&GTZ.

Argumedo, A., & Hall, I. (n.d.). Sumaq Kawsay and Biocultural Conservation. The Experience of the Potato Park (Cusco, Peru). In C. Scott, E. Silva, & K. Sinclair (Eds.), *Indigenous Stewardship of Environment and Alternative Development*. Toronto University Press.

Argumedo, A., & Pimbert, M. (2008). *Protecting Farmers' Rights with Indigenous Biocultural Heritage Territories: The Experience of the Potato Park* (p. 12). International Institute for Environment and Development (IIED), Asociacion ANDES.

Argumedo, A., & Wong, B. (2011). *The Ayllu System of the Potato Park, Cusco, Peru*. https://www.cbd.int/traditional/doc/8jcu-01/2011-05-31-argumedo-en.pdf

Baer, A. (1989). Maintaining Biocultural Diversity. *Conservation Biology*, *3*(1), 97–98. https://doi.org/10.1111/j.1523-1739.1989.tb00233.x

Banerjee, S. B. (2003). Who Sustains Whose Development? Sustainable Development and the Reinvention of Nature. *Organization Studies*, *24*(1), 143–180.

Bannister, K. (2009). Non-legal Instruments for Protection of Intangible Cultural Heritage: Key Roles for Ethical Codes and Community Protocols. In C. Bell & V. Napoleon (Eds.), *Protection of First Nations' Cultural Heritage: Laws, Policy and Reform* (pp. 278–308). UBC Press.

Barelli, M. (2018). Free, Prior, and Informed Consent in the UNDRIP. In J. Hohmann & M. Weller (Eds.), *The UN Declaration on the Rights of Indigenous Peoples: A Commentary* (pp. 247–269). Oxford University Press.

Barton, J. H., & Christensen, E. (1988). Diversity Compensation Systems: Ways to Compensate Developing Nations for Providing Genetic Materials. In J. R. Kloppenburg (Ed.), *Seeds and sovereignty: The Use and Control of Plant Genetic Resources* (pp. 338–355). Duke University Press.

Bavikatte, K., & Jonas, H. (2009). Community Protocols in the Negotiations of the International Regime on Access and Benefit Sharing. In K. Bavikatte & H. Jonas (Eds.), *Biocultural Community Protocols. A Community Approach to Ensuring the Integrity of Environmental Law and Policy* (pp. 37–41). UNEP, Natural Justice.

Bavikatte, K., Jonas, H., & von Braun, J. (2010). Traditional Knowledge and Economic Development: The Biocultural Dimension. In S. M. Subramanian & B. Pisupati (Eds.), *Traditional Knowledge in Policy and Practice: Approaches to Development and Human Well-Being* (pp. 294–326). United Nations University Press.

Bavikatte, K. S. (2014). *Stewarding the Earth: Rethinking Property and the Emergence of Biocultural Rights*. Oxford University Press.

Bavikatte, K. S., & Bennett, T. (2015). Community Stewardship: The Foundation of Biocultural Rights. *Journal of Human Rights and the Environment*, *6*(1), 7–29. https://doi.org/10.4337/jhre.2015.01.01

Bavikatte, K. S., & Robinson, D. F. (2011). Towards a Peoples History of the Law: Biocultural Jurisprudence and the Nagoya Protocol on Access and Benefit Sharing. *Law, Environment and Development Journal (LEAD Journal)*, 7(1), 35–51.

Berkes, F. (2012). *Sacred ecology: Traditional Ecological Knowledge and Resource Management* (3rd ed.). Routledge. http://www.dawsonera.com/depp/reader/protected/external/AbstractView/S9780203123843

Berkes, F., & Berkes, M. K. (2009). Ecological Complexity, Fuzzy Logic, and Holism in Indigenous Knowledge. *Futures*, *41*(1), 6–12. https://doi.org/10.1016/j.futures.2008.07.003

Bessa, A. (2015). Traditional Local Communities: What Lessons Can Be Learnt at the International Level from the Experiences of Brazil and Scotland?: Traditional Local Communities. *Review of European, Comparative & International Environmental Law*, *24*(3), 330–340. https://doi.org/10.1111/reel.12140

Blake, J. E. (2015). *International Cultural Heritage Law*. Oxford University Press. https://ebookcentral.proquest.com/lib/adtca/detail.action?docID=4842246

Blaser, M. (2009a). The Threat of the Yrmo: The Political Ontology of a Sustainable Hunting Program. *American Anthropologist*, *111*(1), 10–20. https://doi.org/10.1111/j.1548-1433.2009.01073.x

Blaser, M. (2009b). Political Ontology: Cultural Studies without 'Cultures'? *Cultural Studies*, *23*(5–6), 873–896.

Blaser, M. (2010). *Storytelling Globalization from the Chaco and Beyond*. Duke University Press. http://site.ebrary.com/id/10412030

Blaser, M. (2013). Ontological Conflicts and the Stories of Peoples in Spite of Europe: Toward a Conversation on Political Ontology. *Current Anthropology*, *54*(5), 547–568. https://doi.org/10.1086/672270

Boer, B. (2020). Part II Substantive Aspects, Ch.14 The Environment and Cultural Heritage. In F. Francioni & A. F. Vrdoljak (Eds.), *The Oxford Handbook of International Cultural Heritage Law* (pp. 318–388). Oxford University Press. https://doi.org/10.1093/law/9780198859871.003.0014

Boisvert, V., & Vivien, F.-D. (2012). Towards a Political Economy Approach to the Convention on Biological Diversity. *Cambridge Journal of Economics*, *36*(5), 1163–1179. JSTOR.

Bonneuil, C. (2019). Seeing Nature as a 'Universal Store of Genes': How Biological Diversity Became 'Genetic Resources', 1890–1940. *Studies in History and Philosophy of Science Part C: Studies in History and Philosophy of Biological and Biomedical Sciences*, *75*, 1–14.

Bowrey, K. (2006). Alternative Intellectual Property?: Indigenous Protocols, Copyleft and New Juridifications of Customary Practices. *Maquarie Law Journal*, *6*, 65.

Bridgewater, P., & Rotherham, I. D. (2019). A Critical Perspective on the Concept of Biocultural Diversity and Its Emerging Role in Nature and Heritage Conservation. *People and Nature*, *1*(3), 291–304.

Brondizio, E. S., & Tourneau, F.-M. L. (2016). Environmental Governance for All. *Science*, *352*(6291), 1272–1273. https://doi.org/10.1126/science.aaf5122

Brosius, J. P. (1999). Green Dots, Pink Hearts: Displacing Politics from the Malaysian Rain Forest. *American Anthropologist*, *101*(1), 36–57.

Brosius, J. P., & Hitchner, S. L. (2010). Cultural Diversity and Conservation. *International Social Science Journal*, *61*(199), 141–168. https://doi.org/10.1111/j.1468-2451.2010.01753.x

Brundtland, G. H. & World Commission on Environment and Development. (1987). *Report of the World Commission on Environment and Development: 'Our Common Future'*. United Nations.

Brush, S. B. (1993). Indigenous Knowledge of Biological Resources and Intellectual Property Rights: The Role of Anthropology. *American Anthropologist, 95*(3), 653–671.
Brush, S. B. (2001). Protectors, Prospectors, and Pirates of Biological Resources. In L. Maffi (Ed.), *On Biocultural Diversity. Linking Language, Knowledge, and the Environment* (pp. 517–530). Smithsonian Institution Press.
Buergin, R. (2013). *Contested Rights of Local Communities and Indigenous Peoples in the Context of the Biocultural Turn in Environment and Development Discourses* (SEFUT Working Paper No. 16). Working Group Socio-Economics of Forest Use in the Tropics and Subtropics, Albert-Ludwigs-Universität Freiburg.
Cámara-Leret, R., & Bascompte, J. (2021). Language Extinction Triggers the Loss of Unique Medicinal Knowledge. *Proceedings of the National Academy of Sciences, 118*(24). https://doi.org/10.1073/pnas.2103683118
Castro, E. V. de. (1992). *From the Enemy's Point of View: Humanity and Divinity in an Amazonian Society* (C. V. Howard, Trans.). University of Chicago Press.
Charters, C. (2018). Indigenous Peoples' Rights to Lands, Territories, and Resources in the UNDRIP. In J. Hohmann & M. Weller (Eds.), *The UN Declaration on the Rights of Indigenous Peoples: A Commentary* (pp. 395–424). Oxford University Press.
Colchester, M. (2003). *Salvaging Nature: Indigenous Peoples, Protected Areas and Biodiversity Conservation*. World Rainforest Movement; Forest Peoples Programme.
Coombe, R. J. (1998). *The Cultural Life of Intellectual Properties: Authorship, Appropriation, and the law*. Duke University Press.
Coombe, R. J. (2011). Cultural Agencies: 'Constructing' Community Subjects and their Rights. In M. Biagioli, P. Jaszi, & M. Woodmansee (Eds.), *Making and Unmaking Intellectual Property* (pp. 79–98). University of Chicago Press.
Coomes, O. T., McGuire, S. J., Garine, E., Caillon, S., McKey, D., Demeulenaere, E., Jarvis, D., Aistara, G., Barnaud, A., Clouvel, P., Emperaire, L., Louafi, S., Martin, P., Massol, F., Pautasso, M., Violon, C., & Wencélius, J. (2015). Farmer Seed Networks Make a Limited Contribution to Agriculture? Four Common Misconceptions. *Food Policy, 56*, 41–50. https://doi.org/10.1016/j.foodpol.2015.07.008
Correa, C. (2016). Sui Generis Protection for Farmers' Varieties. In M. Halewood (Ed.), *Farmers' Crop Varieties and Farmers' Rights: Challenges in Taxonomy and Law* (pp. 115–183). Routledge.
De Obaldia, I. (2005). Western Intellectual Property and the Indigenous Cultures: The Case of the Panamanian Indigenous Intellectual Property Law. *Boston University International Law Journal, 23*(2), 337–394.
De Schutter, O. (2010). The Emerging Human Right to Land. *International Community Law Review, 12*(3), 303–334. https://doi.org/10.1163/187197310X513725
De Schutter, O., & Rajagopal, B. (2019). Conclusion. The Revival of the "commons" and the Redefinition of Property Rights. In O. De Schutter & B. Rajagopal (Eds.), *Property Rights from Below. Commodification of Land and the Counter-Movement* (pp. 203–231). Routledge. https://doi.org/10.1201/9781315621463-1
Descola, P. (1986). *In the Society of Nature: A Native Ecology in Amazonia* (N. Scott, Trans.). Cambridge University Press.
Descola, P., & Sahlins, M. D. (2014). *Beyond Nature and Culture* (J. Lloyd, Trans.; Paperback edition). The University of Chicago Press.
Doolittle, A. A. (2007). Fortress Conservation. In P. Robbins (Ed.), *Encyclopedia of Environment and Society* (pp. 704–705). Sage.
Edelman, M., Weis, T., Baviskar, A., Jr., S. M. B., Holt-Giménez, E., Kandiyoti, D., & Wolford, W. (2014). Introduction: Critical Perspectives on Food Sovereignty. *The Journal of Peasant Studies, 41*(6), 911–931. https://doi.org/10.1080/03066150.2014.963568

Egzibher, G. (1996). A Case of Community Rights. In S. Tilahun & S. Edwards (Eds.), *The Movement for Collective Intellectual Rights*. The Institute for Sustainable Foundation/ The Gaia Foundation.

Eisner, T. (1989). Prospecting for Nature's Chemical Riches. *Issues in Science and Technology, 6*(2), 31–34.

Ellison, N., Martínez Mauri, M., & International Congress of Americanists. (2009). *Paisaje, espacio y territorio: Reelaboraciones simbólicas y reconstrucciones identitarias en América Latina*. Ed. Abya-Yala.

Errico, S. (2018). Control over Natural Resources and Protection of the Environment of Indigenous Territories. In J. Hohmann & M. Weller (Eds.), *The UN Declaration on the Rights of Indigenous Peoples: A Commentary* (pp. 425–457). Oxford University Press.

Errico, S., & Claeys, P. (2020). Human Rights and the Commons: Exploring Approaches to the Governance of Land and Natural Resources beyond Indigenous Peoples' Rights. The Case of Peasants. *International Journal on Minority and Group Rights, 27*(1), 1–33. https://doi.org/10.1163/15718115-02604123

Escobar, A. (1995). *Encountering Development: The Making and Unmaking of the Third World*. Princeton University Press.

Escobar, A. (1997). Cultural Politics and Biological Diversity: State, Capital and Social Movements in the Pacific Coast of Colombia. In R. Fox & O. Starn (Eds.), *Between Resistance and Revolution* (pp. 40–64). Rutgers University Press.

Escobar, A. (1998). Whose Knowledge, Whose Nature? Biodiversity, Conservation, and the Political Ecology of Social Movements. *Journal of Political Ecology, 5*(1), 53. https://doi.org/10.2458/v5i1.21397

FAO. (2016). *Governing Tenure Rights to Commons. A Guide to Support the Implementation of the Voluntary Guidelines on the Responsible Governance of Tenure of Land, Fisheries and Forests in the Context of National Food Security*. FAO.

FAO, Girardin, A., Bradley, A., & Pullar, N. (2017). *FAO Land Tenure Manuals 4. Community Recording of Tenure Relationships Using Open Tenure*. FAO.

Fisher, D. (1988). Boundary Work: A Model of the Relation between Power and Knowledge. *Knowledge, 10*(2), 156–176. https://doi.org/10.1177/0164025988010002004

Freeman, M. R. (1989). Graphs and Gaffs: A Cautionary Tale in the Common-Property Resources Debate. In F. Berkes (Ed.), *Common Property Resources: Ecology and Community-Based Sustainable Development* (pp. 92–109). Belhaven Press.

Garnett, S. T., Burgess, N. D., Fa, J. E., Fernández-Llamazares, Á., Molnár, Z., Robinson, C. J., Watson, J. E. M., Zander, K. K., Austin, B., Brondizio, E. S., Collier, N. F., Duncan, T., Ellis, E., Geyle, H., Jackson, M. V., Jonas, H., Malmer, P., McGowan, B., Sivongxay, A., & Leiper, I. (2018). A Spatial Overview of the Global Importance of Indigenous Lands for Conservation. *Nature Sustainability, 1*(7), 369–374. https://doi.org/10.1038/s41893-018-0100-6

Gibson, J. (2011). Community Rights to Culture: The UN Declaration on the Rights of Indigenous Peoples. In S. Allen & A. Xanthaki (Eds.), *Reflections on the UN Declaration on the Rights of Indigenous Peoples* (pp. 433–453). Hart.

Gilbert, J. (2016). *Indigenous Peoples' Land Rights Under International Law: From Victims to Actors*. Brill Nijhoff.

Gilbert, J. (2018). *Natural Resources and Human Rights: An Appraisal*. Oxford University Press.

Glowka, L., Burhenne-Guilmin, F., Synge, H., & IUCN--The World Conservation Union. (1996). *A Guide to the Convention on Biological Diversity*. IUNC.

Graddy, T. G. (2013). Regarding Biocultural Heritage: In Situ Political Ecology of Agricultural Biodiversity in the Peruvian Andes. *Agriculture and Human Values, 30*(4), 587–604.

Gray, S. (2004). Imagination, Fraud and the Cultural Protocols Debate: A Question of Free Speech or Pornography? *Media and Arts Law Review, 9*(3), 23–37.
Greaves, T. (1996). Tribal Rights. In S. B. Brush & B. Stabinsky (Eds.), *Valuing Local Knowledge: Indigenous Peoples and Intellectual Property Rights* (pp. 25–40). Island Press.
Gudeman, S. F. (2001). *The Anthropology of the Economy: Community, Market, and Culture.* Blackwell.
Halewood, M. (1999). Indigenous and Local Knowledge in International Law: A Preface to Sui Generis Intellectual Property Protection. *McGill Law Journal, 44*(2), 953–996.
Halewood, M., Cherfas, J. J., Engels, J. M. M., Hazekamp, Th., Hodgkin, T., & Robinson, J. (2006). Farmers, Landraces and Property Rights: Challenges to Allocating Sui Generis Intellectual Property Rights to Communities over their Varieties. In S. Biber-Klemm & T. Cottier (Eds.), *Rights to Plant Genetic Resources and Traditional Knowledge Basic Issues and Perspectives* (pp. 173–202). CABI.
Hall, G. H., & Patrinos, H. A. (Eds.). (2012). *Indigenous Peoples, Poverty, and Development.* Cambridge University Press. https://doi.org/10.1017/CBO9781139105729
Hamilton, C. (2008). Intellectual Property Rights, the Bioeconomy and the Challenge of Biopiracy. *Genomics, Society, and Policy, 4*(3), 26–45. https://doi.org/10.1186/1746-5354-4-3-26
Harrop, S. R. (2009). Globally Important Agricultural Heritage Systems: An Examination of Their Context in Existing Multilateral Instruments Dealing with Conservation and Land Tenure. *Journal of International Wildlife Law & Policy, 12*(3), 127–165. https://doi.org/10.1080/13880290903202286
Harvey, D., & Haraway, D. (2016). Nature, Politics, and Possibilities: A Debate and Discussion with David Harvey and Donna Haraway. *Environment and Planning D: Society and Space, 13*, 507–527.
Haugen, H. M. (2020). The UN Declaration on Peasants' Rights (UNDROP): Is Article 19 on Seed Rights Adequately Balancing Intellectual Property Rights and the Right to Food? *The Journal of World Intellectual Property, 23*(3–4), 288–309. https://doi.org/10.1111/jwip.12152
Hayden, C. (2007). Taking as Giving: Bioscience, Exchange, and the Politics of Benefit-sharing. *Social Studies of Science, 37*(5), 729–758.
Heinze, B., & German MAB National Committee (Eds.). (2005). *Full of Life: UNESCO Biosphere Reserves - Model Regions for Sustainable Development* (1st ed). Springer.
Heywood, V. H., Watson, R. T., & United Nations Environment Programme. (1995). *Global biodiversity assessment.* Cambridge University Press.
Huarcaya, S. M. (2018). Land Reform, Historical Consciousness and Indigenous Activism in Late Twentieth-Century Ecuador. *Journal of Latin American Studies, 50*(2), 411–440. https://doi.org/10.1017/S0022216X17001146
iisd Reporting Services. (2009). Summary of the Eighth Meeting of the Working Group on Access and Benefit-Sharing of the Convention on Biological Diversity: 9–15 November 2009. ABS 8 Final. *Earth Negotiations Bulletin, 9*(489).
ILO. (2021). *Understanding the ILO Indigenous and Tribal Peoples Convention, 1989 (No. 169). A Tool for Judges and Legal Practitioners.* International Labour Organization.
IPBES. (2019). *Summary for Policymakers of the Global Assessment Report on Biodiversity and Ecosystem Services* (Summary for Policy Makers). Zenodo. https://doi.org/10.5281/ZENODO.3553579
IPCC (Ed.). (2015). *Climate Change 2014: Impacts, Adaptation, and Vulnerability. Part A.* Cambridge University Press.
IUCN, & The World Conservation Union. (2005). *Benefits Beyond Boundaries Proceedings of the Vth IUCN World Parks Congress, Durban, South Africa, 8–17 September 2003.* IUCN.

IUCN, United Nations Environment Programme, World Wildlife Fund, Food and Agriculture Organization of the United Nations, & UNESCO. (1980). *World Conservation Strategy: Living Resource Conservation for Sustainable Development*. IUCN. http://books.google.com/books?id=_V2HY5N_IHQC

IUCN, & World Commission on Protected Areas (WCPA). (2019). *Recognising and Reporting Other Effective Area-Based Conservation Measures*. IUCN. https://portals.iucn.org/library/node/48773

IUCN, & World Conservation Congress. (2005). *Resolutions and Recommendations: World Conservation Congress, Bangkok, Thailand, 17–25 November 2004*. IUCN.

Janke, T. (2005). *Managing Indigenous Knowledge and Indigenous Cultural and Intellectual Property* (M. Nakata & M. Langton, Eds.; pp. 83–93). UTSePress.

Jonas, H. D., Barbuto, V., Jonas, Kothari, A., & Nelson, F. (2014). New Steps of Change: Looking Beyond Protected Areas to Consider Other Effective Area-Based Conservation Measures. *Parks*, *20*(2), 111–128.

Kania, M. (2016). The Rights of Indigenous Peoples in Peru: From Socio-Political Marginalization to the Modern Principles of Multiculturalism. *Ad Americam*, *17*, 11–32. https://doi.org/10.12797/AdAmericam.17.2016.17.02

Kari, S., & Rössler, M. (2017). A World Heritage Perspective on Culture and Nature—Beyond a Shared Platform. *The George Wright Forum*, *34*(2), 134–141.

Knox, J. H. (2017). *Report of the Special Rapporteur on the Issue of Human Rights Obligations Relating to the Enjoyment of a Safe, Clean, Healthy and Sustainable Environment* (A/HRC/34/49). Human Rights Council, United Nations, General Assembly.

Knox, J. H. (2018). *Report of the Special Rapporteur on the Issue of Human Rights Obligations Relating to the Enjoyment of a Safe, Clean, Healthy and Sustainable Environment* (A/HRC/37/59). Human Rights Council, United Nations, General Assembly.

Kohn, E. (2015). Anthropology of Ontologies. *Annual Review of Anthropology*, *44*(1), 311–327. https://doi.org/10.1146/annurev-anthro-102214-014127

Kothari, A., Corrigan, C., Jonas, H., Neumann, A., Shrumm, H., & Secretariat of the Convention on Biological Diversity. (2012). *Recognising and Supporting Territories and Areas Conserved by Indigenous Peoples and Local Communities: Global Overview and National Case Studies*. Secretariat of the Convention on Biological Diversity, ICCA Consortium, Kalpavriksh, and Natural Justice. http://www.deslibris.ca/ID/242837

Laird, S. A. (Ed.). (2002). *Biodiversity and Traditional Knowledge: Equitable Partnerships in Practice*. Earthscan Publications.

Larsen, P. B., & Wijesuriya, G. (2017). Nature–Culture Interlinkages in World Heritage: Bridging the Gap. *The George Wright Forum*, *34*(2), 142–153.

Latour, B. (2004). Whose Cosmos? Which Cosmopolitics? Comments on the Peace Terms of Ulrich Beck. *Common Knowledge*, *10*(3), 450–462.

Le Teno, S., Cogolati, S., & Frison, C. (Forthcoming). The Right to Seeds: Using the Commons as a Sustainable Governance Scheme to Implement Peasants' Rights? In M. Alabrese, A. Bessa, M. Brunori, & P. F. Giuggioli (Eds.), *The United Nations Declaration on the Rights of Peasants and Other People Working in Rural Areas in Perspective. New International Standards on Natural Resources and Food Systems*. Routledge.

Lenzerini, F. (2006). Sovereignty Revisited: International Law and Parallel Sovereignty of Indigenous Peoples. *Texas International Law Journal*, *42*(1), 155–189.

Liggins, L., Hudson, M., & Anderson, J. (2021). Creating Space for Indigenous Perspectives on Access and Benefit-sharing: Encouraging Researcher Use of the Local Contexts Notices. *MEC Molecular Ecology*, *30*(11), 2477–2482.

Liljeblad, J., & Verschuuren, B. (Eds.). (2019). *Indigenous Perspectives on Sacred Natural Sites: Culture, Governance and Conservation*. Routledge.

Lixinski, L. (2013). *Intangible Cultural Heritage in International Law.* Oxford University Press.
Löfmarck, E., & Lidskog, R. (2017). Bumping against the Boundary: IPBES and the Knowledge Divide. *Environmental Science & Policy, 69*, 22–28. https://doi.org/10.1016/j.envsci.2016.12.008
MacCormack, C. P., & Strathern, M. (Eds.). (1978). *Nature, Culture, and Gender.* Cambridge University Press.
Macpherson, E., Torres Ventura, J., & Clavijo Ospina, F. (2020). Constitutional Law, Ecosystems, and Indigenous Peoples in Colombia: Biocultural Rights and Legal Subjects. *Transnational Environmental Law, 9*(3), 1–20.
Maffi, L. (1999). Linguistic Diversity. In D. A. Posey, Oxford Centre for the Environment, Ethics and Society, & Mansfield College, University of Oxford, UK (Eds.), *Cultural and Spiritual Values of Biodiversity* (pp. 19–57). Intermediate Technology Publications, United Nations Environment Programme.
Maffi, L. (Ed.). (2001a). *On Biocultural Diversity: Linking Language, Knowledge, and the Environment.* Smithsonian Institution Press.
Maffi, L. (2001b). On the Interdependence of Biological and Cultural Diversity. In L. Maffi (Ed.), *On Biocultural Diversity: Linking Language, Knowledge and the Environment* (pp. 1–50). Smithsonian Institution Press.
Maffi, L., & Woodley, E. (2010). *Biocultural Diversity Conservation: A Global Sourcebook.* Earthscan.
Makagon, J., von Braun, J., Cocchiaro, G., & James, S. (2016). *Balancing the Scales. Community Protocols and Extractive Industries.* Natural Justice. https://naturaljustice.org/wp-content/uploads/2017/02/Balancing_the_Scales.pdf
Martinez Luna, J. (2010). The Fourth Principle. In L. Meyer & B. Maldonada Alvarado (Eds.), *New World of Indigenous Resistance. Noam Chomsky and Voices from North, South, and Central America* (pp. 85–100). City Lights.
McAfee, K. (1999). Selling Nature to Save It? Biodiversity and Green Developmentalism. *Environment and Planning D: Society and Space, 17*(2), 133–154. https://doi.org/10.1068/d170133
McGraw, D. M. (2017). The Story of Biodiversity Convention: From Negotiation to Implementation. In P. G. Le Prestre (Ed.), *Governing Global Biodiversity: The Evolution and Implementation of the Convention on Biological Diversity* (pp. 7–38). Routledge. https://www.taylorfrancis.com/books/e/9781315253930
McGuire, S., & Sperling, L. (2016). Seed Systems Smallholder Farmers Use. *Food Security, 8*(1), 179–195. https://doi.org/10.1007/s12571-015-0528-8
McIntyre-Tamwoy, S. (2018). Travelling on the Culture-Nature Journey to New Delhi, India. *Historic Environment, 30*(3), 142–153.
Mgbeoji, I. (2003). Beyond Rhetoric: State Sovereignty, Common Concern, and the Inapplicability of the Common Heritage Concept to Plant Genetic Resources. *Leiden Journal of International Law, 16*(4), 821–837. https://doi.org/10.1017/S092215650300147X
Mgbeoji, I. (2006). *Global Biopiracy Patents, Plants, and Indigenous Knowledge.* Cornell University Press.
Mollinga, P. (2010). Boundary Work and the Complexity of Natural Resources Management. *Crop Science, 50*(1), S1–S9.
Moran, K. (1993). The Healing Forest Conservancy. *Environmental Conservation, 20*(2), 175–175. https://doi.org/10.1017/S0376892900037784
Morgera, E. (2018). Dawn of a New Day? The Evolving Relationship between the Convention on Biological Diversity and International Human Rights Law. *Wake Forest Law Review, 53*(4), 691–712.

Morgera, E., Tsioumani, E., & Buck, M. (2014). *Unraveling the Nagoya Protocol: A Commentary on the Nagoya Protocol on Access and Benefit-sharing to the Convention on Biological Diversity*. Brill; Nijhoff.

Mulrennan, M. E., & Bussières, V. (2020). Indigenous Environmental Stewardship: Do Mechanisms of Biodiversity Conservation Align with or Undermine It? In N. J. Turner (Ed.), *Plants, People, and Places: The Roles of Ethnobotany and Ethnoecology in Indigenous Peoples' Land Rights in Canada and Beyond* (pp. 282–312). McGill-Queen's University Press.

Mwangi, C. K. (2019). *Protecting Traditional Knowledge in Kenya through African Customary Law: An Analysis of Inclusive Subordination*. PhD Dissertation Strathmore University.

Nakashima, D., Krupnik, I., & Rubis, J. T. (Eds.). (2018). *Indigenous Knowledge for Climate Change Assessment and Adaptation*. Cambridge University Press.

Nelson, M. P. (2009). Holism. In J. B. Callicott & R. Frodeman (Eds.), *Encyclopedia of Environmental Ethics and Philosophy* (Vol. 1, pp. 490–493). Macmillan Reference USA.

Nemogá, G. (2019). Indigenous Agrobiodiversity and Governance. In K. S. Zimmerer & S. de Haan (Eds.), *Agrobiodiversity: Integrating Knowledge for a Sustainable Future* (pp. 241–263). The MIT Press. https://ebookcentral.proquest.com/lib/ulaval/detail.action?docID=5748564

Nemogá, G., Domicó, J., & Molina, A. (2018). Designing Biocultural Protocols with the Embera People of Colombia. *Terralingua*, 7(1), 20–24.

Nemogá-Soto, G. R. (2013). *Registro de Conocimientos Colectivos Asociados a la Biodiversidad en Perú* (p. 18). GEF, PNUMA, and UICN.

Nicholas, G. P., & Bannister, K. P. (2004). Copyrighting the Past? Emerging Intellectual Property Rights Issues in Archaeology. *Current Anthropology (Chicago)*, *45*(3), 327–350.

Nietschmann, B. (1992). *The Interdependence of Biological and Cultural Diversity*. Center for World Indigenous Studies.

Nwauche, E. (2017). The Positive Protection of Traditional Cultural Expressions by Intellectual Property Rights in Africa. In E. Nwauche (Ed.), *The Protection of Traditional Cultural Expressions in Africa* (pp. 97–150). Springer International Publishing.

Nzomo, V. (2015, December 16). Comments on the Protection of Traditional Knowledge and Traditional Cultural Expressions Bill, 2015. *IP Kenya*. https://ipkenya.wordpress.com/2015/12/16/comments-on-the-protection-of-traditional-knowledge-and-traditional-cultural-expressions-bill-2015/

Oldfield, M. L., & Alcorn, J. B. (1987). Conservation of Traditional Agroecosystems. *BioScience*, *37*(3), 199–208. https://doi.org/10.2307/1310519

Oxfam, International Land Coalition, & Rights and Resources Initiative. (2016). *Common Ground. Securing Land Rights and Safeguarding the Earth*. Oxfam.

Peet, R., & Watts, M. (1996). Liberation Ecology. Development, Sustainability, and Environment in an Age of Market Triumphalism. In R. Peets & M. Watts (Eds.), *Liberation Ecologies. Environment, Development and Social Movements* (pp. 1–45). Routledge. https://doi.org/10.4324/9780203235096

Pencek, W. (2017). Reflections on the Nature–Culture Journey. *The George Wright Forum*, *34*(2), 211–215.

Pimbert, M. P. (2018). *Food Sovereignty, Agroecology and Biocultural Diversity: Constructing and Contesting Knowledge*. Routledge.

Polo, C. S., & Danielson, M. S. (2013). Political Subsystems in Oaxaca's Usos y Costumbres municipalities. In T. A. Eisenstadt, M. S. Danielson, M. J. Bailon Corres, & C. Sorroza Polo (Eds.), *Latin America's Multicultural Movements* (pp. 169–191). Oxford University Press.

Posey, D. (Ed.). (1999). *Cultural and Spiritual Values of Biodiversity*. Intermediate Technology Publications, United Nations Environment Programme.

Posey, D. A. (1995). *Indigenous Peoples and Traditional Resource Rights: A Basis for Equitable Relationships?* Green College Centre for Environmental Policy and Understanding, University of Oxford. https://www.ecolex.org/fr/details/literature/indigenous-peoples-and-traditional-resource-rights-a-basis-for-equitable-relationships-mon-057188/

Posey, D. A. (1998). Diachronic Ecotones and Anthropogenic Landscapes in Amazonia. Contesting the Consciousness of Conservation. In W. L. Balée (Ed.), *Advances in Historical Ecology* (pp. 104–118). Columbia University Press.

Posey, D. A. (1999). Introduction: Culture and Nature – The Inextricable Link. In D. A. Posey (Ed.), *Cultural and Spiritual Values of Biodiversity* (pp. 1–19). UNEP.

Posey, D. A. (2001). Biological and Cultural Diversity. The Inextricable, Linked by Language and Politics. In L. Maffi (Ed.), *On Biocultural Diversity: Linking Language, Knowledge, and the Environment* (pp. 379–396). Smithsonian Institution Press.

Posey, D. A. (2004). Traditional Resources Rights – De facto Self-determination for Indigenous Peoples. In D. A. Posey & K. Plenderleith (Eds.), *Indigenous Knowledge and Ethics: A Darrell Posey Reader* (pp. 155–168). Routledge.

Posey, D. A., & Dutfield, G. (1996). *Beyond Intellectual Property: Toward Traditional Resource Rights for Indigenous Peoples and Local Communities.* International Development Research Centre.

Posey, D. A., & Plenderleith, K. (2004). *Indigenous Knowledge and Ethics: A Darrell Posey Reader.* Routledge.

Pritchard, J., Lesniewska, F., Lomax, T., Ozinga, S., & Morel, C. (2013). *Securing Community Land and Resource Rights in Africa: A Guide to Legal Reform and Best Practices.* FERN, FPP, ClientEarth and CED.

Puig de la Bellacasa, M. (2017). *Matters of Care: Speculative Ethics in More than Human Worlds.* University of Minnesota Press.

Ramos, A. R. (1994). The Hyperreal Indian. *Critique of Anthropology, 14*(2), 153–171.

Raven, M. (2006). Protocols & ABS. Recognising Indigenous Rights to Knowledge in Australian Bureaucratic Organisations. *Indigenous Law Bulletin, 6*(20), 13–16.

Raven, M. (2010). Protocols: Devices for Translating Moralities, Controlling Knowledge, Defining Actors in Indigenous Research, and Critical Ethical Reflections. *Australian Aboriginal Studies, 2*, 33–47.

Raymond, H. (2007). The Ecologically Noble Savage Debate. *Annual Review of Anthropology, 36*(1), 177–190. https://doi.org/10.1146/annurev.anthro.35.081705.123321

Reed, M. G., & Price, M. F. (2020). *UNESCO Biosphere Reserves: Supporting Biocultural Diversity, Sustainability and Society.* Routledge. https://doi.org/10.4324/9780429428746

Reimerson, E. (2013). Between Nature and Culture: Exploring Space for Indigenous Agency in the Convention on Biological Diversity. *Environmental Politics, 22*(6), 992–1009.

Reyes-García, V., Fernández-Llamazares, Á., Aumeeruddy-Thomas, Y., Benyei, P., Bussmann, R. W., Diamond, S. K., García-del-Amo, D., Guadilla-Sáez, S., Hanazaki, N., Kosoy, N., Lavides, M., Luz, A. C., McElwee, P., Meretsky, V. J., Newberry, T., Molnár, Z., Ruiz-Mallén, I., Salpeteur, M., Wyndham, F. S., … Brondizio, E. S. (2021). Recognizing Indigenous Peoples' and Local Communities' Rights and Agency in the Post-2020 Biodiversity Agenda. *Ambio.* https://doi.org/10.1007/s13280-021-01561-7

Rimmer, M. (2015). Introduction: Mapping Indigenous intellectual property. In M. Rimmer (Ed.), *Indigenous Intellectual Property. A Handbook of Contemporary Research* (pp. 1–44). Edward Elgar. https://doi.org/10.4337/9781781955901

Riphagen, M., & Stolte, G. M. (2016). The Functioning of Indigenous Cultural Protocols in Australia's Contemporary Art World. *International Journal of Cultural Property, 23*(3), 295–320. https://doi.org/10.1017/S0940739116000163
Robinson, D. F. (2011). *Confronting Biopiracy: Challenges, Cases and International Debates.* Earthscan.
Robinson, D. F. (2014). *Biodiversity, Access and Benefit-Sharing: Global Case Studies* (1st ed.). Routledge. https://doi.org/10.4324/9781315882819
Romero, T. L. (2005). Sui Generis Systems for the Protection of Traditional Knowledge. *International Law: Revista Colombiana de Derecho Internacional, 3*(6), Article 6. https://revistas.javeriana.edu.co/index.php/internationallaw/article/view/14043
Rosendal, G. K. (1991). *International Conservation of Biological Diversity: The Quest for Effective International Solutions* (R:012–1991). The Fridtjof Nansen Institute.
Sajeva, G. (2015). Rights with Limits: Biocultural rights – Between Self-Determination and Conservation of the Environment. *Journal of Human Rights and the Environment, 6*(1), 30–54. https://doi.org/10.4337/jhre.2015.01.02
Sajeva, G. (2018). *When Rights Embrace Responsibilities: Biocultural Rights and the Conservation of Environment* (1st ed.). Oxford University Press.
Sajeva, G. (2021). Environmentally Conditioned Human Rights. A Good Idea? In D. P. Corrigan & M. Oksanen (Eds.), *Rights of Nature. A Re-examination* (pp. 87–100). Routledge.
Santilli, J. (2016). *Agrobiodiversity and the Law: Regulating Genetic Resources, Food Security and Cultural Diversity.* Routledge.
Sarrazin, M.-L., & Scott, C. (2020). Livelihoods, Food Sovereignty, and Well-Being. *Policy Brief 2: Nurturing Biocultural Diversity.* https://cicada.world/files/pb/CICADA_PolicyBrief2_Livelihoods.pdf
Saul, B. (2016). *Indigenous Peoples and Human Rights: International and Regional Jurisprudence.* Hart Publishing.
Savaresi, A. (2018). Traditional Knowledge and Climate Change: A New Legal Frontier? *Journal of Human Rights and the Environment, 9*(1), 32–50.
Secretariat of the Convention on Biological Diversity. (2020). *Global Biodiversity Outlook 5.* UNEP, CDB. https://edepot.wur.nl/531233
Sedjo, R. A. (1992). Property Rights, Genetic Resources, and Biotechnological Change. *The Journal of Law & Economics, 35*(1), 199–213.
Shaheed, F. (2011). *Report of the Independent Expert in the Field of Cultural Rights* (A/HRC/17/38). Human Rights Council, United Nations, General Assembly.
Shaheed, F. (2012). *Report of the Special Rapporteur in the Field of Cultural Rights, Farida Shaheed. The Right to Enjoy the Benefits of Scientific Progress and its Applications* (A/HRC/20/26). Human Rights Council, United Nations, General Assembly.
Shaheed, F. (2015). *Report of the Special Rapporteur in the Field of Cultural Rights, Farida Shaheed. Report on Patent Policy and the Right to Science and Culture* (A/70/279). Human Rights Council, United Nations, General Assembly.
Shrumm, H., & Jonas, H. (2012). *Biocultural Community Protocols: A Toolkit for Community Facilitators.* Natural Justice. https://naturaljustice.org/wp-content/uploads/2015/09/Biocultural-Community-Protocols-Toolkit.pdf
Smith, L. T. (2012). *Decolonizing Methodologies: Research and Indigenous Peoples.* Zed Books.
Srinivasan, C. S. (2016). Institutional Capacity and Implementation Issues in Farmers' Rights. In M. Halewood (Ed.), *Farmers' Crop Varieties and Farmers' Rights: Challenges in Taxonomy and Law. Issues in Agricultural Biodiversity* (pp. 249–282). Routledge.
Stengers, I. (2005). The Cosmopolitical Proposal. In B. Latour & P. Weibel (Eds.), *Making Things Public* (pp. 994–1003). MIT Press. http://mnissen.psy.ku.dk/Undervisning/Stengers05.pdf

Stevens, S. (2014). *Indigenous Peoples, National Parks, and Protected Areas: A New Paradigm Linking Conservation, Culture, and Rights*. University of Arizona Press. http://site.ebrary.com/id/10896830

Stoll, T. (2018). Intellectual Property and Technologies. In J. Hohmann & M. Weller (Eds.), *The UN Declaration on the Rights of Indigenous Peoples: A Commentary* (pp. 299–327). Oxford University Press.

Strathern, M. (1996). Potential Property: Intellectual Rights and Property in Persons. *Social Anthropology : The Journal of the European Association of Social Anthropologists*, *4*, 17–32.

Suagee, D. (1999). Human Rights and the Cultural Heritage of Indian Tribes in the United States. *International Journal of Cultural Property*, *8*(1), 48–76. https://doi.org/10.1017/S094073919977061X

Swiderska, K., Mead, A., Dutfield, G., & Argumedo, A. (2016). Designing a Labelling System for Biocultural Heritage-Based Product. In C. Pavel & D. Leaman (Eds.), *Policy Matters: Certification and Biodiversity: How Voluntary Certification Standards Impact Biodiversity and Human Livelihoods: Vol. Issue 21* (pp. 140–148). CEESP and IUCN. https://doi.org/10.2305/IUCN.CH.2014.PolicyMatters-21.en

Tauli-Corpuz, V. (2016). *Report of the Special Rapporteur of the Human Rights Council on the Rights of Indigenous Peoples, Victoria Tauli-Corpuz. Conservation Measures and their Impact on Indigenous Peoples' Rights* (A/71/229). Human Rights Council, United Nations, General Assembly.

Tocancipá Falla, J., & Ramírez Castrillón, C. A. (2018). Las nuevas dinámicas rurales en las Zonas de Reserva Campesina en Colombia. *Perspectiva Geográfica*, *23*(1). https://doi.org/10.19053/01233769.5796

Toledo, V. M. (1990). The Ecological Rationality of Peasant Production. In M. A. Altieri & S. B. Hecht (Eds.), *Agroecology and Small Farm Development* (pp. 53–60). CRC Press.

Toledo, V. M. (2001). Indigenous Peoples and Biodiversity. In S. A. Levin (Ed.), *Encyclopedia of Biodiversity* (pp. 451–463). Elsevier. https://doi.org/10.1016/B978-0-12-384719-5.00299-9

Tsing, A. (2013). More-than-Human Sociality: A Call for Critical Description. In K. Hastrup (Ed.), *Anthropology and Nature* (pp. 27–42). Routledge. https://doi.org/10.4324/9780203795361-8

Tsosie, R. (2007). Cultural Challenges to Biotechnology: Native American Genetic Resources and the Concept of Cultural Harm. *The Journal of Law, Medicine & Ethics*, *35*(3), 396–411.

Tyrrell, T. D., & Alcorn, J. B. (2011). *Analysis of Possible Indicators to Measure Impacts of REDD+ on Biodiversity and on Indigenous and Local Communities: A Report to the Convention on Biological Diversity*. Tentera.

Uddin, N., Gerahz, E., & Chakkarath, P. (2018). Exploring Indigeneity. Introductory Remarks on a Contested Concept. In E. Gerahz, N. Uddin, & P. Chakkarath (Eds.), *Indigeneity on the Move: Varying Manifestations of a Contested Concept* (pp. 1–25). BERGHAHN Books.

Ulloa, A. (2005). *The Ecological Native: Indigenous Peoples' Movements and Eco-governmentality in Colombia*. Routledge.

UNCED (Ed.). (1992). *Agenda 21, Rio Declaration, Forest Principles*. United Nations.

UNESCO & WIPO. (1985). *Model Provisions for National Laws on the Protection of Expressions of Folklore Against Illicit Exploitation and other Prejudicial Actions*. UNESCO, OMPI/WIPO.

UNESCO, World Wildlife Fund for Nature, & Terralingua. (2003). *Sharing a World of Difference: The Earth's Linguistic, Cultural and Biological Diversity*. UNESCO, WWF, Terralingua.

Viveiros De Castro. (2004). Perspectival Anthropology and the Method of Controlled Equivocation. *Tipití: Journal of the Society for the Anthropologu of Lowland America, 2*(1), 3–22.

Westra, L. (2013). *Environmental Justice and the Rights of Indigenous Peoples: International and Domestic Legal Perspectives*. Earthscan.

Wily, L. A. (2011). 'The Law is to Blame': The Vulnerable Status of Common Property Rights in Sub-Saharan Africa. *Development and Change, 42*(3), 733–757. https://doi.org/10.1111/j.1467-7660.2011.01712.x

WIPO. (2001). *Declaration of Shamans on Intellectual Property and Protection of Traditional Knowledge and Genetic Resources. Intergovernmental Committee on Intellectual Property and Genetic Resources, Traditional Knowledge and Folklore, Second Session. Geneva, December 10 to 14, 2001* (WIPO/GRTKF/IC/2/14). WIPO.

Wood, D. (1988). Crop Germplasm: Common Heritage or Farmers' Heritage. In J. R. Kloppenburg (Ed.), *Seeds and Sovereignty. The Use and Control of Plant Genetic Resources* (pp. 274–289). Duke University Press.

World Bank, T. (1989). *Sub-Saharan Africa. From Crisis to Sustainable Growth*. The International Bank for Reconstruction and Development/The World Bank.

World Summit on Sustainable Development & United Nations (Eds.). (2002). *Report of the World Summit on Sustainable Development: Johannesburg, South Africa, 26 August–4 September 2002*. United Nations.

Xanthaki, A. (2014). *Indigenous Rights and United Nations Standards: Self-determination, Culture and Land*. Cambridge University Press.

Younging, G. (2010). *Gnaritas Nullius (No Ones' Knowledge): The Public Domain and Colonization of Traditional Knowledge. WIPO Indigenous Panel on the Role of the Public Domain Concept: Experiences in the Fields of Genetic Resources, Traditional Knowledge and Traditional Cultural Expressions: Experiences from Canada* (WIPO/GRTKF/IC/17/INF/5(A)). WIPO Intergovernmental Committee on Intellectual Property and Genetic Resources, Traditional Knowledge and Folklore (IGC).

PART 1

Conceptual Insights

Biocultural Diversity, Biocultural Rights and Space Making

2

A BIOCULTURAL ETHICS APPROACH TO BIOCULTURAL RIGHTS

Exploring Rights, Responsibilities and Relationships through Ethics Initiatives in Canada

Kelly Bannister[1]

Introduction

> Indigenous peoples have the right to maintain and strengthen their distinctive spiritual relationship with their traditionally owned or otherwise occupied and used lands, territories, waters and coastal seas and other resources and to uphold their responsibilities to future generations in this regard.
>
> (*United Nations Declaration on the Rights of Indigenous Peoples 2007, Article 25*)

It is often said that rights and responsibilities are two sides of the same coin, meaning rights and responsibilities are in a reciprocal or correlative relationship. Rights-based arguments understandably prevail in political, cultural and environmental conflicts involving the biocultural knowledge and heritage of Indigenous peoples. While necessary, having too narrow a focus on rights (and/or limiting interpretations of responsibilities to "duties") can render cultural responsibilities and important relational considerations invisible if attention is primarily directed to legal and policy processes, mechanisms and prescriptions for "doing the right thing".

As indicated in Article 25 of the United Nations Declaration on the Rights of Indigenous Peoples (UNDRIP), within most Indigenous cultures, not only are rights and responsibilities intertwined, they emerge from relationships with, and relational accountability to, the natural world and All Our Relations. *Relational accountability* is a dependence upon, relationship to and responsibility of care for everything and everyone around us (Reo, 2019, p. 66; Wilson, 2001, p. 177). *All Our Relations* is a term that acknowledges human beings as connected to one

DOI: 10.4324/9781003172642-3

another and interdependent with all forms of life within creation. According to Blackfoot scholar Dr. Leroy Little Bear, "In the Indigenous world, everything is animate and has spirit. *All my relations* refers to relationships with everything in creation" (Little Bear, 2009, p. 7).

In the video *How UNDRIP Recognizes the Sacred Relationship with Nibi (Water)*, Anishinaabe-Métis lawyer Aimée Craft explains that Article 25 of UNDRIP ensures protection of a longer-term "spiritual relationship between Indigenous people and their waters, lands, territories, in a way that allows for protection of that water, or that land or that territory, for future generations" (Craft, 2019). She goes on to say:

> UNDRIP is not recognizing water as a resource to be owned or managed, or controlled like most Western legal systems do, but it is actually saying, as Indigenous people, there is something further to that relationship, there is something that is deeper that emanates from your relationship with water, and that you have a right to preserve and maintain that.
>
> *(Craft, 2019)*

To complement the general biocultural rights-based focus of this book for protecting culture and environment, this chapter offers a biocultural ethics perspective, where biological and cultural rights and responsibilities are understood as inextricably interlinked and situated within relationships across generations and species in ways that honor past, present and future. It highlights the important role that ethics have played in supporting biocultural rights of Indigenous peoples in Canada[2] over the last couple of decades, focusing on ways that national research ethics policies and guidelines have evolved to address historical injustices to Indigenous peoples by prioritizing relational dimensions of research involving Indigenous peoples. Notably, these ethics initiatives took place years in advance of the Canadian government's commitments of this nature, which have been articulated recently through new principles, laws and policies as part of a national agenda of reconciliation with Indigenous peoples.

The chapter also points to hard-won existing efforts – and ongoing efforts still needed – to evolve our institutions, policies and practices to secure Indigenous rights more wholly as they extend to cultural responsibilities and spiritual relationships with the natural world. This involves awareness and respect for Indigenous worldviews, customs and legal systems, particularly with regard to how Indigenous knowledge is accessed and used outside of the Indigenous communities and traditions where the knowledge originated. As Anishinaabe legal scholar Dr. John Borrows points out in the Preface to Genetic Resources, Justice and Reconciliation (Borrows, 2018):

> Today, our knowledge and relationships with the genetic diversity of our territories is threatened by governments, corporations, scientists and other bodies and individuals. Biopiracy and cultural appropriation abound. [...]

> Many Anishinaabe people enjoy sharing their knowledge *except* if it is obtained in improper ways or used for inappropriate purposes. [...] [P]ropriety and appropriateness are best defined in harmony with the systems which generate these insights.
>
> *(Borrows, 2018, p. xv)*

An ongoing question, then, is how our Western institutions, policies and practices that aim to reconcile with Indigenous peoples and appropriately include their Indigenous knowledge can evolve *insightfully* and more *harmoniously* with Indigenous knowledge systems, in contrast to becoming more mechanistic and rationalistic in implementing Indigenous rights. In other words, how do we tend to the aspects of biocultural rights that involve righting wrongs by fostering "right relationships", which are not likely to be addressed by directing our gaze too narrowly on rights-based solutions?

A key part of righting past harms has included creating space within Western academic institutions for enacting Indigenous community protocols[3] (e.g. ethical codes, codes of practice and cultural protocols) that have been developed in recent years by many First Nations, Métis and Inuit communities in Canada. However, the formulaic approaches often taken by universities (and governments) to meet what are (in part) inherently relational needs can manifest as transactional processes and outcomes. Because transactions are built primarily on expectations for reciprocal exchange of goods, services or actions, these processes chronically fall short of mutual goals for respectful and equitable interactions and relationships involving Indigenous peoples and their biocultural knowledge.

For example, important considerations such as consent, ownership and mutual benefit feature prominently in discussions of biocultural rights. These are key elements in Canadian ethics policy and in most Indigenous community protocols, as well as in ethical guidelines and toolkits developed by Indigenous and other organizations in Canada. There are deeply held ethical principles underlying consent, ownership and mutual benefit, based in respect for human dignity, concern for well-being and justice. However, the enactment of these ethical principles is often limited to transactional activities that are required by institutionalized procedures and mechanisms (e.g. consent forms, copyright agreements or intellectual property rights agreements). Furthermore, the common use of convenient acronyms, such as FPIC (i.e. free, prior informed consent), OCAP (i.e. ownership, control, access and possession) and MAT (i.e. mutually agreed terms), can diminish our awareness and sensibility to situated meanings and understandings that the acronyms are supposed to represent – especially those of a relational dimension that necessitate respectful, meaningful and sometimes nuanced processes to enact ethically.

With this awareness in mind, the principle of consent is explored throughout this chapter as one of several elements essential for ethical research relationships and for respecting the individual and collective rights of Indigenous peoples, including biocultural rights that relate to protection of culture and environment.[4]

The evolution of consent as an ethical principle in Canada is examined and elaborated from a biocultural ethics perspective to offer a glimpse of how the quality and meaningfulness of principles can shift in powerful ways when understood and embraced relationally (between and beyond humans) rather than only implemented transactionally or mechanistically (as a right, duty or procedural requirement). Extensive quotes are used through this chapter to draw attention to specific nuances within ethics policies that tend to be based on highly negotiated language, and to accurately convey the well-considered words of Indigenous and other authors who emphasize the relational nature of Indigenous rights and responsibilities.

A Biocultural Ethics Approach

The word *ethics* derives from the ancient Greek root *ethos*, which refers to character, disposition, habit or custom. There are many understandings of ethics today, but generally ethics refers to moral character, and a way of living and conducting oneself according to moral principles and values. Ethics is also a formal branch of philosophy that is concerned with the behavior of individuals in society, and the nature of what is considered right/wrong, just/unjust or virtuous/vicious. Ethics may be understood implicitly or made explicit in various ways and forms. In this chapter, ethics generally refers to how we choose to relate to one another as individuals and as part of groups and systems, which can include the natural world. In essence, ethics is about our capacity to know what harms or enhances the well-being of sentient creatures.

Biocultural ethics is an emerging term that has inspired this author's approach to ethics for many years (Bannister, 2000, 2018, 2020; Bannister & Thomas, 2016). The term was first formally described in the literature by environmental philosopher Dr. Ricardo Rozzi (2012) who makes the compelling case for need of a biocultural ethic that acknowledges "the vital links between biological and cultural diversity" (Rozzi, 2012, p. 27), as well as "the diversity of existing sustainable forms of ecological knowledge, practices, and worldviews that have co-evolved within specific ecoregions" (Rozzi, 2012, p. 28), and "the dynamic, reciprocal interrelationships between the well-being and identity of the inhabitants, their habits, and the habitats they inhabit" (Rozzi, 2012, p. 39).

Biocultural ethics can be understood as a form of applied, relational ethics that is vitally informed by Indigenous ethics, and underscores the inextricable interconnections between biological and cultural diversities, including linguistic diversity. Biocultural ethics is the way we relate to one another, the more-than-human world and Mother Earth herself, recognizing that we are all interconnected in the web of life.

Informally and formally, ethics play a fundamental role in guiding our interactions with others, personally and professionally. At organizational levels and within many institutions, ethics are codified into ethical guidelines, which are sometimes voluntary and sometimes implemented through policies or within regulatory frameworks. Since the 1960s and 1970s, professionally and within many academic disciplines, ethical guidance has been formalized into codes of ethics, codes of conduct or codes of practice that prescribe what one should and should not "do" to be ethical.

A biocultural ethics approach draws particular attention to relational dimensions and raises the question of how we should "be" with one another, as pre-requisite to (and co-requisite with) the questions of what we should "do" together or what we should "produce" together. The former question is particularly important amid a flurry of current interest in knowledge co-production, a popularized term (Armitage et al., 2011; Norström et al., 2020; Reed et al., 2014) referring to a collaborative process of bringing a plurality of knowledge sources and types together to address a defined problem and build an integrated or systems-oriented understanding of that problem (Armitage et al., 2011). However, knowledge co-production cannot be pursued without careful attention to ethics of knowledge sharing, lest the co-production process unwittingly reproduces colonial structures and inequities that have historically misrepresented, marginalized and misappropriated the knowledge of Indigenous peoples.

The question of how to "be" with one another is vital but often overlooked in intercultural fields of inquiry, including settings where Western trained scientists, scholars and legal practitioners are seeking to collaborate with Indigenous peoples to address cultural and ecological concerns and protect the biocultural rights of Indigenous peoples. Biocultural ethics invites us to deepen our awareness and cultivate the capacities needed for understanding, creating and enacting ethical processes between and across different worldviews and traditions. In contrast, too heavy reliance on rights-based mechanisms, instruments and tools derived from a Western legal framework, along with a focus on Western notions of ownership, may reproduce a colonial paradigm.

Cree scholar Dr. Greg Younging (2018) notes this phenomenon at work along with a contemporary response by Indigenous peoples:

> Cultural rights are part of contemporary Indigenous cultural realities. Understanding these rights, including how they evolved, is key to working in a culturally appropriate and respectful way. Indigenous Peoples think of Creation as something that includes and sustains all living things. People are part of it and responsible for caring for it. The question of "who owns it" has no context.
>
> By contrast, "who owns it" preoccupies European notions of the world. [...] If something isn't "owned" – air, for example – European notions consider it either free for the taking (mostly without value) or not yet owned. Indigenous Peoples have formulated a new idea of ownership – Indigenous cultural property – to assert their place in a post-contact world of owned things.
>
> *(Younging, 2018, p. 25)*

Central to a biocultural ethics approach is digging deeply into the choice and meaning of the words we choose as our guiding values and principles related to Indigenous rights and responsibilities. Bringing this intentionality into our *ethos* invites conversation and curiosity, rather than assumption that everyone shares the same ethical disposition and understanding of meanings across different disciplines, cultures, languages and worldviews. The latter can lead to unintentional imposition of Western institutionalised values and assumptions on Indigenous peoples.

An example is "scientific integrity", which is defined in the *Model Policy on Scientific Integrity* (Government of Canada, n.d.) as the condition resulting from adherence to the ethical principles and professional standards deemed essential for the responsible practice of Western science. These ethical principles and professional standards include concepts such as transparency, openness, high-quality work, avoidance of conflict of interest, due credit, ensuring high standards of impartiality and abiding by research ethics.

For comparison, Tewa author and professor Dr. Gregory Cajete (2015, p. 137) describes Indigenous science as "founded on a body of practical environmental knowledge that has been learned and transferred over generations of a People through forms of environmental and cultural education that is unique to the People". Indigenous science "perceives and reasons from a 'high context' relational worldview that reflects a 'multiverse of expression and possibility'. As a result, all relational connections are considered". It is a process for exploring, understanding and explaining the natural world based on lived experiences. Indigenous science is "based on observation" and guided by additional values to deepen knowledge, including "spirituality, ethical relationships, mutualism, reciprocity, respect, restraint, a focus on harmony, and an emphasis on interdependence. These values integrate knowledge about a particular place for the purpose of living there in a sustainable way" (Cajete, 2015, pp. 137–138).

At its core, "integrity" is the quality of one's character related to being whole or complete, and requires acting in accordance with the values, beliefs and principles of a particular system or worldview. A key question from a biocultural ethics lens is *which system* and *whose worldview*? Integrity from an Indigenous science perspective (Cajete, 2015) is based on upholding the cultural systems and institutions of knowledge, values, customs, governance, laws, beliefs and practices that are specific to a given Indigenous community and their own scientific systems rooted in their homelands, culture, cosmology and laws. These depend on the particular people, their place, their ways of knowing, being and doing over generations, and honoring of kinship ties with and within the natural world. What constitutes "scientific integrity" from a Western science perspective, therefore, may not be the same as from an Indigenous science perspective.

Intentional and thoughtful unearthing and uprooting of entrenched Western ethical traditions and assumptions, and how these are institutionalized in scientific, legal and educational systems, can make space and create more hospitable ground for co-creating a shared biocultural ethic that respects the different values, worldviews and systems involved. The following sections explore how national research ethics policies and guidelines in Canada have evolved over the last two decades – and continue to evolve – to create more hospitable and productive space for respectful and equitable research relations with Indigenous peoples in Canada. The principle of consent is explored from this approach, to offer an illustrative example of how a biocultural ethics lens can complement biocultural rights-based approaches in protecting culture and environment. The objective of this chapter is to encourage deeper consideration of what it means to be in "right relationships", which Mohawk ethics policy expert Dr. Marlene Brant-Castellano describes as an ethos that is "embodied in First Nation, Inuit and Métis traditional teachings" (Brant Castellano, 2008, p. 23).

Brief History of Canadian Ethics Policies and Guidelines with Focus on Consent

Consent is an ethical and legal principle that is rooted in respect for individual autonomy and autonomous choice – the notion that individuals have a right to make informed choices of their own free will, consistent with their values and preferences. Consent is embedded in both law and ethics, having evolved from untenable and inhumane historical medical treatment, medical research and social science research involving humans. The concept of "voluntary consent" emerged in biomedical research ethics in the 1940s (i.e. Nuremburg Code, 1947)[5] in response to human atrocities committed against inmates of concentration camps by Nazi scientists in the name of research. As a result, consent was originally conceived in research ethics in reference to protection of individuals from physical harm, but has increasingly been extended to collective and non-physical contexts.

Consent as FPIC

Consent has been articulated in ethical and legal realms in a diversity of nuanced ways over the years, such as "voluntary consent", "informed consent", "prior informed consent" and most notably in UNDRIP as "free, prior, informed consent" drawing attention to three important qualitative dimensions:

Free – The choice must be freely and voluntarily made, without any inducement, coercion, intimidation or manipulation involved in giving consent, and without penalty for not giving consent.

Prior – The choice to participate is made sufficiently prior to starting the proposed activity. Adequate time and opportunity to understand the information provided and enable processes to decide on participation are required.

Informed – To be meaningful and valid, the choice must be informed. An informed choice is one that is based on as complete an understanding as is reasonably possible of the nature, purpose, duration, procedures and

> consequences of the activity and its impacts on individuals, communities and the environment. This understanding includes foreseeable risks, harms, costs and potential benefits for all parties involved (UN Permanent Forum on Indigenous Issues, 2005).

FPIC is upheld as "the standard" of consent in international and human rights law regarding the right of Indigenous peoples to self-determination. It is so widely known and used today that the acronym FPIC rarely requires explanation or elaboration. However, FPIC has been a significant source of debate within Canada, related to differing interpretations and specific concerns by government and industry about enabling Indigenous community veto power over development affecting lands, territories and resources (Abouchar et al., 2021; Papillon & Rodon, 2017).

FPIC has started to be referred to as part of implementation of federal laws, but with a disclaimer about veto rights. For example, the *Impact Assessment Act* (2019) for prevention of significant adverse environmental effects in projects carried out in Canada is described vis-à-vis FPIC as follows:

> The process set out in the *Impact Assessment Act* (2019) aligns fundamentally with the objectives of Free, Prior and Informed Consent as set out in the United Nations Declaration on the Rights of Indigenous Peoples (UNDRIP). Free, Prior and Informed Consent does not affirm the ability for Indigenous peoples to veto a government action, decision or project.
>
> *(Government of Canada, 2020)*

One can speculate that Canada's reluctance to the outright adoption of FPIC as a legal principle in past years has been a catalyst for more innovative consideration and articulation of the principle of consent in Canadian ethics policy circles, described below. FPIC was a major sticking point in Canada's initial refusal to support UNDRIP, which was eventually endorsed in 2010 as an aspirational document, and officially adopted in 2016 with a commitment to implementation. It took until 2021 for UNDRIP to come into law through *Bill C-15:* United Nations Declaration on the Rights of Indigenous Peoples Act (2021). The ensuing development and implementation of an action plan to achieve the objectives of UNDRIP will undoubtedly stimulate heightened debate about what FPIC really looks like regarding self-determination of Indigenous peoples in Canada.

Coincident with the timing of the development of UNDRIP, Canada was undergoing a major revision to its national research ethics policy, called the *Tri-Council Policy Statement: Ethical Conduct for Research Involving Humans* (TCPS). The TCPS was developed in 1998 after earlier attempts at creating guidelines for social sciences (Social Sciences and Humanities Research Council of Canada (SSHRC), 1977) and medical research (Medical Research Council of Canada (MRC), 1987) garnered little compliance by researchers who were largely unaware that the guidelines existed (Rocher, 1999 as cited in McDonald, 2009, p. 11). The TCPS aimed to ensure the protection of human participants in any research funded by Canada's three national granting councils, i.e. the Social Sciences and Humanities Research Council (SSHRC), the Natural Sciences and Engineering Research Council (NSERC) and Canadian Institutes for Health Research (CIHR). However, from 1998 to 2010, the section of the TCPS on research involving Indigenous peoples was left in abeyance with explicit recognition that the section was based largely on publicly available information and insufficient discussions had taken place with representatives of Indigenous peoples or researchers involved in such work (McDonald, 2009).

To address the pressing need for national ethical guidance during this interval, CIHR (one of the three granting councils) established a working group from 2004 to 2007 and a process to undertake extensive consultation with Indigenous communities and researchers in Canada to inform development of new *CIHR Guidelines for Health Research Involving Aboriginal People* (Canadian Institutes for Health (CIHR), 2007).[6] The *CIHR Guidelines* offered the first national policy in Canada with detailed philosophical grounding and practical guidance for academic researchers working with Indigenous peoples.

CIHR Guidelines and FPIC

FPIC was explicitly named in the *CIHR Guidelines* as the requirement for one of two levels of consent. As Section 2.5: Community and Individual Consent, Article 4 states:

> A researcher who proposes to carry out research that touches on traditional or sacred knowledge of an Aboriginal community, or on community members as Aboriginal people, should consult the community leaders to obtain their consent before approaching community members individually. Once community consent has been obtained, the researcher will still need the free, prior and informed consent of the individual participants. Community consent is distinct from, and additional to, individual consent from each research participant.
>
> *(Canadian Institutes for Health (CIHR), 2007, p. 20)*

The *CIHR Guidelines* explained the need for two levels of consent as follows:

> Although individual consent is essential, Aboriginal social norms and values tend to be organized around an operative principle of collective Aboriginal knowledge, ownership and decision-making. This is one of the reasons why the notion of community consent is so important in research involving Aboriginal people. Thus, an Aboriginal community is entitled to decide whether a research project is in the best interest of the community

> (community consent) as a precondition to the researcher seeking individual consent from community members.
>
> *(Canadian Institutes for Health (CIHR), 2007, p. 11)*

The *CIHR Guidelines* pointed out that "meaningful consultation and participation are crucial components of a consent process", which "should be undertaken in good faith and with relational accountability" and that "parties should establish a dialogue allowing them to find solutions in an atmosphere of mutual respect in good faith, with full and equitable participation" (Canadian Institutes for Health (CIHR), 2007, pp. 19–20). The consent process should also "take into account the community's own legitimate decision-making processes regarding all phases of planning, implementation, monitoring, assessment, evaluation and wind-up of a research project" and recognize "that consent is an ongoing process and should be reaffirmed periodically, as appropriate to the research project" (Canadian Institutes for Health (CIHR), 2007, p. 21).

Additionally, the *CIHR Guidelines* highlighted some unique cultural considerations related to consent:

> Aboriginal societies are traditionally oral societies and written consent may be seen as contrary to respecting Aboriginal approaches to research initiatives. Oral consent is an appropriate alternative to obtaining written consent. A researcher, however, should document the date, time and place in which the oral consent of a participant was received. Language may be an important consideration as well, and it may be appropriate to have a written consent form translated into the community's language.
>
> *(Canadian Institutes for Health (CIHR), 2007, p. 21)*

The *CIHR Guidelines* may have been one of the first policies of a Canadian federal agency to explicitly incorporate FPIC, somewhat remarkable given the federal government's aversion to the term at the time. The *CIHR Guidelines* were seen as part of an evolving culture of research and ethics that was redefining research relationships as part of a "broader movement transforming the relationship between Aboriginal peoples and Canadian society", in which research is understood as having "a critical role to play in creating the knowledge base for mutually respectful relationships and full participation in Canadian life, with all its responsibilities and benefits" (Aboriginal Research Ethics Initiative (AREI), 2008, p. 5).

TCPS2 Core Principles

The *CIHR Guidelines* remained in place until 2010 when they were superseded by a new and completely revised version of the TCPS, called TCPS2, which was further revised in 2014 and 2018. TCPS2 (Canadian Institutes for Health (CIHR) et al., 2018) is premised on the underlying value of *Respect for Human*

Dignity, requiring that research involving humans be conducted in a way that is sensitive to the inherent worth of all human beings and gives every person due respect and consideration (Canadian Institutes for Health (CIHR) et al., 2018, p. 7). Respect for Human Dignity is expressed through three complementary and interdependent core principles, including *Respect for Persons, Concern for Welfare* and *Justice*, described as follows:

Respect for Persons – recognizes the intrinsic value of human beings and offers the respect and consideration they are due. It incorporates dual moral obligations to respect autonomy and to protect those with developing, impaired or diminished autonomy. An important mechanism for respecting participants' autonomy in research is the requirement to seek their "free, informed and ongoing consent", described in more detail subsequently (Canadian Institutes for Health (CIHR) et al., 2018, p. 7).

Concern for Welfare – refers to concern about the quality of a person's experience of life in all its aspects, including the impact of factors such as physical, mental and spiritual health, as well as physical, economic and social circumstances. It involves preventing or minimizing potential harms and finding an acceptable balance of risks and benefits involved in participation (Canadian Institutes for Health (CIHR) et al., 2018, pp. 7–8).

Justice – refers to the obligation to treat people fairly and equitably. Fairness entails treating all people with equal respect and concern. Equity requires distributing the benefits and burdens of research participation in such a way that no segment of the population is unduly burdened by the harms of research or denied the benefits of the knowledge generated from it (Canadian Institutes for Health (CIHR) et al., 2018, p. 8).

TCPS2 and The Consent Process

TCPS2 (Canadian Institutes for Health (CIHR) et al., 2018, pp. 27–48, Chapter 3) includes a 21-page chapter devoted to "The Consent Process" that applies to all research involving humans (not just research involving Indigenous peoples), and is based on the underlying principle of Respect for Persons, interpreted as respect for an individual's agreement or refusal to participate. TCPS2 Chapter 3 (Canadian Institutes for Health (CIHR) et al., 2018, pp. 27, 35) avoids FPIC as a term and instead describes consent as "free, informed and ongoing", noting that "free" and "voluntary" are used interchangeably. The consent of participants is required "prior" to engaging in research, except under certain circumstances (Canadian Institutes for Health (CIHR) et al., 2018, pp. 27, 35) but the word "prior" is not directly linked with "free, informed and ongoing". The guidance on consent gives substantial detail about the expectations inherent in the three elements of voluntary, informed and ongoing as follows:

Voluntary – TCPS2, Article 3.1 (Canadian Institutes for Health (CIHR) et al., 2018, p. 28) requires that consent be given *freely* or *voluntarily* and be able to withdraw at any time. In withdrawing consent, the participant can also request his or her data be withdrawn, to the extent possible, given the methods of data collection and storage that were used (e.g. individual data may be impossible to identify if anonymity was used or if information was collected from a group without attribution to individuals). The consent process is required to make clear at the onset whether or not withdrawal of data is possible and individuals "should not suffer any disadvantage or reprisal for withdrawing" (Canadian Institutes for Health (CIHR) et al., 2018, p. 29). Participants should also be made aware that "it is impracticable, if not impossible, to withdraw results once they have been published or otherwise disseminated" (Canadian Institutes for Health (CIHR) et al., 2018, p. 29).

Methods of participant recruitment are seen as "important elements in assuring (or undermining) voluntariness" since "undue influence, coercion or the offer of incentives may undermine the voluntariness of a participant's consent" (Canadian Institutes for Health (CIHR) et al., 2018, p. 28). Undue influence and manipulation are seen as potentially arising in situations related to power relationships and positions of authority where individuals "may feel constrained to follow the wishes of those who have some form of control over them" and where "control may be physical, psychological, financial or professional". Likewise, paying attention to existing relationships of trust and dependency is advised as "any relationship of dependency, even a nurturing one, may give rise to undue influence even if it is not applied overtly" (Canadian Institutes for Health (CIHR) et al., 2018, p. 28).

Informed – TCPS2, Article 3.2 (Canadian Institutes for Health (CIHR) et al., 2018, p. 30) requires "full disclosure of all information necessary for making an *informed* decision to participate" and offers a lengthy list of elements to include in plain language, such as:

- Purpose, nature and expected duration of research activities, as well as the identity and contact information of a qualified designated representative of the research institution "who can explain scientific or scholarly aspects of the research";
- Assurance that there is no obligation to participate, that participants may withdraw at any time without penalty or prejudice to pre-existing entitlements, and that information relevant to continuing or withdrawing will be provided throughout the project;
- Identities of research proponents (individuals, institutions, partners) and funders of the research;
- Responsibilities of all parties involved, including participants;
- What information will be collected from or about participants and for what purposes;
- Anticipated uses of data;
- Who will have access to information collected and the identity of participants;
- How privacy and confidentiality will be protected;

- Measures for disseminating results and whether participants will be identified directly or indirectly;
- All reasonably foreseeable risks and potential benefits that may arise from participation;
- Any possibility of findings being commercialized;
- Any real, potential or perceived conflicts of interest on the part of the researchers, their institutions or the research sponsors;
- Information about any payments, including incentives for participants, reimbursement for participation-related expenses and compensation for injury;
- A statement that, by consenting, participants have not waived any rights to legal recourse in the event of research-related harm; and
- The identity and contact information of an appropriate arms-length individual who may be contacted regarding ethical issues arising (e.g. ombudsperson).

TCPS2 requires that *language* not be a barrier to the consent process, which may require "an intermediary who has the necessary language skills to ensure effective communication" (Canadian Institutes for Health (CIHR) et al., 2018, p. 31).[7] *Adequate time and opportunity* are required for this initial phase of the consent process to ensure mutual understanding, which will depend on many factors, such as complexity of information and likelihood of harms.

Ongoing – TCPS2, Article 3.3 (Canadian Institutes for Health (CIHR) et al., 2018, p. 33) speaks to the ongoing nature of consent as *encompassing a process* that begins with initial contact and carries through to the end of the participants' involvement in a project. It involves "an ongoing duty" to provide participants with all information relevant to their ongoing consent to participate. It highlights the researcher's "ongoing ethical and legal obligation to bring to participants' attention any changes to the research project that may affect them", including "changes to the risks or potential benefits of the research" to give participants "the opportunity to reconsider the basis for their consent in light of the new information".

TCPS2 (Canadian Institutes for Health (CIHR) et al., 2018, p. 47) also requires that consent be *documented*, noting that "written consent in a signed statement from the participant is a common means of demonstrating consent, and in some instances, is mandatory" but there are many other ethically acceptable means of providing consent. These might include oral consent or other verbal agreements, intentional physical gestures such as a handshake, or exchange of gifts, which can symbolize "the establishment of a relationship comparable to consent". These alternative processes for demonstrating consent are intended for circumstances where there are valid reasons for not documenting consent in writing (e.g. in a signed consent form), not as options for convenience of the proponents of research. The procedures that are used to seek consent must be documented, and it is recommended that a "written statement of the information conveyed in the consent process" be left with the participant as "evidence that they have agreed to participate" and a reminder of what they agreed to at that moment in time.

TCPS2 and Indigenous Research

TCPS2 (Canadian Institutes for Health (CIHR) et al., 2018) also contains a 28-page chapter on Indigenous research (Chapter 9: Research Involving the First Nations, Inuit and Métis Peoples of Canada) that was informed by a technical advisory committee called the Panel on Research Ethics-Technical Advisory Committee on Aboriginal Research (PRE-TACAR).[8] PRE-TACAR did not explicitly recommend using the term FPIC in TCPS2 Chapter 9 but framed the requirement for collective and individual consent as one of several fundamental elements necessary for ethical research involving Indigenous peoples, especially noting the importance of "group interests and the risk of doing harm due to misunderstanding of cultural norms" (Aboriginal Research Ethics Initiative (AREI), 2008, p. 6). Ethical research was situated within the underlying value of Respect for Human Dignity, which was understood as also implying "respect for diversity within and among [Indigenous] communities, for Indigenous knowledge systems, and for [Indigenous] cultural heritage" (Aboriginal Research Ethics Initiative (AREI), 2008, p. 9).

TCPS2 Chapter 9 (Canadian Institutes for Health (CIHR) et al., 2018) interprets the three core principles of TCPS2 in Indigenous contexts and tunes the reader's understanding to the concerns and considerations in research involving Indigenous peoples, as follows:

Respect for Persons – is primarily expressed and implemented through a free, informed and ongoing consent process, and ethical protections are needed beyond the scope of the individual to "address concerns of First Nations, Inuit and Métis for their continuity as peoples with distinctive cultures and identities" (Canadian Institutes for Health (CIHR) et al., 2018, p. 111). Thus, ethical protections must "extend to the interconnection between humans and the natural world, and include obligations to maintain, and pass on to future generations, knowledge received from ancestors as well as innovations devised in the present generation" (Canadian Institutes for Health (CIHR) et al., 2018, p. 111). Abiding by codes of research practice developed by Indigenous communities themselves is indicated as a primary tool for this purpose.

Concern for Welfare – is understood as broader than the well-being of individuals, "requiring consideration of participants [...] in their physical, social, economic and cultural environments [...] as well as concern for the community to which participants belong", and recognizing the "important role of Indigenous communities in promoting collective rights, interests and responsibilities that also serve the welfare of individuals" (Canadian

Institutes for Health (CIHR) et al., 2018, p. 111). The interpretation of this principle underscores the need for research involving Indigenous peoples to "enhance their capacity to maintain their cultures, languages and identities as First Nations, Inuit or Métis peoples, and to support their full participation in, and contributions to, Canadian society" (Canadian Institutes for Health (CIHR) et al., 2018, p. 111).

Justice – raises awareness of historical imbalances of power between researchers and participants, and harms that have resulted. For many Indigenous peoples in Canada, "abuses stemming from research have included: misappropriation of sacred songs, stories and artefacts; devaluation of Indigenous peoples' knowledge as primitive or superstitious; violation of community norms regarding the use of human tissue and remains; failure to share data and resulting benefits; and dissemination of information that has misrepresented or stigmatized entire communities" (Canadian Institutes for Health (CIHR) et al., 2018, p. 111).

Dr. Marlene Brant-Castellano, who chaired PRE-TACAR and had a significant role in the development of TCPS2 Chapter 9, acknowledged existing tensions in the colonial nature of the TCPS 2 language and the attempts to codify Indigenous ethical principles. She offered some ways to reframe the TCPS2 language to help bridge the words with applications in Indigenous contexts. For example, she suggested that where the TCPS2 uses "respect for human dignity", an Indigenous articulation may include "spiritual responsibilities to maintain right relationships" (Brant Castellano, 2008, p. 23). She shared the following translation offered by Anishinabek Elder George Courchene, in the context of relationships between researchers and Indigenous participants in research:

- Kindness implies respect for the dignity of the others involved, not dominating or pressing our own agenda at the others' expense;
- Honesty involves communicating our principles and intentions as the basis for relationship and ensuring free, informed consent for actions taken;
- Sharing recognizes that the common good requires give and take by all, with respect for the different gifts that each party brings; and
- Strength is courage to stand firm for our principles; in some cases, strength is resilience, as in the capacity to bend to circumstance while holding on to important values (Brant Castellano, 2008, p. 23).

Brant-Castellano (2008) concluded:

> Together, these virtues balance one another to maintain respect for self and others. All parties to a relationship are responsible for maintaining this ethical balance. While words to describe relationships differ, it is possible to see the harmony between the ethics of "respect for human dignity"

> endorsed by researchers and the ethics of "right relationships" embodied in First Nation, Inuit and Métis traditional teachings.
>
> *(Brant Castellano, 2008, p. 23)*

TCPS2 and Consent in Indigenous Research

TCPS2 Chapter 9 is described as a "framework for the ethical conduct of research involving Indigenous Peoples" and "is not intended to override or replace ethical guidance offered by Indigenous peoples themselves" but "to ensure, to the extent possible, that research involving Indigenous peoples is premised on respectful relationships" (Canadian Institutes for Health (CIHR) et al., 2018, p. 107). TCPS2 Chapter 9 focuses on four primary means to implement the core principles and to address requirements for free, informed and ongoing consent. These include community engagement, collaboration, research agreements and community customs and codes of research practice.

Community engagement – TCPS2 Articles 9.1 through 9.6 provide detailed guidance on requirements for community engagement, which is defined as "a process that establishes an interaction" that "signifies the intent of forming a collaborative relationship between researchers and communities" (Canadian Institutes for Health (CIHR) et al., 2018, p. 110). Community engagement is required when "research is likely to affect the welfare of an Indigenous community, or communities to which prospective participants belong" (Canadian Institutes for Health (CIHR) et al., 2018, p. 112), for example, if research takes place on Indigenous lands, includes Indigenous identity as criteria or involves documentation or interpretation of Indigenous cultural heritage, artifacts, traditional knowledge, language or history. Community engagement is required for both primary collection of research data and secondary use of information that was originally collected for another purpose.

Community engagement is understood as an upfront process that may take a variety of forms, depending on the circumstances, the nature of the research and those involved. For example, community engagement may include:

> review and approval from formal leadership to conduct research in the community, joint planning with a responsible agency, commitment to a partnership formalized in a research agreement, or dialogue with an advisory group expert in the customs governing the knowledge being sought.

The degree and type of collaboration may vary, from "information sharing to active participation and collaboration, to empowerment and shared leadership of the research project", including the possibility of a community choosing not to engage but to indicate no objection to a research project (Canadian Institutes for Health (CIHR) et al., 2018, p. 110).

TCPS2 Chapter 9 clarifies that neither "engagement with formal leadership" (Canadian Institutes for Health (CIHR) et al., 2018, p. 115) of an Indigenous community nor "community agreement that a research project may proceed" (Canadian Institutes for Health (CIHR) et al., 2018, p. 123) are substitutes for seeking the consent of individual participants. As well, community engagement requires researchers be "informed about formal rules or oral customs that may apply in accordance with a particular First Nations, Inuit or Métis authority" (Canadian Institutes for Health (CIHR) et al., 2018, p. 112).

Collaboration – TCPS2 Article 9.12 (Canadian Institutes for Health (CIHR) et al., 2018, p. 123) advises researchers and communities to "consider applying a collaborative and participatory approach as appropriate to the nature of the research, and the level of ongoing engagement desired by the community" where "the nature and degree of collaboration between the researcher and the community will depend on the nature of the research, and the community context". Collaboration is described as based on mutually respectful relationships between colleagues who contribute their different expertises, capacities and perspectives in productive ways for mutual benefit. Collaboration is not seen as equal contributions; rather, different partners may take primary responsibility for different aspects of the research.

A participatory approach is described as "action-oriented" and "based on respect, relevance, reciprocity and mutual responsibility", where "those involved in the research process collaborate to define the research project, collect and analyze the data, produce a final product and act on the results" (Canadian Institutes for Health (CIHR) et al., 2018, p. 124). The TCPS2 advises that the terms and conditions of collaborative and participatory research be set out in a research agreement to clearly define the research relationship.

Research agreements – After formal community engagement and before initiating participant recruitment or research activities, TCPS Article 9.11 specifies that the terms and activities of researchers and Indigenous communities, including consent processes, should be set out in a research agreement. Research agreements are described as the "primary means of clarifying and confirming mutual expectations and, where appropriate, commitments between researchers and communities". At minimum, research agreements are seen to achieve the following (Canadian Institutes for Health (CIHR) et al., 2018):

- Address ethical protections related to securing individual consent;
- Specify commitments regarding collective community participation and decision making, sharing of benefits, and review or updating of the research agreement;
- Set out the purpose of the research;
- Detail mutual responsibilities in project design, data collection and management, analysis and interpretation, credit due to knowledge holders, protection (and non-disclosure) of restricted knowledge, sharing of benefits, sharing of royalties flowing from intellectual property where applicable;

production of reports, co-authorship, dissemination of results and a conflict resolution process;

- Include provisions for any anticipated secondary use of information;
- Set out responsibilities and requirements in accordance with any code of research practice that has been put in place by an Indigenous community (Canadian Institutes for Health (CIHR) et al., 2018, pp. 122–123).

TCPS2 Chapter 9 indicates that research agreements should adhere to codes of research practice that have been adopted by, or are already in place within Indigenous communities. It also recognizes the time and resources needed for community engagement, building relationships, agreeing on research goals and developing research agreements, and suggests that these costs should be factored into research funding proposals, as possible.

Indigenous customs and codes of research practice – Respect for Indigenous community "cultural traditions, customs and codes of practice" (Canadian Institutes for Health (CIHR) et al., 2018, p. 108) is a prevalent theme throughout TCPS2 Chapter 9, and is stated as an explicit obligation of researchers seeking to work with Indigenous communities in Article 9.8:

> Researchers have an obligation to become informed about, and to respect, the relevant customs and codes of research practice that apply in the particular community or communities affected by their research. Inconsistencies between community custom and [the TCPS2] should be identified and addressed in advance of initiating the research, or as they arise.
>
> *(Canadian Institutes for Health (CIHR) et al., 2018, pp. 118–119)*

Moreover, TCPS2 Chapter 9 states:

> The absence, or perceived absence, of a formal local [Indigenous community] research code or guidelines does not relieve the researcher of the obligation to seek community engagement in order to identify local customs and codes of research practice.
>
> *(Canadian Institutes for Health (CIHR) et al., 2018, p. 119)*

TCPS2 Chapter 9 recognizes that First Nations, Inuit and Métis communities have customs and procedures derived from oral traditions that may or may not be in written form but are integral to respecting First Nations, Inuit and Métis identities, cultures and knowledge systems and to protecting the integrity of Indigenous knowledge that may be shared outside traditional forms of cultural transmission. These customs and codes will vary depending on the community and culture involved. They are seen as vital to determine "what information may be shared, and with whom" and to "distinguish among knowledge that can be publicly disclosed, disclosed to a specific audience, or disclosed under certain conditions". As well, "custom may restrict the observation, recording, or reporting of ceremonies or certain performances and require approval of appropriate individuals" (Canadian Institutes for Health (CIHR) et al., 2018, p. 119).

TCPS2 Chapter 9 (Canadian Institutes for Health (CIHR) et al., 2018, p. 121) notes that while Indigenous community codes of practice, institutional research ethics policies and research agreements developed between communities and researchers may share similar goals, "the approaches to achieving those goals may differ significantly". It is seen as inappropriate to "insist on uniformity between community practices and institutional policies". An example offered is when Indigenous Elders are willing to share their knowledge according to traditional customs of consent, other processes and language that may feel culturally inappropriate or awkward for the Elders should not be imposed.

Like the *CIHR Guidelines*, TCPS2 essentially offers insights on how to uphold the key elements of FPIC while deepening and strengthening the principle of consent through relational processes that are needed to *enact* rather than simply *transact* consent, namely within collaborative relationships based on participatory approaches and guided by the Indigenous communities involved. These understandings are preferably represented in a written research agreement, not necessarily in the spirit of trying to create an enforceable contractual arrangement, but as a shared documentation of what is agreed between those who are in relationship together.

It is important to acknowledge that the implementation of TCPS2 as national research ethics policy varies widely across universities and other research institutions in Canada. Thus, the relational potential of consent (and other principles) also depends on interpretations of research ethics boards and specific institutional research policies. For better and worse, conflation between the goals of research ethics (i.e. protection of humans) and managing general risk of liability in research at the institutional level is common in Canadian universities, influencing the form and nature of research agreements.

Indigenous Community Protocols

Across Canada, many First Nations, Métis and Inuit communities have developed Indigenous community protocols (e.g. community research codes and guidelines referred to in TCPS2 Chapter 9) as a governance tool to articulate their expectations and requirements regarding conditions such as access, use, ownership and protection of their Indigenous knowledge and biocultural heritage. Indigenous community protocols can be thought of as expressions of Indigenous self-determination that also fill legal, policy and educative voids in Canada amid a nation-wide moral and legal commitment to reconciliation with First Nations, Métis and Inuit peoples.

Canada's legal commitment to reconciliation stems from the Truth and Reconciliation Commission of Canada's 6-volume final report and 94 Calls to

Action (Truth and Reconciliation Commission of Canada, 2015), adoption of UNDRIP in 2010 and passing of *Bill C-15:* United Nations Declaration on the Rights of Indigenous Peoples Act in 2021. Even prior to UNDRIP becoming law, the federal government articulated a clear moral commitment to achieving reconciliation with Indigenous peoples through renewed Indigenous-crown relationships "based on recognition of rights, respect, co-operation, and partnership as the foundation for transformative change" by adopting *Principles respecting the Government of Canada's relationship with Indigenous peoples* (Government of Canada, 2018).

These principles are rooted in existing Aboriginal and treaty rights in Section 35 of Canada's Constitution Act 1982 (Part II: Rights of the Aboriginal Peoples of Canada), guided by UNDRIP (2007), and informed by both the TRC Calls to Action (Truth and Reconciliation Commission of Canada, 2015) and the Report of the Royal Commission on Aboriginal Peoples (Canada & Royal Commission on Aboriginal Peoples (RCAP), 1996). RCAP was established by the federal government in 1991 and concluded in 1996 with a 5-volume report consisting of 4,000 pages and 440 recommendations, including *Ethical Guidelines for Research* that was sponsored by RCAP (Canada & Royal Commission on Aboriginal Peoples (RCAP), 1993). The RCAP report called for extensive changes to the relationship between Indigenous and non-Indigenous people and governments in Canada, and focused on the vision of a "new relationship", founded on the recognition of Indigenous peoples as self-governing Nations within Canada (Canada & Royal Commission on Aboriginal Peoples (RCAP), 1996).

Essentializing Indigenous Community Protocols

Amid Canada's current "age of reconciliation" and especially given new legislation that enshrines UNDRIP in law, Indigenous community protocols serve a vital role in informing government, academic, educational, non-profit and industry sectors about specific cultural and place-based rights, responsibilities and relationships of Indigenous peoples with respect to their traditional lands, airs and waters, including the biodiversity within these places. Indigenous community protocols are increasingly recognized as the appropriate starting place for establishing respectful research, environmental monitoring and resource development projects and partnerships with Indigenous communities.

Federal government policies are pointing to Indigenous community protocols as essential tools in respecting Indigenous self-determination and fostering collaborative working relations. For example, the *Policy on Scientific and Indigenous Knowledge Integrity* of Crown-Indigenous Relations and Northern Affairs Canada (2019) states:

> Crown-Indigenous Relations and Northern Affairs Canada will adhere to Indigenous Knowledge protocols that have been developed by Indigenous

> governments, organizations, or communities, in order to foster respectful, collaborative, and productive working relationships with knowledge holders.
>
> *(Crown-Indigenous Relations and Northern Affairs Canada, 2019, Article 7.2.2.2)*

National Indigenous organizations in Canada have developed guides, toolkits, templates and reports to support Indigenous communities in understanding key issues and developing their own public-facing protocols for sharing and protecting their cultural knowledge, for example:

Ethics in First Nations Research by the Assembly of First Nations (2009), *Aboriginal Women and Aboriginal Traditional Knowledge: Input and Insight on Aboriginal Traditional Knowledge* by Native Women's Association of Canada (n.d.), *Considerations and Templates for Ethical Research Practices* by the National Aboriginal Health Organization (2007) and *Ownership, Control, Access, and Possession (OCAP) or Self-Determination Applied to Research: A Critical Analysis of Contemporary First Nations Research and Some Options for First Nations Communities* by the National Aboriginal Health Organization, First Nations Centre (National Aboriginal Health Organization (NAHO) & First Nations Centre, 2007). In addition to National Indigenous organizations, many guides, templates and toolkits have been developed by Indigenous entrepreneurs (e.g. a *Template Traditional Knowledge Confidentiality and Consent Form* by Indigenous Corporate Training Inc.)[9] and allies working under the guidance of Indigenous experts (e.g. *the Indigenous Guardians Toolkit* developed by Nature United).[10]

There are also recent efforts by federal government agencies in Canada to assist Indigenous communities in developing written protocols. One example is the *Guidance Document for Community Knowledge Protocols and Data Sharing Agreements* commissioned by the Indigenous Community-Based Climate Monitoring Program of Crown-Indigenous Relations and Northern Affairs Canada to advance Indigenous community-based climate data management and data sharing (Bannister et al., 2021). The *Guidance Document* is intended

> to assist First Nations, Inuit and Métis Peoples in Canada who wish to develop written protocols for sharing their knowledge, information and/or data with the goal to ensure Indigenous knowledge, and community information and data are shared in ways and forms that respect and protect the integrity of the community's Indigenous knowledge and protect the community's interests and rights.
>
> *(Bannister et al., 2021, p. 2)*

Clearly, Indigenous community protocols have become widely accepted within Canada. Moreover, they increasingly are deemed *essential* to inform and guide culturally appropriate ways for academic, government, industry and non-profit sectors to work with Indigenous communities and respectfully include

Indigenous knowledge in all manner of projects, programs and initiatives, including but not limited to those of an environmental nature. However, while some Indigenous community protocols are well known and publicly available, many are not. In some cases, versions of these documents have circulated online in the past and may be currently posted by third parties, but are no longer available from the original source community for various reasons (e.g. Bannister, 2004, 2009 include examples formerly available online, only some of which are still available). Documents or processes may have changed or be under revision, web links may no longer be active or content may be considered proprietary and only shared within a collaborative relationship directly involving the Indigenous community. Many Indigenous communities across Canada are at various stages of considering or developing their own written articulations, sometimes borrowing elements from existing examples developed by other Indigenous communities. Some Indigenous communities may prefer not to have their protocols codified in writing at all.

The diversity of examples of Indigenous community protocols publicly available in Canada reveal that some Indigenous communities have been in a position to take the lead on articulating their expectations and requirements on their own terms, in their own words and languages, according to their own principles, values and laws to meet their own goals. Other Indigenous communities may have developed their written protocols to comply with external processes and templates, presumably as a pragmatic reaction to externally imposed research, monitoring or development pressures.

Indigenous Community Protocols and Consent

Consent is an element of most (but not all) publicly available Indigenous community protocols. When included, consent is articulated in a wide variety of ways, sometimes explicitly as FPIC, but more often as a variation of "informed consent". The description of consent may be accompanied by a detailed list of required criteria and processes similar to what is summarized in a previous section of this chapter, based on TCPS2 Chapter 3 (Canadian Institutes for Health (CIHR) et al., 2018), along with a template consent form. Or consent may be referred to only briefly, for example, as an expected part of a research agreement.

The conception of consent as a principle emerging from Western systems of ethics and law can lead to articulations of a transactional nature within written Indigenous community protocols. This is especially the case when a community's requirements of consent are examined in isolation of that community's broader principles and values, which situate consent within the community's specific understandings and worldview related to what is being consented to, by who and how the consenting process should happen in culturally appropriate ways. Four examples are provided below to illustrate the diversity of how consent is articulated by different Indigenous communities in Canada. Also included is a brief summary of that community's written expression of principles and values, which place consent within a relational approach to protecting Indigenous knowledge and biocultural rights.

Sambaa K'e Dene Band Policy Regarding the Gathering, Use and Distribution of *Yúndíit'õh* (Traditional Knowledge) (Sambaa K'e Dene Band, 2003) – Consent is not mentioned specifically but procedures are outlined for the use, gathering and distribution of *yúndíit'õh* (traditional knowledge) in a step-by-step way that is based on developing a research agreement to detail all terms and conditions, including usage, copyright, storage and ownership (Sambaa K'e Dene Band, 2003). *Yúndíit'õh* is translated as "the past time of the land" or as "our heritage" which "includes all of the stories, legends, experiences, practices, beliefs, etc. of the Sambaa K'e Dene people from time immemorial" (Sambaa K'e Dene Band, 2003, p. 1).

The gathering, use and distribution of Sambaa K'e *yúndíit'õh* are guided by several principles that make clear the quality and type of interactions required for sharing to take place, for example:

- *Yúndíit'õh* is derived from a traditional process of intuition, observation, testing and validation and is of equal value to Western scientific processes;
- *Yúndíit'õh* belongs to the Sambaa K'e Dene as a whole and is therefore a collective responsibility. Decisions concerning what *Yúndíit'õh* information to share with outside agencies must be made through a community process, with the full and active involvement of the elders;
- *Yúndíit'õh* is closely linked with, and dependent on, the language in which it is rooted and must therefore be documented and shared to the greatest extent possible in the Sambaa K'e Dene Yatie dialect (Sambaa K'e Dene Band, 2003, p. 2).

Tl'azt'en Nation Guidelines for Research in Tl'azt'en Territory (Tl'azt'en Nation, 1998) – "Informed consent" is briefly described as being required of individuals and groups, in writing or recorded, without pressure to participate, after being provided with information about the purpose and nature of the research activities, expected benefits and risks, and the degree of confidentiality that will be maintained (Tl'azt'en Nation, 1998, p. 2).

Some of the accompanying principles require that:

- Research must reflect "the distinctive perspectives and understandings, deriving from [Tl'azt'enne] culture and history and, embodied in Tl'azt'en language";
- There must be an "opportunity to correct misinformation or to challenge ethnocentric and racist interpretations" that may be part of existing research before it is taken as a basis for new research;
- The "authenticity of orally transmitted knowledge" should be validated by appropriate means within the particular Indigenous traditions;

- Researchers must "understand and observe the protocol concerning communications within any Tl'azt'en community" and "observe ethical and professional practices relevant to their respective disciplines" (Tl'azt'en Nation, 1998, p. 1).

Mi'kmaq Ecological Knowledge Study Protocol (2nd Edition) of the Assembly of Nova Scotia Mi'kmaq Chiefs (Assembly of Nova Scotia Mi'kmaq Chiefs, n.d.) – There is an expectation for "free, informed consent" noting that "education and agreement" are key elements, and a "written Consent and Release Form" that specifies the "agreement factors" is required. It must be made clear to participants that the Mi'kmaq Ecological Knowledge Study "is not Consultation for the purpose of justifying an infringement of Aboriginal and Treaty Rights" (Assembly of Nova Scotia Mi'kmaq Chiefs, n.d., pp. 17–18).

The document also includes a lengthy description of what Mi'kmaq Ecological Knowledge is and the principles that must be included in Mi'kmaq Ecological Knowledge Studies:

> Mi'kmaq maintain a deep and profound relationship with their traditional lands, waters and resources. This longstanding relationship has given rise to Mi'kmaq Ecological Knowledge (MEK) – a body of knowledge that the Mi'kmaq maintain regarding the natural environment.
>
> *(Assembly of Nova Scotia Mi'kmaq Chiefs, n.d., p. 3)*

"Mi'gmaq ways of knowing" include the principles of *Kepmite'tmnej ta'n wettapeksulti'k* and *netukulimk*:

- *Kepmite'tmnej ta'n wettapeksulti'k* translates to "Let us greatly respect our Mi'gmaq roots" and references that Mi'gmaq acknowledge themselves as being born from and rooted in the traditional lands of Mi'gma'qi (Assembly of Nova Scotia Mi'kmaq Chiefs, n.d., p. 5);
- *Nekutulimk* is "the use of natural bounty provided by the Creator for self-support and well-being of the individual and community. *Nekutulimk* is achieving adequate standards of community nutrition and economic well-being without jeopardizing the integrity, diversity, or productivity of our environment" (Assembly of Nova Scotia Mi'kmaq Chiefs, n.d., p. 6).

The Foreword to the document ends with *M'st No'gmaq*, which is the Mi'gmaq way to indicate the end of a prayer "as an acknowledgement to [Mi'gmaq] ancestors and to all creation for giving us the lives we have". *M'st No'gmaq* translates to "All Our Relations", which expresses the Mi'gmaq

> social/spiritual concept of understanding that each and every life depends on all other beings (animate and inanimate) for survival here on Mother Earth. Embedded within this concept is the reality that all creation, and all it encompasses, are interconnected and interdependent upon one another

> as a collective. The collective includes future generations as well as the present and past so this concept of *M'st No'gmaq* has relevance throughout the temporal and spatial dimensions of our spirituality.
>
> *(Assembly of Nova Scotia Mi'kmaq Chiefs, n.d., p. 4)*

The document also makes a clear link between Indigenous rights and responsibilities: "In addition to court affirmed rights in Canada and rights under International Law, the inherent rights of the Mi'gmaq include corresponding responsibilities bestowed upon them by the Creator" (Assembly of Nova Scotia Mi'kmaq Chiefs, n.d., p. 10).

Negotiating Research relationships with Inuit communities: A Guide for Researchers (Inuit Tapiriit Kanatami (Organization) & Nunavut Research Institute, 2007) – General elements of "informed consent" are listed, and discussion with the appropriate local authorities is advised to determine specific requirements and protocols for consent when interviewing people. Clear format and wording are emphasized for written or verbal evidence of consent. It is noted that broader consent should be gained from umbrella organizations or community representatives in addition to individual consent (Inuit Tapiriit Kanatami (Organization) & Nunavut Research Institute, 2007, pp. 8–9).

The document also offers extensive guidance on negotiating "a research relationship" whereby researchers and Inuit communities "jointly define their respective roles and responsibilities, outlining mutual benefits and expectations" (Inuit Tapiriit Kanatami (Organization) & Nunavut Research Institute, 2007, p. 7) to "ensure more responsible, reciprocal, and mutually beneficial research" (Inuit Tapiriit Kanatami (Organization) & Nunavut Research Institute, 2007, p. 21). The document recognizes that research relationships may involve creating a formal agreement or be based on informal arrangements, depending on the community's preference (Inuit Tapiriit Kanatami (Organization) & Nunavut Research Institute, 2007, p. 7).

Practical suggestions to include in the negotiation are provided, such as:

- Be honest and straightforward about community participation and how the project can address community goals;
- Be humble about academic credentials;
- Be willing to learn from local people and foster a reciprocal relationship;
- Be informed about community demographics and socio-economic characteristics;
- Be familiar with the local Inuktitut dialect and/or cultural practices;
- Be open about plans;
- Be patient in getting on with research, and understand that "Inuktitut dialects are not always easy to translate into English" so take time to clarify;
- Keep lines of communication open with the community;
- Respect local cultures, customs and authority;
- Use the local language whenever possible (Inuit Tapiriit Kanatami (Organization) & Nunavut Research Institute, 2007, pp. 7–8).

As the brief summaries of these four examples imply, consent as a principle or element of any Indigenous community protocol is necessary to set out specific expectations and requirements of an Indigenous community. Understandably, in Canada and elsewhere, there has been a trend toward homogenizing and streamlining Indigenous articulations and approaches to consent, partly influenced by widespread awareness of FPIC as a key element of UNDRIP, as well as a need to fit with Western institutional consent processes of academe, government and industry partners. But contemporary articulations of consent (e.g. as lists of criteria and templates) tend to be limited in conveying the relational dimension that is important from a biocultural ethics perspective. What is needed additionally is careful attention to articulations of Indigenous principles and values, whether expressed in writing within Indigenous community protocols or through verbal exchanges, to shift consent from a process of *transaction* based primarily on rights and duties, to one of *interaction* that also fosters relational accountability and right relationships. How to facilitate this shift is a particularly relevant question in Canada, in light of government commitments to reconciliation that promote renewed relationships with Indigenous peoples.

Indigenous Principles and Values as a Foundation for Right Relationships

Criminal law and legal ethics scholar Professor John Humbach (2001) offers a helpful distinction between different paths in the pursuit of justice through legal rights compared with justice through right relationships:

> The justice of rights and relational justice differ enormously in their approach to the conflicts and clashes of human social life. In particular, to pursue relational justice is to seek to actually resolve disputes rather than just subliminalize them, pushing them beneath the surface. The idea that the enforcement of rights actually resolves conflicts is, in most cases, nothing but illusion. The actual "resolution" of disputes means mending tears in the fabric of relationships and, if possible, fostering the establishment of right relationships in the place of manifestly wrong ones. This can only happen, however, when all of those concerned come away feeling they have reached an accord that is respectful of the equal dignity and the legitimate claims and needs of everyone concerned. It is only then that a right relationship can begin to form and build.
>
> *(Humbach, 2001, p. 14)*

Humbach (2001, p. 17) creatively depicts and contrasts rights-based relationships as "paint-by-numbers" with right relationships as "genuine works of art", and points out how rights-based approaches alone have the potential to lead to profoundly wrong relationships. This observation is consistent with the sentiments of the recent *10 Calls to Action to Natural Scientists Working in Canada* by Wong et al. (2020,

p. 780) who express concern about the potential of scientists, funding bodies and research institutions to underestimate what reconciliation means for Canada: "Reconciliation will necessarily be long and deep, and must not devolve into a shallow series of box-ticking exercises". Wong et al. (2020, p. 779) urge that "true reconciliation requires going beyond what is required" and underscore the "need to resist the urge to colonize the process of reconciliation itself" (2020, p. 780).

Emerging Ethical Guidance for Sharing Knowledge across Knowledge Systems

From a biocultural ethics perspective, something that will help prevent the pursuit of biocultural rights for protection of culture and environment from becoming paint-by-numbers or box-ticking exercises is enabling Indigenous values and principles to be living and embodied (rather than merely aspirational) through co-creation of intentional and meaningful relationships and fostering concomitant relational accountability. An example of ethical guidance that embraces this objective has recently emerged in Canada. Rather than starting from Western ethical principles that prioritize the rights of autonomous individuals and duties to one another, this *Ethical Guidance for Knowledge Sharing Across Indigenous and Western Scientific Knowledge Systems* (Ethics Circle, 2021) builds upon foundational Indigenous principles and values that uphold broader interconnected relationships of care and responsibility.

Inspired by Indigenous ethics, relational ethics and ethics of care, and informed by existing national and international ethical norms, this *Ethical Guidance* extends ethical thought and action beyond "research" and beyond "humans", to offer far-reaching voluntary guidance for respectful knowledge sharing across diverse knowledge systems (Ethics Circle, 2021).[11] Development of the *Ethical Guidance* was supported indirectly by government and academic institutions but primarily led by Indigenous and non-Indigenous experts in a voluntary capacity, outside the constraints of these Western systems. The *Ethical Guidance* is intended to complement, not replace or contradict, other existing forms of voluntary and mandatory ethical guidance and regulatory measures that are in place.

Instead of offering a prescription of ethical conduct, the *Ethical Guidance* fosters "working together to create capacities that transform ethical thought, sensibility and action through increasing awareness, understanding and connection" and "seeks to elevate and encourage ethics as a core awareness and practice within research, monitoring, education, curriculum development and other endeavours in a facilitative way" (Ethics Circle, 2021, p. 6). It is meant to support

> all those who seek better ways to work together respectfully and equitably across our diversities by drawing upon those diverse ways of knowing to reach and share new understandings, insights and approaches to action for the wellbeing of all beings and Mother Earth.
>
> *(Ethics Circle, 2021, p. 6)*

The *Ethical Guidance* comprises nine interrelated principles that are intended to be considered simultaneously to nurture "ethical understandings and capacities to share knowledge in a good way" (Ethics Circle, 2021, p. 9), meaning in respectful ways that uphold the integrity of all knowledge systems involved and with "full awareness of interconnectedness between the spiritual and physical realms" (Flicker et al., 2015). The nine guiding principles "address a pressing need for contemporary ethical guidance that considers the intercultural, interspecies and intergenerational aspects of reconciling diverse ways of knowing, being and doing" (Ethics Circle, 2021, p. 9).

The principles are intentionally written as verbs rather than nouns, to emphasize the ongoing action-oriented nature of ethical relations and relational accountability. The principles include:

- Respecting the inherent dignity and interconnectedness of All Our Relations;
- Caring about well-being;
- Embracing humility and precaution;
- Honoring Indigenous languages;
- Nurturing mutual trust and respect;
- Reciprocating positively with generosity;
- Upholding the Integrity of Indigenous knowledge systems;
- Consenting meaningfully;
- Co-protecting Indigenous knowledge.

The first named principle "Respecting the inherent dignity and interconnectedness of All Our Relations" (Ethics Circle, 2021, pp. 13–15) clearly establishes the relational context for all the principles by extending "inherent dignity" (a principle and right enshrined in ethics and law) beyond humankind to embrace all living beings in a way that honors and respects kinship understandings that all things in the universe are related and interconnected in the web of life, and all forms of life, including Mother Earth, exist with an inherent right of being. It means acting in ways that equally value all humans and accepting responsibilities to coexist respectfully with one another and All Our Relations.

The other eight principles are similarly founded in Indigenous principles, values and teachings, and bridged with Western ethical traditions, including the principle of consent, which is articulated as "Consenting meaningfully" (Ethics Circle, 2021, p. 27). This principle draws upon and elaborates elements of FPIC within a broader and deeper relational context. It was partially inspired by the Office of the Privacy Commissioner of Canada's *Guidelines For Obtaining Meaningful Consent* (2018), and draws on the key elements of consent (noted previously) that are described in the *CIHR Guidelines* (Canadian Institutes for Health (CIHR), 2007) and TCPS2 (Canadian Institutes for Health (CIHR) et al., 2018), as well as "educated prior informed consent" described in the *International Society of Ethnobiology's Code of Ethics* (International Society of Ethnobiology (ISE), 2006).

"Consenting meaningfully" (Ethics Circle, 2021, pp. 27–31) is an ongoing and dynamic process of choosing to participate in an activity. The choice is based on fully understanding the nature, purpose and consequences of all aspects of the activity and its physical, emotional, mental and spiritual impacts on individuals, communities and All Our Relations. To be given access to Indigenous knowledge as part of knowledge sharing requires valid and meaningful processes of consenting by individual Indigenous knowledge holders who may extend their responsibility for consent to All Our Relations, for example, by considering the well-being and rights of plants, animals and the local ecology. It may additionally require collective processes of consent by the community, as determined by the Indigenous knowledge holders, community protocols and community governance structures. It must be an iterative and interactive process to maintain the element of voluntariness, especially when the circumstances that were consented to change or when new choices are faced, which requires that information relevant to ongoing consent is continuously communicated. Meaningfully consenting requires mechanisms to extend over long timeframes "when what is shared and agreed to now has implications for future generations (*e.g.,* recordings of Elders)" (Ethics Circle, 2021, p. 31).

Meaningful consenting also extends to secondary use of information that is in the public domain and requires careful consideration about what constitutes ethical access for secondary use of documented Indigenous knowledge that was published by others if "how the knowledge came to be in the public domain is not known, consent is not evident and/or if there is insufficient attribution to the Indigenous Knowledge Holders and Indigenous communities of origin" (Ethics Circle, 2021, p. 31). It is further noted that "intentional secondary use of Indigenous knowledge that was not published with due credit and attribution to the Indigenous community sources, and/or lacks evidence of consent for the Indigenous knowledge to be made public is unethical" (Ethics Circle, 2021, p. 31).

In essence, according to this *Ethical Guidance*, "meaningfully consenting" is not possible outside of relationships that recognize physical and spiritual responsibilities of care and accountability to past, present and future Ancestors and other sentient beings through ongoing stewardship of lands, airways and waterways.

Envisioning Biocultural Rights and Responsibilities within Relationships of Care and Accountability

From a biocultural ethics perspective, the biocultural rights of Indigenous peoples are inextricably linked with responsibilities based in caring relationships and relational accountability that extend across species and generations. The intentional twinning of rights with responsibilities that are derived from Indigenous worldviews and Indigenous articulations of natural law (Battiste, 2016; Borrows, 2002, 2010; Napoleon, 2013) points us toward right relationships among humans, and between humans and the natural world. This approach can deepen our awareness and capacities to work well together across our diversities, in shared goals of protecting culture and environment and caring for the well-being of our earth and All Our Relations.

This chapter offers an ethical lens to complement a tendency toward primarily rights-based approaches, which can lean us away from Indigenous worldviews based in interrelationships and favor Western philosophical, legal and scientific paradigms that are based primarily on the autonomous, rational individual. The legal and ethical principle of consent was chosen here as an example, in the context of Canadian ethics policy initiatives, to explore important relational dimensions of *consenting* – representing more of an interactive verb rather than a transactional noun. In this understanding, consent cannot be reduced to an acronym, and implementation of consent cannot be understood only as a signature on a consent form indicating that a checklist of criteria has been met. The quality and meaningfulness of consent as an element for protecting Indigenous biocultural rights (along with others, such as ownership and mutual benefit) can shift when intentionally situated within relationships. An understanding of how rights and responsibilities are inextricably linked has the potential to not just bring about justice of rights, but also relational justice through right relationships.

Health researcher Julie Bull (2018) points out that in health research involving northern Indigenous peoples in Canada, "[r]esearchers are ethically and culturally obligated to build relationships on a foundation of respect, relevance, reciprocity, and responsibility in an authentic way", stating that these qualities are "imperative" in research involving Indigenous peoples as "a means to respect and practice self-determination in research".

In his book, *Law's Indigenous Ethics*, Dr. John Borrows (2019) explores ethics in relation to Indigenous rights and other legal issues through the Anishinaabe Seven Grandmother and Grandfather Teachings. He concludes:

> There is no guarantee that the application of good principles will deliver us from the challenges we face. [...] Yet [...] our lives could be much better if we applied concepts of love, truth, bravery, humility, wisdom, honesty and respect more fully in our law.
>
> *(Borrows, 2019, p. 240)*

In simplistic terms, a biocultural ethics approach speaks to our choice of values and principles that underlie our quality of *being* together in whatever it is that we are *doing* together. Biocultural ethics inspires more than just good intentions and fulfilling obligations, but invites deeper consideration and deliberation of the ethical opportunities in our work, inviting us to build our awareness and capacities for developing meaningful relationships of care and accountability, as well as respect, humility, reciprocity and integrity, from which will come ethical actions and greater capacity to address the biocultural rights of Indigenous peoples as they relate to protection of culture and environment.

Notes

1 *All URLs retrieved on 1 September 2021.
2 Indigenous peoples in Canada refers to First Nations, Métis and Inuit peoples as the original inhabitants of Canada, each with unique worldviews, histories, laws, languages, cultures, traditions, customs and knowledge systems. The term "Aboriginal people" used in the *CIHR Guidelines for Health Research Involving Aboriginal People* (2007) means the same as Indigenous peoples in Canada.
3 The term "Indigenous community protocols" is used in this chapter as a general description of a variety of instruments and mechanisms developed by many First Nations, Inuit and Métis communities in Canada to convey the information, expectations, policies and processes that they wish to make known regarding culturally appropriate engagement with their community or Nation. Each community has their own name or way to refer to these kinds of instruments and mechanisms, not all of which are articulated in writing. This diversity of expression and articulation that exists across First Nations, Inuit and Métis communities to fulfill their own specific purposes is important to keep in mind.
4 In part, this chapter offers an update and elaboration on the evolution and relevance of consent as an ethical principle in Canada in addressing Indigenous biocultural rights, originally described in Bannister (2004).
5 See "The Nuremberg Code (1947)", 1996.
6 For transparency, the author was a member of the Aboriginal Ethics Working Group (AEWG) from 2004 to 2007, which developed the *CIHR Guidelines for Health Research Involving Aboriginal People* (2007). See http://www.cihr-irsc.gc.ca/e/29134.html
7 It is noteworthy that Canada passed *Bill C-91 An Act* respecting *Indigenous languages* in 2019 and created a new Office of the Commissioner of Indigenous Languages in 2021 to support implementation and creation of effective measure to uphold Indigenous language-related rights found in UNDRIP Articles 13, 14 and 15.
8 For transparency, the author was a member of the Panel on Research Ethics-Technical Advisory Committee on Aboriginal Research (PRE-TACAR) from 2005 to 2008.
9 https://cdn2.hubspot.net/hubfs/374848/docs/TKConsent_Form_.pdf?t=1521485539033
10 https://www.indigenousguardianstoolkit.ca/
11 For transparency, the author is a member of the Ethics Circle who developed the *Ethical Guidance for Knowledge Sharing Across Indigenous and Western Scientific Knowledge Systems* (Ethics Circle, 2021).

References

Aboriginal Research Ethics Initiative (AREI). (2008). *Issues and Options for Revisions to the Tri-Council Policy Statement on Ethical Conduct of Research Involving Humans (TCPS): Section 6: Research Involving Aboriginal Peoples. Report prepared by the Aboriginal Research Ethics Initiative (AREI) of the Interagency Advisory Panel on Research Ethics (PRE) with the Assistance of PRE's Technical Advisory Committee on Aboriginal Research (PRE-TACAR) and the Guiding Consortium for the Development of TCPS Guidelines for Research Involving Aboriginal Peoples*. Interagency Advisory Panel and Secretariat on Research Ethics. http://icwrn.uvic.ca/wp-content/uploads/2013/10/ethics_aboriginal_guidelines_metis_e.pdf

Abouchar, J., Birchall, C. J., & Wortsman, L. (2021, June 21). Bill C-15 Passed: Federal Government to Align Federal Laws with the United Nations Declaration on the Rights of Indigenous Peoples. *Willms & Shier Environment Indigenous Energy Law*. https://www.

willmsshier.com/docs/default-source/articles/article---bill-c-15-passed---federal-government-to-align-federal-laws-with-the-united-nations-declaration-on-the-rights-of-indigenous-peoples---june-21-2021ac4c37826168616da574ff000044313a.pdf?sfvrsn=382456d5_0

Armitage, D., Berkes, F., Dale, A., Kocho-Schellenberg, E., & Patton, E. (2011). Co-management and the Co-production of Knowledge: Learning to Adapt in Canada's Arctic. *Global Environmental Change, 21*(3), 995–1004. https://doi.org/10.1016/j.gloenvcha.2011.04.006

Assembly of First Nations & Environmental Stewardship Unit. (2009). *Ethics in First Nations Research*. Assembly of First Nations, Environmental Stewardship Unit. https://www.afn.ca/uploads/files/rp-research_ethics_final.pdf

Assembly of Nova Scotia Mi'kmaq Chiefs. (n.d.). *Mi'kmaq Ecological Knowledge Study Protocol* (2nd ed.). Kwilmu'kw Maw-klusuaqn Negotiation Office. https://novascotia.ca/abor/aborlearn/docs/MEK%20Protocol%20Second%20Edition.pdf

Bannister, K. (2000). *Chemistry Rooted in Cultural Knowledge: Unearthing the Links between Antimicrobial Properties and Traditional Knowledge in Food and Medicinal Plant Resources of the Secwepemc (Shuswap) Aboriginal Nation*. University of British Columbia.

Bannister, K. (2004). *Lessons for ABS: Academic Policies, Community Protocols and Community-level PIC*. International Expert Workshop on Access to Genetic Resources and Benefit Sharing, Cuernavaca, Mexico. https://polisproject.org/PDFs/Bannister%202004_ABS%20Lessons.pdf

Bannister, K. (2009). Non-legal Instruments for Protection of Intangible Cultural Heritage: Key Roles for Ethical Codes and Community Protocols. In C. Bell & V. Napoleon (Eds.), *Protection of First Nations' Cultural Heritage: Laws, Policy and Reform* (pp. 278–308). UBC Press.

Bannister, K. (2018). Ethical Guidance for Access and Benefit-Sharing: Implications for Reconciliation. In C. Oguamanam (Ed.), *Genetic Resources, Justice and Reconciliation* (1st ed., pp. 212–230). Cambridge University Press. https://doi.org/10.1017/9781108557122.013

Bannister, K. (2020). Right Relationships – Legal and Ethical Context for Indigenous Peoples' Land Rights and Responsibilities. In N. J. Turner (Ed.), *Plants, People, and Places: The Roles of Ethnobotany and Ethnoecology in Indigenous Peoples' land Rights in Canada and Beyond* (pp. 254–268). McGill-Queen's University Press. http://search.ebscohost.com/login.aspx?direct=true&scope=site&db=nlebk&db=nlabk&AN=2406450

Bannister, K., de Koninck, V., Munson, M., Cerpnjak, N., & Doyle, M. (2021). *Guidance Document for Community Knowledge Protocols (CKP) and Data Sharing Agreements (DSA)*. Commissioned by Indigenous Community-Based Climate Monitoring Program, Crown-Indigenous Relations and Northern Affairs Canada. Unpublished.

Bannister, K., & Thomas, M. (2016). The Importance of Cultural Knowledge in Understanding the Antimicrobial Chemistry of Balsamroot. In M. Ignace, N. J. Turner, & S. L. Peacock (Eds.), *Secwepemc People and Plants: Research Papers in Shuswap Ethnobotany* (pp. 336–365). Society of Ethnobiology.

Battiste, M. (Ed.). (2016). *Living Treaties: Narrating Mi'kmaw Treaty Relations*. Cape Breton University Press. http://public.ebookcentral.proquest.com/choice/publicfullrecord.aspx?p=4562571

Borrows, J. (2002). *Recovering Canada: The Resurgence of Indigenous Law*. University of Toronto Press.

Borrows, J. (2010). *Canada's Indigenous Constitution*. University of Toronto Press.

Borrows, J. (2018). Preface: "Mashkikiikwe". In C. Oguamanam (Ed.), *Genetic Resources, Justice and Reconciliation* (pp. xiii–xvi). Cambridge University Press. https://doi.org/10.1017/9781108557122.001

Borrows, J. (2019). *Law's Indigenous Ethics*. University of Toronto Press.

Brant Castellano, M. (2008). Aboriginal Ethics Guide Ethical Research. *Visions Journal, 5*(1), 22–23.

Bull, J. (2018). Chapter 23: A Northern and Indigenous Ethical Imperative for Working with Communities. In H. Exner-Pirot, B. Norbye, & L. Butler (Eds.), *Northern and Indigenous Health and Health Care* (pp. 285–295). University of Saskatchewan. openpress.usask.ca/northernhealthcare

Cajete, G. (2015). *Indigenous Community Rekindling the Teachings of the Seventh Fire*. Living Justice Press.

Canada & Royal Commission on Aboriginal Peoples (RCAP). (1993). *Integrated Research Plan: Ethical Guidelines for Research, Appendix B, 37–40*. Royal Commission on Aboriginal Peoples. https://publications.gc.ca/collections/collection_2016/bcp-pco/Z1-1991-1-41-17-eng.pdf

Canada & Royal Commission on Aboriginal Peoples (RCAP). (1996). *Report of the Royal Commission on Aboriginal Peoples*. Royal Commission on Aboriginal Peoples. https://www.bac-lac.gc.ca/eng/discover/aboriginal-heritage/royal-commission-aboriginal-peoples/Pages/final-report.aspx

Canadian Institutes for Health (CIHR). (2007). *CIHR Guidelines for Health Research Involving Aboriginal People*. CIHR. http://icwrn.uvic.ca/wp-content/uploads/2013/10/ethics_aboriginal_guidelines_metis_e.pdf

Canadian Institutes for Health (CIHR), Natural Sciences and Engineering Research Council of Canada (NSERC), & Social Sciences and Humanities Research Council (SSHRC). (2018). *Tri-Council Policy Statement: Ethical Conduct for Research Involving Humans (TCPS2)*. Government of Canada. http://www.pre.ethics.gc.ca/eng/documents/tcps2-2018-en-interactive-final.pdf

Craft, A. (2019). *How UNDRIP Recognizes the Sacred Relationship with Nibi (Water)*. Centre for International Governance Innovation. https://www.cigionline.org/multimedia/how-undrip-recognizes-sacred-relationship-nibi-water/

Crown-Indigenous Relations and Northern Affairs Canada. (2019). *Policy on Scientific and Indigenous Knowledge Integrity* [Policy]. https://www.rcaanc-cirnac.gc.ca/eng/1575567784632/1575567805298

Ethics Circle. (2021). *Ethical Guidance for Knowledge Sharing Across Indigenous and Western Scientific Knowledge Systems*. July 1 draft version, under review. On file with author. Final version forthcoming in 2022. https://polisproject.org/biocultural-ethics/

Flicker, S., O'Campo, P., Monchalin, R., Thistle, J., Worthington, C., Masching, R., Guta, A., Pooyak, S., Whitebird, W., & Thomas, C. (2015). Research Done in "A Good Way": The Importance of Indigenous Elder Involvement in HIV Community-Based Research. *American Journal of Public Health, 105*(6), 1149–1154. https://doi.org/10.2105/AJPH.2014.302522

Government of Canada. (n.d.). *Model Policy on Scientific Integrity*. Office of the Chief Science Advisor. https://www.ic.gc.ca/eic/site/063.nsf/eng/h_97643.html

Government of Canada. (2020). *Impact Assessments in Canada: Frequently Asked Questions. Fact Sheet*. As of 2020-09-28. Government of Canada. https://www.canada.ca/en/impact-assessment-agency/services/policy-guidance/impact-assessments-canada-faq.html

Government of Canada, Department of Justice. (2018). *Principles Respecting the Government of Canada's Relationship with Indigenous Peoples*. Department of Justice, Canada. https://www.justice.gc.ca/eng/csj-sjc/principles.pdf

Humbach, J. A. (2001). Towards a Natural Justice of Right Relationships. *SSRN Electronic Journal*. https://doi.org/10.2139/ssrn.287485

International Society of Ethnobiology (ISE). (2006). *ISE Code of Ethics (with 2008 Additions)*. ISE. https://www.ethnobiology.net/wp-content/uploads/ISE-COE_Eng_rev_24Nov08.pdf

Inuit Tapiriit Kanatami (Organization) & Nunavut Research Institute. (2007). *Negotiating Research Relationships with Inuit Communities: A Guide for Researchers*. Inuit Tapiriit Kanatami; Nunavut Research Institute. https://www.itk.ca/wp-content/uploads/2016/07/Negotitiating-Research-Relationships-Researchers-Guide_0.pdf

Little Bear, L. (2009). *Naturalizing Indigenous knowledge: A Synthesis Paper*. Aboriginal Learning Knowledge Center; University of Saskatchewan. https://www.afn.ca/uploads/files/education/21._2009_july_ccl-alkc_leroy_littlebear_naturalizing_indigenous_knowledge-report.pdf

McDonald, M. (2009). From Code to Policy Statement: Creating Canadian Policy for Ethical Research Involving Humans. *Health Law Review*, *17*(2–3), 12–25.

Medical Research Council of Canada (MRC). (1987). *Guidelines for Research Involving Human Subjects*. Medical Research Council of Canada. http://catalog.hathitrust.org/api/volumes/oclc/20007254.html

Napoleon, V. (2013). Thinking About Indigenous Legal Orders. In R. Provost & C. Sheppard (Eds.), *Dialogues on Human Rights and Legal Pluralism* (pp. 229–245). Springer Netherlands. https://doi.org/10.1007/978-94-007-4710-4_11

National Aboriginal Health Organization (NAHO), & First Nations Centre. (2007). *Considerations and Templates for Ethical Research Practices*. National Aboriginal Health Organization (NAHO); First Nations Centre. https://achh.ca/wp-content/uploads/2018/07/Guide_Community_FNC_ConsiderationsandTemplates.pdf

Native Women's Association of Canada. (n.d.). *Aboriginal Women and Aboriginal Traditional Knowlege (ATK): Input and Insight on Aboriginal Traditional Knowledge*. Native Women's Association of Canada. Retrieved 2 September 2021, from https://www.nwac.ca/wp-content/uploads/2015/05/2014-NWAC-Aborignal-Women-and-Aborignal-Traditional-Knowledge-Report1.pdf

Norström, A. V., Cvitanovic, C., Löf, M. F., West, S., Wyborn, C., Balvanera, P., Bednarek, A. T., Bennett, E. M., Biggs, R., de Bremond, A., Campbell, B. M., Canadell, J. G., Carpenter, S. R., Folke, C., Fulton, E. A., Gaffney, O., Gelcich, S., Jouffray, J.-B., Leach, M., … Österblom, H. (2020). Principles for Knowledge Co-Production in Sustainability Research. *Nature Sustainability*, *3*(3), 182–190. https://doi.org/10.1038/s41893-019-0448-2

Papillon, M., & Rodon, T. (2017). Indigenous Consent and Natural Resource Extraction Foundations for a Made-in-Canada Approach. *IRPP*, *16*. https://irpp.org/research-studies/insight-no16/

Privacy Commissioner of Canada. (2018). *Guidelines for Online Consent*. Office of the Privacy Commissioner of Canada. https://priv.gc.ca/en/privacy-topics/collecting-personal-information/consent/gl_omc_201805/

Reed, M. S., Stringer, L. C., Fazey, I., Evely, A. C., & Kruijsen, J. H. J. (2014). Five Principles for the Practice of Knowledge Exchange in Environmental Management. *Journal of Environmental Management*, *146*, 337–345. https://doi.org/10.1016/j.jenvman.2014.07.021

Reo, N. J. (2019). Inawendiwin and Relational Accountability in Anishnaabeg Studies: The Crux of the Biscuit. *Journal of Ethnobiology, 39*(1), 65. https://doi.org/10.2993/0278-0771-39.1.65

Rocher, G. (1999). Origin and Development of the Tri-Council Policy Statement on the Ethics of Research Involving Humans. *NCEHR Communique, 9*(2), 4–6.

Rozzi, R. (2012). Biocultural Ethics: Recovering the Vital Links Between the Inhabitants, their Habits, and Habitats. *Environmental Ethics, 34*(1), 27–50.

Sambaa K'e Dene Band. (2003). *Sambaa K'e Dene Band Policy Regarding the Gathering, Use, and Distribution of Yúndíit'õh (Traditional Knowledge).* https://nwtresearch.com/sites/default/files/sambaa-k-e-dene-band.pdf

Social Sciences and Humanities Research Council of Canada (SSHRC). (1977). *Ethics Guidelines for Research Involving Human Subjects.* Social Sciences and Humanities Research Council of Canada.

The Nuremberg Code (1947). (1996). *British Medical Journal, 313*(7070), 1448.

Tl'azt'en Nation. (1998). *Tl'azt'en Nation Guidelines for Research in Tl'azt'en Territory.* http://nafaforestry.org/forest_home/documents/CEM-TlaztenGuidelines.pdf

Truth and Reconciliation Commission of Canada. (2015). *Final Report and Calls to Action.* Truth and Reconciliation Commission of Canada. https://web.archive.org/web/20200505161532/http://trc.ca/about-us/trc-findings.html

UN Permanent Forum on Indigenous Issues. (2005). *Report of the International Workshop on Methodologies regarding Free, Prior and Informed Consent and Indigenous Peoples* (E/C.19/2005/3). United Nations Economic and Social Council. https://undocs.org/pdf?symbol=en/E/C.19/2005/3

Wilson, S. (2001). What is an Indigenous Research Methodology? *Canadian Journal of Native Education, 25*(2), 175–179.

Wong, C., Ballegooyen, K., Ignace, L., Johnson, M. J. (Gùdia), & Swanson, H. (2020). Towards Reconciliation: 10 Calls to Action to Natural Scientists Working in Canada. *FACETS, 5*(1), 769–783. https://doi.org/10.1139/facets-2020-0005

Younging, G. (2018). *Elements of Indigenous Style: A Guide for Writing By and About Indigenous Peoples.* Brush Education.

3

SUMAQ KAWSAY (GOOD LIVING) AND INDIGENOUS POTATOES

On the Delicate Exercise of Ontological Diplomacy

Ingrid Hall[1]

In this chapter, I address the alternative vision of nature conservation proposed by the Parque de la Papa (Potato Park) in Pisac, Peru (Map 3.1). This model has been baptized *Sumaq kawsay* in Quechua, or "good living" (*buen vivir* in Spanish), in the same vein as propositions elaborated in Bolivia and Ecuador. At the same time, it has been adapted according to the conservation objectives of potato biodiversity, which are the Park's *raison d'être*. Although the model is presented as "indigenous" (*indígena* in Spanish) and is generally recognized as such both nationally and internationally, the way it is formulated is actually due to an intermediary, the non-governmental organization ANDES[2]. I will focus on the role of this mediator, especially in framing the principles embedded in *Sumaq kawsay* to make them understandable for a national and international audience.

I propose analyzing the role played by the NGO ANDES in terms of ontological diplomacy, building on the "cosmopolitical proposal" of Stengers (2005) and Latour (2004a, 2004b) who show how taking account of non-humans can (and must) lead to rethinking political ecology in a philosophical sense.[3] For Latour (2004a, pp. 209–215), "diplomats" work for the recognition of hybrid communities, made up of both humans and non-humans, thus disrupting modern collectives and allowing the development of a new "common world" (which is evolving and constantly negotiated). From an ethnographic perspective, we must place this philosophical reflection in the specific context at hand, and set ourselves within a perspective of political ecology in the more classical sense (Benjaminsen & Svarstad, 2009), taking account of the specific configurations of actors, their practices, and their visions. It seems essential to consider how environmental conservation, particularly in Latin America, is closely linked to the recognition of indigenous rights (Dumoulin, 2007; Tsing, 2007). Some ethnographers have studied the actors who have worked

DOI: 10.4324/9781003172642-4

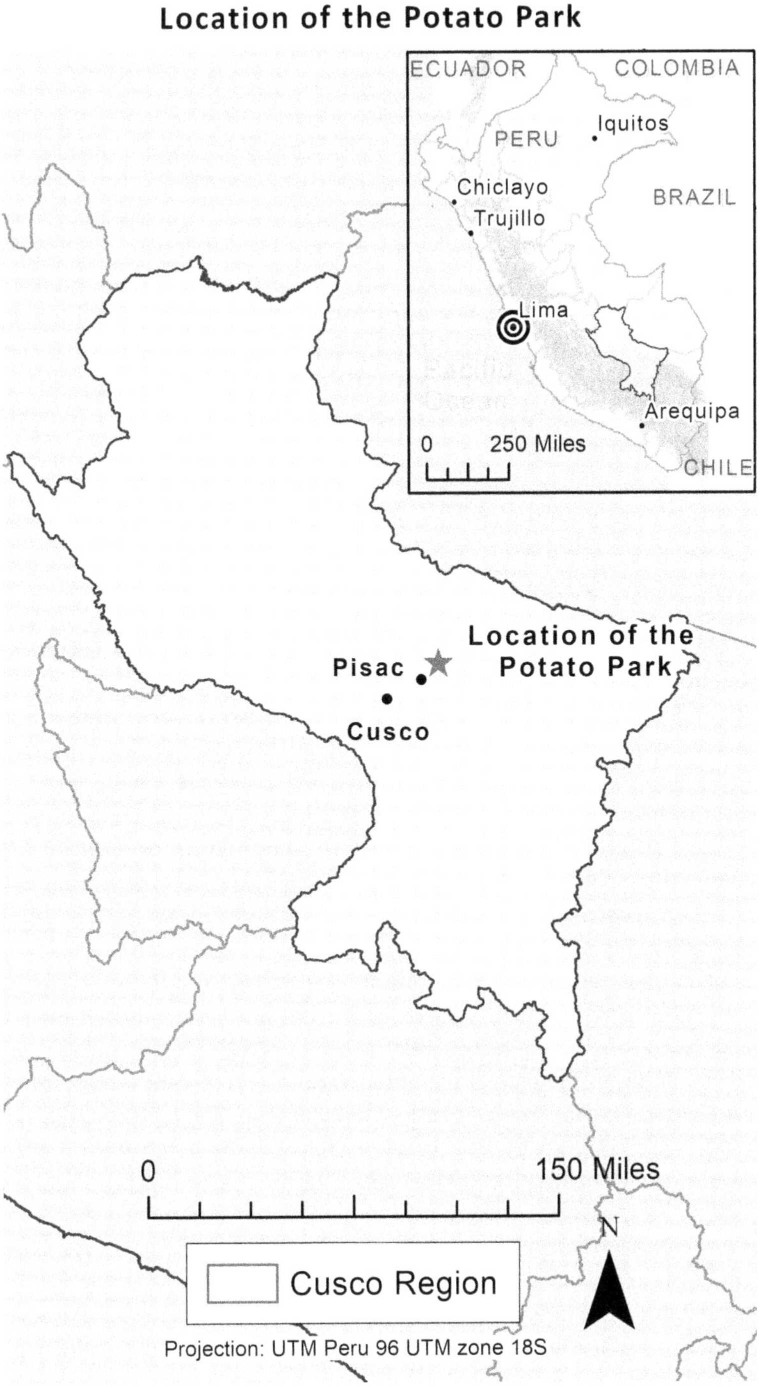

MAP 3.1 Location of the Pisac Potato Park.

Source: Alejandra Uribe Albornoz 2020, used with permission of the author.

toward this *rapprochement* (Albert, 1993; Ramos, 1994; Conklin & Graham, 1995). While showing the diversity of the agents involved (state agencies, NGOs,[4] environmentalists, local communities), they have focused more specifically on actors from the communities with which they work, above all on indigenous leaders, allowing them to make a powerful case for an alternative ontology to naturalism (Blaser, 2009; de la Cadena, 2010, 2015). However, the elaboration of this ontological discourse sometimes results from improbable alliances – for instance, with a priest, as in the Peruvian case described by Li (2013). In the example I will develop here, although the indigenous vision of conservation is upheld mainly by a Peruvian NGO, the discourse of ANDES results from the "intercultural politicization of cosmological categories" (Albert, 1993, p. 369), leading to the translation of local concepts inasmuch as they are compatible with the global ecological imaginary (Conklin & Graham, 1995, p. 697).

This work should be understood in the context of debates on "double conservation" (Dumoulin, 2007), described by ANDES as "biocultural" conservation (ANDES et al., 2012), which are now widely debated in the international governance of biodiversity conservation[5] (see Chapter 1). This vision, as spearheaded by the Potato Park, seems to bode well for furthering mechanisms of access and benefit-sharing linked to the use of biodiversity for the benefit of indigenous and local populations (Thomas & Filoche, 2015). An analysis of ANDES vision may thus clarify the relevance of the tools currently being designed within this paradigm. Within the framework set out by the Convention on Biological Diversity (CBD), we will see that representatives of local and indigenous communities tend to address the biocultural dimension in ontological terms. This observation will allow me to investigate conditions concerning the exercise of such diplomacy in the present context. I take, therefore, an approach that is critical of biocultural visions, following Foyer (2015) and Kohler (2011).

The data upon which this text is based were gathered over several periods of fieldwork carried out in Peru between 2011 and 2018. I conducted interviews on the development of this vision with the head of ANDES, the staff of ANDES and people living in the Park; I consulted various texts produced by ANDES – published or otherwise – as well as internal documents; and I took part in various events where ANDES staff presented their vision. On four occasions, I attended meetings of the working group on recognizing the role of local and indigenous populations for the *in situ* conservation of biodiversity (article 8(j) of the CBD). Together with Alejandro Argumedo, the head of ANDES, I have written on the concept of *Sumaq kawsay* (Argumedo & Hall, forthcoming).

This chapter is organized as follows: after providing an overview of the Potato Park, I will detail its biocultural vision of conservation. I will then analyze the various objectives pursued at local, national, and international levels, in order to situate the framing of this vision within the various streams of logic that give it meaning.

The Potato Park, an "area of indigenous biocultural heritage"[6]

The Potato Park is located within the territory of five peasant communities (*comunidad campesina*)[7] in the high part of the district of Pisac in the southern Peruvian Andes, in the region of Cusco. The Park covers 12,000 hectares at an altitude between 3,500m and 5,000m above sea level. It is home to around 6,000 people, who self-identify as "*campesinos*", which I translate as peasants (despite its awkward connotations in English, on the local terminology see Robin, 2004). The Park covers 12,000 hectares at an altitude between 3,500m and 5,000m above sea level (Figure 3.1). As its name indicates, the Park is devoted to the conservation of potato biodiversity. Potatoes were domesticated in the central Andes 7,000 years ago (Dillehay et al., 2004), and this region has the greatest diversity of varieties in the world. This includes around 180 wild, undomesticated varieties, as identified by the International Potato Center [CIP]) [8], as well as more than 4,000 Andean landraces (Villa et al., 2005) i.e. indigenous Andean potatoes that are the result of the long process of domestication and selection (Figure 3.2). These wild varieties and peasant-domesticated landraces represent a genetic resource of great importance for selecting and creating new varieties that are resistant to certain diseases, or adapted to different environments and certain climatic conditions such as drought or frost. Since the potato is the third-most consumed food worldwide (Campos & Ortiz, 2019), and its importance is growing in the diets of emerging countries,[9] the conservation of this heritage is of worldwide significance. Moreover, as the limits of *ex situ* conservation have become clear,[10] *in situ* conservation of

FIGURE 3.1 Landscape of the Potato Park.
Source: Otárola, 2016, used with permission of the author.

FIGURE 3.2 Indigenous potatoes.
Source: Hall, 2016.

this genetic material has acquired greater urgency. In addition, peasant agriculture in Peru depends heavily on seeds from these landraces: less than 10% of seeds come from the commercial sector (Lapeña, 2012a, 2012b). Production is mainly for self-consumption by peasants themselves and seeds are chosen at harvest time. *In situ* conservation of this genetic material is thus closely linked to food security.

Potato conservation activities in the Park consist of collecting indigenous varieties and breeding them on-site. Three plant collections have been created, of which the two main ones are the local collection and the "repatriated" collection. The first of these was created by the peasants of the Park with the help of the NGO ANDES and it includes more than 600 cultivars. This effort went hand-in-hand with the first phase of establishing the Park itself. The second collection consists of 410 landraces from the germplasm bank of the CIP gathered in the region of Cusco, which have been "repatriated" to the Park under an agreement between the Park and the CIP (on this subject see Hall, 2022). Every year, these two collections are cultivated on three parcels of land of less than a hectare each, located in three communities within the Park. Park peasants, who have formed a collective and are paid for their work, carry out the various aspects of agricultural work, from preparing the soil to collecting the crop. The harvested tubers are kept in cold chamber until they are sown the following year. Alongside these activities there is a project taking place on medicinal plants that includes landscape conservation.

The Potato Park adheres to the three criteria identified by the International Union for Conservation of Nature (IUCN) that define an area or territory conserved by indigenous peoples and local communities (ICCA).[11] First, the population in question must live in close connection with the associated territory. Second, the communities must play a major role in decision-making concerning the management of the territory and the implementation of decisions. Last, the population must ensure that the nature of the area, as well as the cultural values linked to it, is conserved. The Park has been a pioneering example of this type of protected area[12] (Borrini-Feyerabend et al., 2013; Kothari et al., 2014), now considered by the IUCN as a new conservation paradigm (Stevens, 2014). The relevance of this paradigm is discussed in the framework of the CBD and in the working group corresponding to article 8 (j), dedicated to *in situ* conservation and to indigenous and local communities. At the 2008 IUCN Congress, a group was created to promote this new type of protected area, called the ICCA Consortium.

The Potato Park was created in 1996, with support from the Peruvian NGO ANDES, which, the previous year, had begun to document the knowledge of peasants with the help of a group of volunteers (Argumedo, 2008; Argumedo & Stenner, 2008). At this time, six communities were involved in the project. ANDES is run by two brothers originally from the Ayacucho region – Cesar and Alejandro Argumedo – both engineers, one educated in Peru and the other in Canada (Montréal). Coming from a middle-class background, they are concerned by environmental matters and by the precarity in which the Andean peasant population lives. One brother overseas local activities, while the other is in charge of links with international partners and institutions. Since the beginning, conserving agrobiodiversity and recognizing the value of knowledge held by peasants concerning potatoes have been at the heart of the NGO's project. At first, the team was made up mainly of Peruvians, most of them Quechua-speakers engineers like the two directors. ANDES quickly turned into a project for the *in situ* conservation of indigenous domesticated potato biodiversity (Argumedo, 2008). It has organized various activities, using funding from various sources. Despite initial difficulties, a solution has been found for the Park's governance (Argumedo & Stenner, 2008): the Association of Potato Park communities[13] has been constituted and registered with the Peruvian public registry. The directing body ("the office") which presides over this association is now made up of the presidents of the different communities involved, so that the Park's governance is aligned with that of each community[14]. At present, the Park is relatively well-established locally and is recognized as an independent entity by the municipality of Pisac, as well as by various institutions from the local to the international level. Since March 2020, the 25th anniversary of ANDES, it is officially recognized by the Peruvian government as a zone of agrobiodiversity (*zona de agrobiodiversidad*).

In the Potato Park, emphasis is placed not only on the *in situ* conservation of plant material, but also on the cultural dimension of this conservation; this is why the decision was taken to use the term "area of indigenous biocultural heritage" rather than ICCA (Argumedo, 2008). This orientation toward the development of biocultural rights is considered highly important for ANDES. In the Park,

ANDES experimented and promoted different biocultural tools: "biocultural heritage" (Argumedo, 2013; Argumedo & Stenner, 2008), "biocultural territory" (IIED, 2014), biocultural innovations, and biocultural knowledge. The Park is considered to have developed the first biocultural protocols (ANDES et al., 2012).

Sumaq kawsay, an Alternative Vision of Conservation

The Potato Park has a program of valorizing the biocultural heritage of its inhabitants, which frames their specific vision of their environment in terms of *Sumaq kawsay*, translatable as "good living" (or *buen vivir* in Spanish). These terms began to gain ground in Andean countries around the beginning of the present century. They were enshrined in the constitutions of Ecuador in 2008 and Bolivia in 2009. In these two cases, the inspiration for *Sumaq kawsay* is asserted to be indigenous, and the importance of nature is emphasized (see especially Coordinadora Andina de Organizaciones Indígenas & Huanacuni Mamani (2010)). The Ecuadorean constitution, in particular, attributes right to nature by identifying it with Pachamama ("Mother Earth[15]"). However, critical investigation of the concept's development (Bretón Solo de Zaldívar, 2013) and the various schools of thought that make claims on it (Hidalgo-Capitán & Cubillo-Guevara, 2017) shows that it arose from a negotiation among different sectors of society, and that various perspectives on it co-exist. In particular, these studies show that the indigeneity of the concept's origin is relative. Peru has not included "good living" in its constitution, although it is a well-known concept in the Peruvian NGO community. The lack of state-led discourse opens more room for other actors to appropriate the concept and adapt it to their own particular objectives, as in the Potato Park.

What follows is a description and analysis of the concept of conservation, as articulated within *Sumaq kawsay* at the Pisac Potato Park. When I arrived in 2012, the institutional discourse of ANDES had already taken shape for the most part, even though one of the tasks requested of me was to help the NGO in conceptualizing it (Argumedo & Hall, Forthcoming).

On the ANDES website, there is a reference to an "indigenous philosopher" from Peru, namely Javier Lajo, educated in economics and director of the little-known *Instituto del Sumaq Kawsay*. This author, both online and in various published articles (2013), presents *Sumaq kawsay* as a pre-Hispanic model and advocates it as a decolonializing approach (Mignolo, 2001),[16] as an extension of the indigenist current of thought that developed in the Cusco region at the start of the twentieth century.[17] The definition of *Sumaq kawsay* put forward by ANDES draws from this school of thought and takes up the principle of a three-way partition. Yet, it reformulates the local principles.

The first idea upon which *Sumaq kawsay* rests, as emphasized in the Park, is the identification of three distinct communities, each called an *ayllu*. This polysemic term (Masuda & Sato, 1981) is generally used to reference pre-Hispanic communities (Castro Pozo, 1973), small or large family groups (see Sendón, 2016), or the territory of a community. Recently, however, there is a tendency to reposition the term in a way that emphasizes the idea that Andean conceptions of the environment do not rest on a naturalist ontology (de la Cadena, 2015). ANDES uses the term in a similar way, although it does not reference the work of de la Cadena.[18] This discourse identifies three *ayllu* or distinct communities (Argumedo & Wong, 2010). First, the *runa ayllu*, or the "community of human beings", refers specifically the community of peasants, according to the terminology used in communities within the region (*runa* meaning "man" or "person" in Quechua). Second, the *sallqa ayllu*, or "the community of wild natural resources", includes both animal and plant lives. Third, there is the *awki ayllu*, or "the community of spiritual powers", more often called *apu* and *Pachamama* in the Cusco region (Isbell, 1978; Valderrama Fernández & Escalante Gutiérrez, 1988; La Riva González, 2005; Ricard Lanata, 2010).

Until recently, the structure of the NGO's website explicitly followed the distinction of these three communities, around which various ongoing and completed projects were organized. The decolonialist aspect of the vision manifests itself in the use of ethnohistorical sources and, above all, in the use of a drawing that supposedly existed as an altarpiece in the Qoricancha, the former Inca temple upon whose foundations the temple of Santo Domingo was built. This drawing appears in the 1574 chronicle of Juan de Santa Cruz Pachacuti Yamqui Salcamayhua (1879 [1574], pp. 16–17) (Figure 3.3). Although ethnohistorians have shown that this drawing should be interpreted as an evangelizing tool (Duviols, 1997),[19] in certain pro-indigenous circles, it is considered to be "emblematic of Andean cosmology".[20] In this sense, ANDES used the drawing in participatory workshops with peasants in order to identify the various communities (*ayllu*) of humans and non-humans according to Andean cosmology.

Thus, ANDES consulted the local population with respect to the debates taking place in certain intellectual and activist circles. Interviews carried out on site show that peasants recognize each term in Quechua (*runa*, *sallqa*, *ayllu*), but they do not necessarily associate them as different kinds of *ayllu*. The division into three communities makes sense for peasants already familiar with the ANDES vision. Some indicated that it had not been easy to appropriate the idea.

From the perspective of anthropological or philosophical theory, these three *ayllu* place communities of non-humans on the same footing as the collectivity of humans. They form a hybrid collective in the sense of Latour (2004a), which brings into question the opposition of nature and culture which, as Descola has shown (Descola, 2005; Descola & Sahlins, 2014), is not universal and is not applicable to Andean peasants.[21] Furthermore, such ontological arguments need to be understood in a context where their political impact is tangible, that is, from the perspective of political ontology as in cases studied by Blaser (2009) and Martínez Mauri (2007). In the case of Potato Park, these arguments are marshalled in a joint defense of cultural and biological diversity, in continuity with the position developed by Maffi and Woodley (2010), with the participation of Alejandro Argumedo since 1996. Since Argumedo found the linguistic aspect

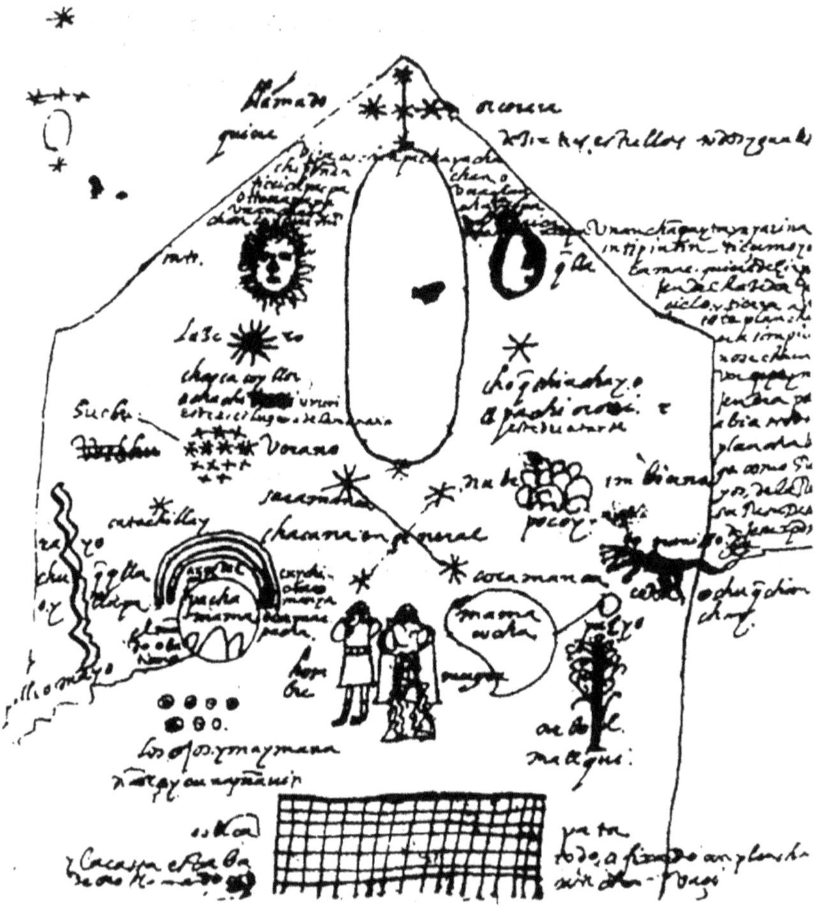

FIGURE 3.3 Relación de las antigüedades deste Reyno del Piru, 1613.
Source: Santa Cruz Pachacuti Yamqui Salcamayhua, Juan de (1613), public domain.

of this position to be over-emphasized, he decided to develop his own approach (personal communication, Cusco, 2014), preserving the importance of cultural factors as in the initial schema, but in a broader sense that includes even the ontological dimension. I should emphasize, however, that the people concerned do not use the term *ontology*, and tend to consider such matters as an aspect of culture.

The second characteristic of the ANDES vision is based upon the close reciprocal relationships that the three communities (*ayllu*) necessarily have with each other, relationships referred to by the Quechua term *ayni*. This reciprocity is an essential condition for *Sumaq kawsay*. Andean peasants mainly use the term *ayni*, as is reported in the ethnographic literature,[22] to refer to a mechanism of mutual help between close relatives that is reciprocal and symmetrical. In order to carry out heavy agricultural work such as plowing, sowing, harvesting, or even building a house, a man will usually ask those close to him (siblings, siblings-in-law, cousins, friends) to come and help him. The people who come receive no payment and are treated with great respect: the *ayni* relationship is privileged and treasured. Subsequently, the person who benefited from this help will have to pay back this day's work to each helper. In rural communities such as those in Potato Park, men keep a detailed mental record of the days of work that are due to them and those that they have to pay back. Dense social networks of individuals are often constructed and reinforced through *ayni*, and as a principle it appears symbolically important. The ability to mobilize a large network of mutual help translates into a social capital of great significance. Even though this practice has declined in the agricultural domain over the last 20 years or so, its underlying ideal of reciprocity is still highly valued.

ANDES uses this term in order to highlight the importance of reciprocity beyond the pragmatic agricultural context. In this conceptualization, *Sumaq kawsay*, results directly from the quality of reciprocal relations (*ayni*) that exist between the communities of humans, natural resources, and spiritual entities. My interlocutors confirmed that they consider rituals to take place within a logic of reciprocity similar to *ayni*, although they primarily conceive *ayni* in terms of mutual help. Thus, the NGO's interpretation stresses a discreet dimension of *ayni*.

In its interpretation of the *ayllu*, ANDES takes up certain local concepts while also proposing something new. In Andean communities, offerings to Pachamama must be made in order to ensure a good harvest. The offering of August 1st, which begins the agricultural cycle, is especially important[23] and livestock farmers observe specific rituals (Ricard Lanata, 2010). Decreased productivity and ominous signs of climate change are often associated with the gradual abandon of these rituals (de la Cadena, 2010; Cometti, 2015), thus putting reciprocal relationships between humans and non-humans at risk. The logic of this reciprocity makes sense within a particular conceptualization of humans, plants, animals, and various elements of the landscape. Ethnographic research has shown clearly that humans share the same vital energy, called *animu* or *samay*, with plants, animals, Pachamama, and the *apus* (Allen, 1982; La Riva González, 2005; Ricard Lanata, 2010). Moreover, entities such as Pachamama and the *apu* (or *awki*) play a fundamental role in the proper circulation and conservation of this energy. The rituals owed to these entities aim to satisfy them so that they will bring the necessary energy for the success of various human endeavors, especially health, livestock breeding, and agricultural production. This is manifest not only in rituals, but also in daily practices related to production. For example, the local classification of potatoes is based on distinct uses for potatoes. One of the three categories of potato grown is the cooking potato (*wayku papa*) that is clearly distinguished from the peeling potato (*monda papa*). To peel a *wayku papa* risks disturbing its state of *animu* – in both its energetic and emotional senses – due to the pain inflicted by the knife cutting into its flesh. As the *animu* of plants is conceived collectively, injuring one potato could mean the end of production of this landrace for the offender. This reflects the fact that in Andean conceptions of the body, whether human, animal, or vegetable, vital energy or *animu* must "go" ("*puriy*") around the body, that is, it must circulate properly, neither too fast nor too slowly, for an individual to be in good health (Bastien, 1985; Hall, 2012).

The way that ANDES represents conservation thus takes inspiration from the concepts held by the local population, which are well-documented ethnographically. The uniqueness of the ANDES vision rests upon its identification of hybrid communities (*ayllu*) and its stated need for reciprocal relationships (*ayni*), and thus, upon an ontological model that is an alternative to modern naturalism. The NGO assigns these aspects to *Sumaq kawsay* while emphasizing an Andean spirituality that manifests through stories and rituals. This vision bypasses the influence of Catholicism, despite the fact that those rituals are carried out by people who self-identify as "Catholics". It excludes some local conceptions, such as local classificatory schemes and agricultural practices that ANDES has documented, but have mostly stayed in the organization's archives as raw data in the form of video recordings or reports. We see that the promotion of this alternative ontological model implies a series of choices, whose logic will be the topic of the following section.

Conservation Issues at Local, National, and International Scales

The formulation of this alternative to the dominant ontology has a particular intention, which takes its meaning from a multi-level governance of biodiversity. This section explores these various levels starting with the international, which is central to the discussion.

As Alejandro Argumedo explains, the work of ANDES must be understood in an international context; it is this context that makes it meaningful. His role in the NGO is to coordinate activities at the international level by taking part in various events, building partnerships and networks. This also allows to raise funds to finance the NGO's activities and the conservation activities at the heart of Potato Park. The Park has a particular importance in matters of worldwide biodiversity governance. As it focuses on *in situ* conservation and the role of local and indigenous populations, the CBD and, more specifically, the working group on article 8(j) are particularly interesting. Since the 1970s, following the Meadows report by the Club of Rome, international institutions have set up organizations to conserve agrobiodiversity, which has drastically declined in the face of mechanized agriculture and the increasing specialization of the seed sector (Bonneuil & Thomas, 2009; Halewood et al., 2013). In this context, *ex situ* germplasm banks were created in order to conserve the biodiversity of the most cultivated and consumed plants; the International Potato Center in Lima, which holds the main germplasm bank for tubers and roots, is one of these. However, these mechanisms are not wholly satisfying, since the genetic diversity collected in this way benefits, at no cost, the selection processes of private seed enterprises with no consultation with, or recompense for, the people who contributed to domesticating and selecting this material and who continue to conserve it *in situ* (Halewood et al., 2013; Thomas & Boisvert, 2015). In the 1990s, the international community began to reflect harder on sharing biotechnology profits in order to compensate the communities possessing the relevant natural resources and associated knowledge, and this reflection led to the creation of the CBD. The Nagoya Protocol adopted in 2010 specifies the mechanisms for the access and benefit sharing (ABS) of the benefits generated by the use of genetic material. Its implementation remains complex, however, and discussion continues within the framework of the International Treaty on Plant Genetic Resources (ITPGRFA).[24] In the case of the potato, which generates considerable profits worldwide, the issues surrounding a "fair and equitable sharing of benefits" are very significant, which makes the example of the Park particularly emblematic. Argumedo, who actively takes part in these various international events, is fully aware of these issues and he has been able to turn the Park into an international model.

With this success have come consequences: it is necessary to adapt the Park's to the concepts and categories that are relevant in the international arena. First of all, the Pisac Potato Park is depicted as led by "Quechua" indigenous communities and Argumedo presents himself as their representative. Seen from Peru – and even more so from Cusco – the reality is more complex. On one hand, for most people, "Quechua" is a language and not an ethnic category (Robin, 2004). On the other, even the peasants do not consider themselves indigenous (*indígenas*). The Park's claim makes sense within the United Nations definition of *indigeneity*, wherein a generic and consensual definition has been adopted (Verdeaux & Roussel, 2006). What is more, at international events, the representatives of local and indigenous communities (to use the established terminology) are in a minority. Their spokespeople are equated with the populations they represent, whether they are indigenous leaders, representatives of community associations, or representatives of pro-indigenous NGOs. Ramos (1994), using Amazonian case studies in Brazil, shows how the recognition of indigenous communities' rights and the growing importance of environmental conservation together have led to the representation of a "hyperreal Indian" by pro-indigenous institutions. This representation is not easy to reconcile with the realities of local people, but helps them to defend certain rights, especially land rights (Albert, 1993). Within the working group on article 8(j) of the CBD, a similar process has taken place: the figure of the hyperreal Indian has established itself on an international scale (see Hall's Chapter 4, in this volume that focuses specifically on the way indigenous people and local communities (IPLCs) participate in the CBD meetings).

The recognition of *Sumaq kawsay* by Bolivia has also had repercussions in this domain. Since the adoption of the new Bolivian constitution in 2005 under the Evo Morales government, the country has used all available opportunities to assert, loudly and clearly, the need to rethink the relationship between humankind and nature. As a "State party" (i.e., signatory country), Bolivia has an official say and a vote at the Conference of the Parties (COP) of the CDB, and makes great use of this. The representatives of local and indigenous communities, for their part, do not vote and only participate marginally in the debates. Bolivia's involvement and its aggressive use of *Sumaq kawsay* is a way to fight against the naturalist convictions that, due to economic issues and the weight of biological sciences, dominate this international setting. Local and indigenous communities, as well as pro-indigenous NGOs, thus have every interest in following the movement led by Bolivia; for these actors, the Bolivian vision in itself has become an essential reference point. At the same time as it takes part in these debates, ANDES extends them by proposing a model specifically adapted to issues of *in situ* conservation; this model is mainly presented at side events and feeds into the deliberations that accompany the official debates. The recognition gained by ANDES in this domain, including in other contexts such as the congress on parks organized by the IUCN or the meetings of the International Society of Ethnobiology, helps to publicize the Potato Park. This is an important source of social capital that in recent years has directly contributed to obtaining funding, all of which has been international (including from Oxfam, the European Union, and the International Institute for Environment and Development).

Furthermore, in the current context of environmental crisis, non-indigenous actors in both the global South and North are now calling for the reformulation of our relationship with nature. This translates into a willingness to rework relationships with nature, as is clear in the title of the United Nations Decade for Biodiversity: "Living in harmony with nature". The (partial) convergence of these demands ensures that claims made by local and indigenous communities gain momentum.

The vision of *Sumaq kawsay* promoted by ANDES thus has meaning within an international context linked to the CBD. In this context, the NGO has worked out how to position itself and Indigenous people strategically. We must consider the conservationist orientation of the Potato Park within this logic: a productivist logic would create confusion. Part of the Park's success, for its funders, rests on its action-research program that consists of showing the particularities and advantages of an alternative vision of conservation, and also reflects on the possibility of upscaling the model in Peru and abroad.

Within Peru, this international recognition is a powerful force; the resulting networks and knowledge of current debates allow ANDES to position itself effectively in discussions. In Peru, recognition of peasants' rights over landraces is still problematic, although some innovative potential solutions have been found. Peru signed up to the agreements of the International Union for the Protection of New Varieties of Plants (UPOV) in 1993 and 2011, thus deciding to favor intensive agriculture that uses improved seeds to feed a growing urban population and stimulate the expanding industry of export-oriented agriculture (Lapeña, 2012a, 2012b). The UPOV system, developed in Europe and North America, has entailed a major loss of agrobiodiversity and is not adapted to the realities of family agriculture as practiced by Andean peasants. As specialists have noted, membership in the UPOV not only endangers the richness of genetic heritage cultivated *in situ*, but is also likely to affect the food security of small-scale farmers in the Andes (Chevarria Lazo et al., 2004; Lapeña, 2012b). The latter have only limited access to commercial seeds and mainly grow landraces which they sow again the following year; yet, these practices are restricted by the UPOV agreements. As such, a whole sector of production is put in danger, in a context where the government does not offer any development policy addressed specifically at community peasants (Castillo Castañeda, 2013, p. 15; Eguren, 2013, p. 11).

ANDES considers that action at several levels is necessary to remedy this. Since options are relatively limited, it is necessary to know how to combine strategies. At the national level, the main objective of ANDES is to work for the recognition and the protection of informal systems of seed production and exchange. To achieve this, it was necessary to first raise the profile of issues relating to the peasant population and to make the urban population, as well as political decision-makers, more aware of such matters. Several important events have enabled ANDES to mark some points, and become well-known. The most impactful of these was the refusal to allow genetically modified potatoes into the Park in 2007 when, with the help of ANDES, the peasants of the Potato Park mobilized against a supreme decree authorizing the introduction of GMO seeds.[25] The movement gained momentum, leading to the adoption of a ten-year moratorium by the Department of Cusco in 2007,[26] which has since been renewed. Then, following the mobilization of other regions of Peru, this moratorium was extended to the whole country. This step was crucial, as Peru put into action a series of measures concerning potato agrobiodiversity (Filoche, 2009). In particular, a register of indigenous potatoes – an initiative in theory beneficial to the cultivars concerned – was put in place. Nonetheless, another event demonstrated that the protective measures relating to indigenous potato varieties were not effective enough. The National Institute of Agrarian Innovation (INIA in Spanish) submitted around 50 requests for seed certification, some of which concerned landraces. Once again, mobilization by ANDES and the population of the Potato Park was fruitful and the requests were withdrawn.

These struggles, which have had successful outcomes, have allowed ANDES to establish itself as an important actor in negotiations concerning landraces in Peru. What is more, a network has been built up around these issues, including the Peruvian Society for Environmental Law (SPDA in Spanish, a Peruvian NGO) and various actors belonging to universities. Thus in 2016, a workshop was organized in the Pisac Potato Park to discuss the implementation of regulations specifically concerning "ancestral seeds" briefly mentioned in an article of the General Law on seeds of 2012 (supreme decree n° 006–2012-AG), which allows for the implementation of the UPOV system in Peru. The aim was to organize a regulatory framework that would allow Andean peasants – such as those of the Park – to continue to use these seeds, to resow them, to exchange them, and even to sell them. In this framework, not only would peasant agricultural practices potentially be made legal, but their seed system would be maintained; at the same time, it would be better supported, which could help to improve the quality of landrace seeds used by peasants and to develop a specific market for them. In these two cases, ANDES managed to situate national debates within international trends and mobilized important alliances that made the government withdraw. For the moment, however, concrete results are few and far between as far as the Park is concerned: the Park was only recognized by the State in 2020 and the regulation on ancestral seeds still has not been ratified. The summer of 2018 seemed a promising time, but the chaos of Peruvian political life in recent years has once again delayed the implementation of these measures. In 2022, we do not know if the new government of Pedro Castillo will be of any help on this subject, despite the agrarian reform it has initiated.

Given that the State does not recognize the concept of *Sumaq kawsay* and that Peruvian society tends not to recognize the Andean population as indigenous,[27] the particular ontologies held by peasants in the Park do not make for a promising argument at the national level, at least not from an institutional perspective. With the arrival of multicultural policies, the concept of "culture" is more significant. Negotiations are under way between the Park and the Peruvian Ministry of Culture for the recognition of a specific "life plan". This translates into a showcasing of stories and rituals linked to the potato, as well as traditional clothing, in other words, the valorization of classic cultural elements. Thanks to the social capital it has gained on the international stage, the NGO ANDES has indirectly become an important actor on the national scene.

At a local level, the Park represents various economic opportunities. Peasants from the communities are marginalized and have limited means, while their needs have increased significantly over the last few decades. They now have to pay for electricity and mobile phones, their children go to school for longer periods and their wants and needs have changed accordingly[28]. These economic opportunities are welcome, especially from women. Emphasizing biodiversity and its conservation attracts a well-informed public, whose presence generates revenues through guided activities, participative rural tourism, the sale of handicrafts, and the sale of meals. All of these activities are organized by the communities, benefiting those who are directly involved, but also contributing to a common fund that serves to finance certain activities and expenditures (up to a maximum of 10%). The "Biocultural protocol" previously mentioned (see Argumedo, 2012 and ANDES et al., 2012)stipulates the rules for the division of profits that results from the various activities.

The emphasis on *Sumaq kawsay* has other kinds of effects locally. In particular, it gives rise to festivities organized for National Potato Day, when a ritual, known as *Papa watay*, dedicated to "tying down" the souls of freshly harvested potatoes, is performed (Hall, 2022). Certain tensions surface on this occasion. On one hand, Protestant converts are uneasy with this ritual that they have abandoned. If they take part, it is because the community takes part as a collective. On the other hand, for the Catholics, the event allows for a revalorization of the ritual and of knowledge and practices that were formerly experienced as stigmatizing. In these ways, valorization of *Sumaq kawsay* feeds the religious tensions at the heart of these communities. Not least, the people who have key roles in the Potato Park are mainly Catholic, which aggravates underlying tensions. This bias of ANDES plays against the NGO since these tensions may manifest themselves within the "bureau" of the Park and lead to the withdrawal of a community. The recognition of the Park by the State will possibly alleviate this problem.

Conclusion

The Peruvian NGO ANDES has formalized an alternative vision of biodiversity based upon the Potato Park. On the surface, it would seem that the issue is simply one of explaining the peasants' worldview, but it turns out that the situation is much more complex. The vision draws upon the specific logic of the working group on article 8(j) of the CBD, and should be understood in a context where Bolivia fiercely promotes a vision with the same name. In this context, opposition to neoliberal conceptualizations of nature is the common denominator that unites local and indigenous communities together with their representatives. In this way, ANDES engages in an effort at ontological diplomacy, trying to make the CDB and the working group on article 8(j) – the main international institution involved in environmental conservation and in recognizing the rights of local and indigenous communities – recognize the existence of alternative relationships to the environment. Thus, there is a certain intertextuality at work that leads to the "intercultural politicization of cosmological categories" (Albert, 1993, p. 369): local cosmological categories – as they are perceived, understood, and explained by the NGO – are mobilized in such a way that they will resonate with those of other actors. In this framework, anchoring local cosmology within the non-modern composition of worlds appears as the most effective rallying point for the actors involved with local and indigenous communities in this domain.

In order to formulate this discourse, the NGO has taken inspiration from the practices and concepts of the local population, but according to a non-local logic. This translates into a tendency to reconstitute a pre-Hispanic Andean worldview. This in turn leads to a hyper-valorization of the "Catholic" perspective (as it is paradoxically considered locally), while the Protestant perspective does not have a voice. Thus, the overall vision remains unfamiliar to most people concerned. The *Sumaq kawsay* that is promoted is the philosophy of a "model Andean".[29]

Although the ontological dimension of the vision is less relevant at the national level,[30] the international recognition of ANDES allows it to mobilize alliances and to successfully position itself in national debates, for example, on GMOs, biopiracy, and laws concerning seed.

Thus, ANDES participates internationally in the recognition of a non-naturalist conceptualization of the environment, a vision which – although it is not completely faithful to the peasants' conceptualization – has "transformative potential", to use Stengers' term (2005), and may contribute to the development of a "common world" (Latour, 2004a). ANDES displays a powerful ontological diplomacy. The philosophical nature of this discussion should not make us forget that the work of diplomacy takes place within relations of power. As such, the biocultural tools and rights shaped by the CBD are developed on the basis of discourses negotiated and formulated according to a logic that is largely foreign or extraneous to local and indigenous communities. This process reflects the weight of the naturalist and capitalist ontology. Such framing of the local views proves to be politically powerful, but entails some accommodations. On one hand, emphasis on this discourse within these communities has consequences that are worth evaluating. On the other hand, one might well ask to what extent the international tools forged according to cases like the Park may need to be more finally tuned in order to defend the interests of specific Indigenous peoples or Local communities.

Notes

1 This chapter was originally published in French in the journal Anthropologie et Sociétés: Hall Ingrid (2019). "Le 'bien-vivre' (sumaq kawsay) et les pommes de terre paysannes. Du délicat exercice de la diplomatie ontologique". *Anthropologie et Sociétés, 43*(3), 217–244, https://doi.org/10.7202/1070155ar. This version has been updated and slightly modified. This research has been possible, thanks to the Fonds de recherche du Québec – société et culture (FRQSC) and the Social Sciences and Humanities Research Council (CRSH) of Canada. *All URLs retrieved on 1 September 2021.

2 *Asociación para la naturaleza y el desarrollo sostenible* [Association for nature and sustainable development].

3 Latour understands political ecology as a refounding of the "modern constitution" that sets human collectives apart from non-human collectives (Latour, 2004a). According to Latour and Stengers (2004), this process has to take place via a new "composition of worlds" based upon hybrid collectives, composed of both humans and non-humans.

4 Other authors discuss the omnipresence of NGOs in the realm of biodiversity conservation (Aubertin, 2005b).

5 On the emergence of international governance concerning biodiversity, see Aubertin (2005a), Halewood et al. (2013), Nazarea and Rhoades (2013), and Thomas and Boisvert (2015).

6 Following the title of an article by Argumedo (2008): "The Potato Park, Peru: Conserving Agrobiodiversity in an Andean Indigenous Biocultural Heritage Area".

7 The peasant community is the established form of organization of rural communities in the Peruvian Andes; nowadays, there are more than 6,000 of them in Peru (Tipula & Alvarado, 2016).

8 See https://cipotato.org/genebankcip/process/potato/potato-cultivated, and https://cipotato.org/crops/potato/.

9 https://cipotato.org/programs/.

10 Namely in the large germplasm banks such as the International Potato Center, which depends on the Food and Agriculture Organization of the United Nations (Halewood et al., 2013).

11 See https://www.iccaconsortium.org/index.php/discover/.

12 An article signed by Alejandro Argumedo (2008) appeared in the first volume of the series "Values of Protected Landscapes and Seascapes" edited by the IUCN.

13 *Asociación de Comunidades del Parque de la Papa.*

14 Since membership of the Park depends upon voluntary membership of the Association of Potato Park communities, this membership can be cancelled. As a matter of fact, one community did quit the Park.

15 Ecuador's Constitution of 2008 (*Registro Oficial* 449 de 20-oct-2008) (English translation: https://pdba.georgetown.edu/Constitutions/Ecuador/english08.html).

16 A group of academics of Latin American origin (often occupying a position in North America) has developed a decolonial approach that is distinct from the post-colonial approaches developed in the Anglo-Saxon world more broadly. Informed by a specific historical, social, political, and demographic context, their approach emphasizes "the permanence of the global coloniality of power, knowledge, and being, after decolonization" (Boidin, 2009) and focuses on the need for emancipation by searching appropriate alternative solutions such as *Sumaq kawsay*.

17 On this school of thought, see the works of Tamayo Herrera (1980).

18 In fact, de la Cadena analyses this kind of proposal, helping to elaborate it and give it academic recognition.

19 Duviols (1997, paragraph 5) concludes that this drawing

> [was meant to] remind all those contemplating it that there is only one God and creator, who is all-powerful and who governs the universe, and that all the other entities of the world, who might appear to have power and whom Andean people have worshipped and continue to worship, in fact neither have power, nor are gods, but are merely God's creatures.
>
> *(author's translation)*

20 See https://donambro.wordpress.com/2010/09/17/pachacuti-yamqui-dibujo-cosmogonico/.

21 In this instance, from an anthropological point of view, the dominant type of ontological identification for this population is analogism, according to Descola's typology (2005; Descola & Sahlins, 2014). I have addressed elsewhere the way analogism manifests itself in agricultural practices in a nearby community of Cusco (Hall, 2012). Although I anticipated working with ANDES would be a continuation of this work, it has not been the case.

22 The classic reference on this topic is Alberti and Mayer (1974).

23 For an analysis of Andean concepts related to agricultural production and the importance of spiritual entities, see Hall (2012), Rivière (1994), and Arnold and de Dios Yapita (1996).

24 As far as the Treaty is concerned, however, the emphasis on the indigeneity of the relevant populations – and on their particular ontologies – seems less promising. The terms "small-scale agriculture" and "locality" seem to be more appropriate.

25 See the blog by Leighton published in 2007, Retrieved 9 July 2021, from https://www.scidev.net/america-latina/news/el-cusco-prohbe-ogm-en-resguardo-de-sus-papas-nat/.

26 See https://grain.org/fr/article/entries/4650-and-now-gm-potatoes-in-.

27 See the debates on this issue that followed the adoption of the consultation law, found in the work edited by Castillo Castañeda et al. (2007). This might change with the new president elected in 2021.

28 In Peru, 83% of the poor population, usually peasants, makes their living from agriculture (https://www.ifad.org/es/web/latest/news-detail/asset/40031910).

29 Here, we refer to the terminology "model Indian" proposed by Ramos (1994, p. 162) in the Brazilian case.

30 Cultural and spiritual aspects, being more in line with multicultural policies, receive more emphasis.

References

Albert, B. (1993). L'Or cannibale et la chute du ciel: Une critique chamanique de l'économie politique de la nature (Yanomami, Brésil). *L'Homme*, *33*(126/128), 349–378.

Alberti, G., & Mayer, E. (Eds.). (1974). *Reciprocidad e intercambio en los Andes peruanos*. Instituto de Estudios Peruanos.

Allen, C. J. (1982). Body and Soul in Quechua Thought. *Journal of Latin American Lore, 8*(2), 179–196.
ANDES, the Potato Park Communities, & IIED. (2012). *Community Biocultural Protocols: Building Mechanisms for Access and Benefit Sharing among the Communities of the Potato Park based on Quechua Customary Norms.* International Institute for Environment and Development (IIED). https://pubs.iied.org/G03168/
Argumedo, A. (2008). The Potato Park, Peru: Conserving Agrobiodiversity in an Andean Indigenous Biocultural Heritage Area. In T. Amend, J. Brown, A. Kothari, A. Phillips, & S. Stolton (Eds.), *Protected Landscapes and Agrobiodiversity Values* (Vol. 1, pp. 45–58). Kasparek Verlag.
Argumedo, A. (2012). Descolonizando la investigación-acción: El protocolo biocultural del Parque de la Papa para la distribución de beneficios. In K. Swiderska, A. Milligan, K. Kohli, H. Jonas, H. Shrumm, W. Hiemstra, & M. J. Oliva (Eds.), *Biodiversidad y cultura: Exploración de protocolos comunitarios, derechos y consentimiento (PLA 65)* (pp. 99–108). IIED.
Argumedo, A. (2013). *Collective Trademarks and Biocultural Heritage. Towards New Indications of Distinction for Indigenous Peoples in the Potato Park, Peru.* IIED. https://pubs.iied.org/sites/default/files/pdfs/migrate/16528IIED.pdf
Argumedo, A., & Hall, I. (Forthcoming). The Potato Park of Pisac. Building Tools to Strengthen the Local Life Project, the Sumaq kawsay. In C. Scott, E. Silva-Rivera, & K. Sinclair (Eds.), *Indigenous Stewardship of Environment and Alternative Development.* University of Toronto Press.
Argumedo, A., & Stenner, T. (2008). Association ANDES: Conserving Indigenous Biocultural Heritage in Peru. *IIED – Gatekeeper, 137a.* https://pubs.iied.org/sites/default/files/pdfs/migrate/14567IIED.pdf
Argumedo, A., & Wong, B. Y. L. (2010). The Ayllu System of the Potato Park, Cusco, Peru | International Partnership for the Satoyama Initiative. *The satoyama initiative.* https://satoyama-initiative.org/case_studies/the-ayllu-system-of-the-potato-park-cusco-peru/
Arnold, D. Y., & de Dios Yapita, J. (Eds.). (1996). *Madre Melliza y sus crías: Ispall Mama Wawampi.* Hisbol: ILCA.
Aubertin, C. (2005a). La biodiversité: Une notion en quête de stabilité. In C. Aubertin (Ed.), *Représenter la nature? ONG et biodiversité* (pp. 99–122). IRD Éditions.
Aubertin, C. (Ed.). (2005b). *Représenter la nature? ONG et biodiversité.* IRD Éditions.
Bastien, J. W. (1985). Qollahuaya-Andean Body Concepts: A Topographical-Hydraulic Model of Physiology. *American Anthropologist, 87*(3), 595–611.
Benjaminsen, T. A., & Svarstad, H. (2009). Qu'est-ce que la 'political ecology'? *Natures Sciences Sociétés, 17*(1), 3–11.
Blaser, M. (2009). The Threat of the Yrmo: The Political Ontology of a Sustainable Hunting Program. *American Anthropologist, 111*(1), 10–20.
Boidin, C. (2009). Études décoloniales et postcoloniales dans les débats français. *Cahiers des Amériques latines, 62*, 129–140. https://doi.org/10.4000/cal.1620
Bonneuil, C., & Thomas, F. (2009). *Gènes, pouvoirs et profits: Recherche publique et régimes de production des savoirs de Mendel aux OGM.* Quae.
Borrini-Feyerabend, G., Dudley, N., Jaeger, T., Lassen, B., Broome, N. P., Phillips, A., & Sandwith, T. (2013). *Governance of Protected Areas: From Understanding to Action.* IUCN.
Bretón Solo de Zaldívar, V. (2013). Etnicidad, desarrollo y 'Buen Vivir': Reflexiones críticas en perspectiva histórica. *ERLACS European Review of Latin American and Caribbean Studies | Revista Europea de Estudios Latinoamericanos y del Caribe, 0*(95), 71–95.
Campos, H., & Ortiz, O. (Eds.). (2019). *The Potato Crop: Its Agricultural, Nutritional and Social Contribution to Humankind.* Springer Nature.

Castillo Castañeda, P. (2013). Las comunidades campesinas en el siglo XXI : balance jurid-ico. In P. Castillo Castañeda, A. Diez Hurtado, B. Zulema, J. Urrutia Ceruti, & P. Del Valle (Eds.), *¿Que sabemos de las comunidades campesinas?* (pp. 15–106). Grupo allpa.

Castillo Castañeda, P. J., Diez Hurtado, Burneo, J., Urrutia Ceruti, J., & de Valle, P. (Eds.). (2007). *¿Qué sabemos de las comunidades campesinas?* Allpa.

Castro Pozo, H. (1973). *Del ayllu al cooperativismo socialista*. Peisa. http://books.google.com/books?id=0wk-AAAAYAAJ

Chevarria Lazo, M., Rosales Benitez, M., & Sigüeñas Saavedra, M. (2004). *Reporte Tecnico: Conveniencia de La Adhesión Del Perú Al Convenio UPOV-1991 Dentro Del Marco de Negociaciones Del Tratado de Libre Comercio (TLC) Entre El Perú y Los Estados Unidos de Norteamerica – USA*. INRENA, INIEA, CONCYTEC.

Cometti, G. (2015). *Lorsque le brouillard a cessé de nous écouter: Changement climatique et migrations chez les Q'eros des Andes péruviennes*. Peter Lang.

Conklin, B. A., & Graham, L. R. (1995). The Shifting Middle Ground: Amazonian Indians and Eco-Politics. *American Anthropologist*, *97*(4), 695–710. https://doi.org/10.1525/aa.1995.97.4.02a00120

Coordinadora Andina de Organizaciones Indígenas, & Huanacuni Mamani, F. (Eds.). (2010). *Buen vivir / vivir bien: Filosofía, políticas, estrategias y experiencias regionales andinas*. CAOI.

de la Cadena, M. (2010). Indigenous Cosmopolitics in the Andes: Conceptual Reflections beyond "Politics". *Cultural Anthropology*, *25*(2), 334–370.

de la Cadena, M. (2015). *Earth Beings: Ecologies of Practice Across Andean Worlds*. Duke University Press.

Descola, P. (2005). *Par-delà nature et culture*. Gallimard.

Descola, P., & Sahlins, M. D. (2014). *Beyond Nature and Culture* (J. Lloyd, Trans.; Paperback edition). The University of Chicago Press.

Dillehay, T. D., Bonavia, D., & Kaulicke, P. (2004). The First Settlers. In H. Silverman (Ed.), *Andean Archaeology* (pp. 16–34). Blackwell Pub. http://catalog.hathitrust.org/api/volumes/oclc/52819673.html

Dumoulin, D. (2007). Grandeur et décadence de la double conservation dans les arènes internationales. *Quaderni*, *64*(1), 23–35.

Duviols, P. (1997). La interpretación del dibujo de Pachacuti-Yamqui. In T. Bouysse Cassagne (Ed.), *Saberes y memorias en los Andes* (pp. 101–123). Éditions de l'IHEAL. https://doi.org/10.4000/books.iheal.806

Eguren, F. (2013). Prólogo. In P. Castillo Castañeda, A. Diez Hurtado, B. Zulema, J. Urrutia Ceruti, & P. Del Valle (Eds.), *¿Que sabemos de las comunidades campesinas?* (pp. 11–14). Grupo allpa.

Filoche, G. (2009). Les connaissances, innovations et pratiques traditionnelles en matière de biodiversité: Un kaléidoscope juridique. *Droit et société*, *72*(2), 433–456. Cairn.info. https://doi.org/10.3917/drs.072.0433

Foyer, J. (2015). Chapitre 4. Défendre les maïs au Mexique, entre ressources génétiques et ressources politiques. In F. Thomas & Boisvert Valérie (Eds.), *Le pouvoir de la biodiversité* (pp. 119–130). IRD Éditions. https://doi.org/10.4000/books.irdeditions.21849

Halewood, M., Lopez Noriega, I., & Louafi, S. (Eds.). (2013). *Crop Genetic Resources as a Global Commons: Challenges in International Law and Governance*. Routledge.

Hall, I. (2012). Labourer la terre, tisser la vie. Éclats d'analogies dans les Andes sud-péruviennes. *Journal de la société des américanistes*, *98*(98–1), 101–131.

Hall, I. (2022). Tying Down the Soul of a Potato in the Southern Peruvian Andes: Performance and Frictions. In F. Dussart & S. Poirier (Eds.), *Sensing Religion in Indigenous Context: Towards New Methods* (pp. 157–186). Press of Toronto.

Hall, I. (Forthcoming). Rapatrier le matériel phytogénétique, en revenir aux semences. *Revue d'études autochtones*, 51(1).
Hidalgo-Capitán, A. L., & Cubillo-Guevara, A. P. (2017). Deconstruction and Genealogy of Latin American Good Living (Buen Vivir). The (Triune) Good Living and its Diverse Intellectual Wellsprings. *International Development Policy | Revue Internationale de Politique de Développement, 9*, 23–50. https://doi.org/10.4000/poldev.2351
IIED. (2014). *Biocultural Heritage Territories*. IIED. http://andes.center/wp-content/uploads/2018/07/G03843.pdf
Isbell, B. J. (1978). *To Defend Ourselves: Ecology and Ritual in an Andean village*. Waveland Press. http://hdl.handle.net/1813/2135
Kohler, F. (2011). Diversité culturelle et diversité biologique: Une approche critique fondée sur l'exemple brésilien. *Natures Sciences Sociétés, 19*(2), 113–124. Cairn.info.
Kothari, A., Corrigan, C., Jonas, H., Neumann, A., & Shrumm, H. (Eds.). (2014). *Recognising and Supporting Territories and Areas Conserved by Indigenous Peoples and Local Communities: Global Overview and National Case Studies*. Secretariat of the Convention on Biological Diversity, ICCA Consortium, Kalpavriksh, and Natural Justice. http://www.deslibris.ca/ID/242837
La Riva González, P. (2005). Las representaciones del animu en los Andes del sur peruano. *Revista andina, 41*, 63–88.
Lajo, J. (2013). Sumaq Kawsay o Esplendida Existencia. *Debates de Ciencias, Tecnologia y Sociedad, 12*, 4–16.
Lapeña, I. (2012a). *La nueva legislación de semillas y sus implicancias para la agricultura familiar en el Perú. Serie de Politica y Derecho Ambiental, no. 26* (pp. 1–20). Sociedad Peruana de Derecho Ambiental. https://orgprints.org/id/eprint/25051/
Lapeña, I. (2012b). *La Propiedad Intelectual sobre las Semillas y sus Implicaciones para la Agricultura Familiar en el Perú. Serie de Politica y Derecho Ambiental, no. 25* (pp. 1–20). Sociedad Peruana de Derecho Ambiental. https://orgprints.org/id/eprint/25051/
Latour, B. (2004a). *Politics of Nature: How to Bring the Sciences into Democracy*. Harvard University Press.
Latour, B. (2004b). Whose Cosmos? Which Cosmopolitics? Comments on the Peace Terms of Ulrich Beck. *Common Knowledge, 10*(3), 450–462.
Li, F. (2013). Relating Divergent Worlds: Mines, Aquifers and Sacred Mountains in Peru. *Anthropologica, 55*(2), 399–411.
Maffi, L., & Woodley, E. (2010). *Biocultural Diversity Conservation: A Global Sourcebook*. Earthscan.
Martínez Mauri, M. (2007). Une réserve de la biosphère non achevée. Les Kunas et la conservation de la nature. *Cahiers d'anthropologie sociale, 3*(1), 97–107. Cairn.info. https://doi.org/10.3917/cas.003.0097
Masuda, S., & Sato, N. (Eds.). (1981). El concepto de ayllu, y Qata/Q'acun: Un estudio de la familia, el parentesco y el ayllu. In *Estudios etnográficos del Peru meridional* (pp. 139–171). Universidad de Tokio. http://books.google.com/books?id=SLgKAQAAIAAJ
Mignolo, W. (2001). Géopolitique de la connaissance, colonialité du pouvoir et différence coloniale. *Multitudes, 6*(3), 56–71. https://doi.org/10.3917/mult.006.0056
Nazarea, V. D., & Rhoades, R. E. (2013). Conservation Beyond Design: An Introduction. In V. D. Nazarea, R. E. Rhoades, & J. Andrews-Swann (Eds.), *Seeds of Resistance, Seeds of Hope: Place and Agency in the Conservation of Biodiversity* (pp. 3–18). University of Arizona Press. http://site.ebrary.com/id/10777429
Ramos, A. R. (1994). The Hyperreal Indian. *Critique of Anthropology, 14*(2), 153–171.

Ricard Lanata, X. (2010). *Les voleurs d'ombre: L'univers religieux des bergers de l'Ausangate (Andes centrales)*. Société d'ethnologie.

Rivière, G. (1994). El sistema de aynuqa. Memoria e historia de la comunidad (comunidades aymara del altiplano boliviano). In D. Hervé, D. Genin, & G. Rivière (Eds.), *Dinámicas del descanso de la tierra en los Andes* (pp. 89–105). IBTA-ORSTOM. https://horizon.documentation.ird.fr/exl-doc/pleins_textes/pleins_textes_7/b_fdi_03_01/41709.pdf

Robin, V. (2004). Indiens, Quechuas ou Paysans ? *Amérique Latine Histoire et Mémoire. Les Cahiers ALHIM. Les Cahiers ALHIM, 10*, Article 10. https://doi.org/10.4000/alhim.98

Santa Cruz Pachacuti Yamqui Salcamayhua, J. de. (1879). *Tres relaciones de antigüedades peruanas*. Impr. de M. Tello. (Original work published 1574)

Sendón, P. F. (2016). *Ayllus del Ausangate: Parentesco y organización social en los Andes del sur peruano*. PUCP, IEP, CBC.

Stengers, I. (2005). The Cosmopolitical Proposal. In B. Latour & P. Weibel (Eds.), *Making Things Public Atmospheres of Democracy* (pp. 994–1003). MIT Press.

Stevens, S. (2014). *Indigenous Peoples, National Parks, and Protected Areas: A New Paradigm Linking Conservation, Culture, and Rights*. University of Arizona Press. http://site.ebrary.com/id/10896830

Tamayo Herrera, J. (1980). *Historia del indigenismo cuzqueño: Siglos XVI–XX*. Instituto nacional de cultura.

Thomas, F., & Boisvert, V. (Eds.). (2015). *Le pouvoir de la biodiversité: Néolibéralisation de la nature dans les pays émergents*. Quae, IRD éditions.

Thomas, F., & Filoche, G. (2015). Chapitre 1. Le partage des avantages, une nouvelle éthique pour la biodiversité ? In F. Thomas & Boisvert Valérie (Eds.), *Le pouvoir de la biodiversité* (pp. 43–64). IRD Éditions. https://doi.org/10.4000/books.irdeditions.21837

Tipula, P., & Alvarado, S. (2016). *Directorio 2016: Comunidades campesinas del Perú*. Sistema de Información sobre Comunidades Campesinas del Perú (SICCAM), Instituto del Bien Común, Centro Peruano de Estudios Sociales (CEPES).

Tsing, A. L. (2007). Indigenous Voice. In M. de la Cadena & O. Starn (Eds.), *Indigenous Experience Today* (pp. 33–68). Berg. http://site.ebrary.com/id/10233374

Valderrama Fernández, R., & Escalante Gutiérrez, C. (1988). *Del Tata Mallku a la Mama Pacha: Riego, sociedad y ritos en los Andes peruanos*. DESCO.

Verdeaux, F., & Roussel, B. (2006). Y a-t-il un autochtone dans l'avion? Des ethnies locales à l'autochtonie globale en passant par la gestion durable de la biodiversité. *Autrepart, 38*(2), 15.

Villa, T. C. C., Maxted, N., Scholten, M., & Ford-Lloyd, B. (2005). Defining and Identifying Crop Landraces. *Plant Genetic Resources, 3*(3), 373–384.

4

UNMAKING THE NATURE/CULTURE DIVIDE

The Ontological Diplomacy of Indigenous Peoples and Local Communities at the CBD

Ingrid Hall

> Could the present rupture in the meaning of the natural lead to a new art of living in society/nature?
>
> *(Escobar, 1999, p. 15)*

A variety of writers agree on the idea that a naturalist ontology (Descola, 2005), resting upon a distinction and opposition between that which pertains to nature and that which pertains to culture, is the dominant paradigm in the realm of global biodiversity governance (Aubertin et al., 2007; Bonneuil Christophe, 2011; Büscher et al., 2012; Escobar, 1999; Ulloa, 2010). This opposition is increasingly criticized, especially by the Indigenous Peoples and Local Communities (IPLCs) who have taken advantage of these spaces in order to formulate broader demands for political recognition (de Sousa Santos et al., 2007; Escobar, 1998; Ulloa, 2010). This phenomenon is particularly noteworthy in the field of biodiversity, where the Convention on Biological Diversity (CBD), signed in 1992, is one of the central institutions. Thus, following the 10th Conference of the Parties (COP) of the CBD (2010), which coincided with the creation of a stronger participation mechanism for IPLCs, the relationship between nature and culture became an important issue, one linked to the recognition of biocultural diversity although not reducible to it (Bavikatte, 2014; Dunbar, 2019; Foyer, 2011). Since the Nagoya Protocol (2010), notably, traditional knowledge has been at the center of debates regarding the rights of IPLCs (Aubertin & Filoche, 2011; Reimerson, 2013, p. 215). This had led to the emergence of "cultural politics" (Escobar, 1999) based on the valorization of culture. Yet, the chain of logic has been extrapolated further still and the very meaning of nature has begun to be debated (Foyer, 2015, p. 15; Parks, 2020, p. 110; Uggla, 2010, p. 87), which gives rise to a "politics of nature" (Escobar, 1999) dependent upon cosmopolitics

DOI: 10.4324/9781003172642-5

(Cadena, 2012; Latour, 2007; Stengers, 2005), otherwise known as "political ontology" (Blaser, 2009; Müller, 2014). Together, these two kinds of politics entail a reflection upon what nature and culture represent, and contribute to a rethinking of the naturalist ontological regime that is predominant. The title chosen for the Strategic Plan for Biodiversity 2011–2020, "Living in Harmony with Nature", partly reflects the great hopes that are held in this regard (although the Nagoya Protocol does not necessarily live up to these, as Harrop (2011) indicates). At the same time, with the recognition of biocultural rights, it seems that a space is being created that might allow for ontologies alternative to naturalism within the CBD (Bavikatte, 2014, p. 6; Hermitte, 2009; Parks, 2017). Parks, who has tracked the main topics addressed within the CBD across time, shows that this opening up is reflected especially in the increasingly frequent use of terms such as "worldviews" or "cosmologies", with a peak in 2007 (Parks, 2020, p. 4). Hopes are high, and not only for IPLCs: what is at stake is the (re)creation of a more "harmonious" world for everyone – indigenous or otherwise – as Escobar explains in the quotation at the beginning of this chapter. Engal Anderson, who represented the UNEP (United Nation Environmental Program) at the opening of the meeting of the Working Group on Article 8(j) (WG8j) in 2019, explained this position in an exemplary manner, stating that the IPLCs should help us "to close the divide between nature and culture", regretting that "we pay too little attention to these voices" (20 November 2019). In this chapter, I focus on how IPLCs call into question the nature-culture dualism at the heart of the CBD, despite significant resistance, which still has plenty of life left in it if the formulation of the Zero draft of the post-2020 Global Biodiversity Framework is anything to go by.[1] In this document, the importance of IPLCs is recognized and their "full and effective participation" is enshrined; yet, the IPLCs themselves judge that their demands have been given little consideration in the negotiations within the group dedicated to the elaboration of this strategic plan (CBD, 2021). Some observers also point out that the consideration of IPLCs is too restricted (Reyes-García et al., 2021). Thus, I concentrate on the meetings that mobilize IPLCs in preparation for the COP of the CBD. During the meetings, the IPLCs have varying degrees of influence, which calls for a qualitative analysis of the spaces that are available for IPLCs to express themselves, and of the way that IPLCs use such spaces.

To this end, I will make use of the concepts of "ontological diplomacy" developed by Latour (2004, pp. 275–285) and "nature regime" developed by Escobar (1999, p. 5). The concept of ontological diplomacy allows us to address the way in which IPLCs negotiate the recognition of a non-naturalist ontological regime – a regime which Escobar calls "organic", distinguishing this from a "capitalist" (i.e. naturalist) nature regime. At stake here is a questioning of the "nature" of both culture and nature, as well as the relationship between the two. Under ontological diplomacy, I include both politics of culture and politics of nature. Nevertheless, as we will see, for IPLCs, the issue is not one of defending their relationships to their world in an unequivocal manner; this diplomacy does not consist of simply proposing to replace one nature regime with another. According to context, IPLCs will in fact defend quite different positions. Escobar states that:

> The nature regimes can be seen as constituting a structured social totality made up of multiple and irreducible relations [...]; there is a double articulation, within each regime and between one and another. The identity of each regime is the result of discursive articulations—with biological, social, and cultural couplings—that take place in an overall field of discursivity wider than any particular regime (Laclau & Mouffe, 1990 [1985]).
>
> *(1999, p. 5)*

This concept thus helps us to understand the way that certain actors conceive nature, culture, and the relationship between them, while at the same time placing this analysis in a broader discursive framework. These tools allow us to contemplate the contribution of IPLCs to the formulation of hybrid nature regimes, together with other actors, taking inspiration from work carried out in science and technology studies (Callon, 2013; Mol, 1999), especially the works of Vadrot (2014), which deals with a similar topic and field-site. This contributes to a reflection on the possibility of taking account of the diversity of worldviews – that some call the "pluriverse" (Cadena & Blaser, 2018; Latour, 2007; Mignolo, 2013) – and of the potential remaking of a "common world" (Latour, 2007; Stengers, 2005), whose architecture we do not know and whose emergence we should be attentive to.[2]

The CBD is a particularly interesting space to investigate this question, even though opinions on it in the literature diverge. If we take account of the measures put in place since 1992 and the creation of the CBD for the recognition of IPLC rights (especially the access and benefit-sharing (ABS) mechanisms of the Nagoya Protocol in 2010), or the specificity of IPLCs' relationship with the world, concrete results have been limited. In fact, a highly economic and naturalist logic still predominates (Bellier, 2015; Foyer, 2015; Reimerson, 2013; Thomas & Boisvert, 2015; Uggla, 2010). Nonetheless, this institution seems to be one of the spaces where a rearticulation of nature and culture is beginning to be worked out (Parks, 2020, p. 4), especially if compared to other legal instruments as Uggla has done (2010). This is linked to various factors. First, the term "biodiversity" has emerged on the public stage as a social issue (Aubertin, 2005; Hannigan, 2014) and indigenous peoples have taken it up to support identity-based demands both nationally and internationally (Tsing, 2007), especially in Latin America (Escobar, 1998; Ramos, 1994; Ulloa, 2010). Moreover, with its Working Group on Article 8j (WG8j), the CBD is one of the United Nations institutions that has a mode of governance particularly favorable to the participation of IPLCs (Friis, 2020). Thus, certain spaces within the CBD seem to be privileged sites for globalized demands from below (de Sousa Santos et al., 2007; Kurasawa, 2004; Ulloa, 2010).

The approach taken here is ethnographic. I have followed (to the extent that it has been possible from institutional and practical points of view) and analyzed the debates that took place in November 2019 during three meetings in Montreal organized by the CBD: (1) the First Global Thematic Dialogue for Indigenous Peoples and Local Communities (18–19 November 2019),[3] (2) the 11th Meeting of the Working Group on Article 8j (20–22 November 2019),[4] and (3) the 23rd Meeting of the Subsidiary Body on Scientific, Technical and Technological Advice (SBSTTA) (25–29 November 2019).[5] The second of these meetings was particularly important; it is within the WG8j that IPLCs are best represented and can best make their opinions known during negotiations with State Parties. Throughout these different meetings, I paid particular attention to discussions around the "link between nature and culture": this was the topic of a specific "item", a preparatory document had been written on it, and it had been debated in a plenary session,[6] all in order to help draw up the Post-2020 Strategic Plan on Biodiversity Conservation. I was attentive to the various contexts in which participants took a stand on their viewpoints, and I asked myself the following questions: which nature regimes are put forward by IPLCs? In what contexts? What are the alliances sought? What strategy, or strategies, do these positions refer to? The aim was to draw out competing and recurrent nature regimes that are mobilized by IPLCs at the same meetings – in the various different spaces where IPLCs express themselves – and sometimes even in the same documents.

This research situates itself in continuity with work carried out on the CBD by social science researchers (Aubertin & Filoche, 2011; Boisvert & Caron, 2002; Campbell et al., 2014; Escobar, 1997; Foyer, 2015; Reimerson, 2013; Suiseeya, 2014), especially the work of Parks (2017; Parks & Schröder, 2018). My approach, however, is different from previous work in its combination of several choices: on the one hand, I concentrate on meetings that prepare the COP, where IPLCs are better represented, and on the other hand I focus on meetings that happened synchronously, in 2019. In addition, I take an ethnographic approach that combines the analysis of documents, speeches, interviews, and ethnographic observations. Lastly, I concentrate on nature regimes and I adopt a political-ecological approach. This research, therefore, helps to inform the way that IPLCs negotiate for the recognition of their rights in international arenas through the exercise of ontological diplomacy.

The text is organized as follows: after having given an overview of the various meetings under study and specified how IPLCs take part in them, I present the various nature regimes put forward by IPLCs according to particular contexts, focusing on the way that IPLCs present the nature-culture relationship. I begin with one of the first nature regimes supposed to "integrate" nature and culture, one which calls naturalism into question, then I develop the second nature regime, which aims to "articulate" nature and culture and which is largely naturalist. At this point, let us note that these terminological choices stem from the choices of non-indigenous actors, whose usage of terms has crystallized over various exchanges and negotiations; this allows us to state that these positions do not reflect a nature regime specific to IPLCs, but instead relate to positions developed in specific contexts according to other non-indigenous actors.

The Spaces for IPLCs' Exercise of Ontological Diplomacy

The analysis presented here basically concerns three meetings that took place in the following order: the First Global Thematic Dialogue for Indigenous Peoples and Local Communities, the 11th Meeting of WG8j, and the 23rd Meeting of the SBSTTA. Each of these meetings has particular characteristics, especially as regards the participation of IPLCs, which decreases in degree from the first meeting to the last. In addition, there is continuity across the three events. The recommendations from the Global Thematic Dialogue feed into the reflections of the WG8j, which, in turn, feed into those of the SBSTTA to further preparation of the next COP 15 (which was supposed to be held in autumn 2021). In this section, I give an overview of the various spaces in which IPLCs participate according to the relevant meetings. As Campbell and her co-authors (2014, p. 6) point out, the rules that regulate when and how IPLCs can speak in the various contexts under investigation are essential in order to understand how IPLCs position themselves in negotiations. I present the various types of meetings according to the creation date of the corresponding working group in order to take account of the gradual opening up of the CBD to IPLCs.

At this point, it is necessary to clarify that the aim of these meetings is to produce various kinds of (non-binding) legal texts that formalize principles and mechanisms concerning a range of biodiversity-related topics (including conceptions of the relationship between nature and culture).[7] The State Parties that are signatories of the Convention negotiate[8] the texts resulting from previous meetings with a view to their validation at the COPs. Throughout the meetings, various kinds of negotiation take place: before the meeting, State Parties examine the texts, and then they may form and share their opinions; during the meeting, the State Parties negotiate with their regional partners (such as the European Union) at parallel sessions; and in plenary sessions all the Parties have to reach a compromise collectively. The ethnographic data that is harder to capture is the informal interactions that take place in the corridors. Since unanimity is required, if the parties do not come to an agreement, then it can be decided to leave a text "under brackets" to be discussed again later, at the COPs that vote on the definitive texts.[9] Certain moments during these meetings, however, are not directly to do with negotiations but aim to contribute information that is judged relevant by the CBD secretariat. This includes presentations made in plenary sessions before the negotiations proper, or "side events" carried out on various themes at midday or in the evening (in other words outside of the time dedicated to negotiations). On these occasions, various kinds of information are put forward concerning an initiative or a piece of research. As we will see, IPLC express themselves according to the kind of access they have to the different spaces.

The Subsidiary Body on Scientific, Technical and Technological Advice (SBSTTA)

The SBSTTA is a scientific advisory body of the COP whose creation relates to Article 25 of the CBD.[10] The first meeting of the SBSTTA took place in 1995. It is open to participation by all Parties and its aims are "providing assessments of the status of biological diversity; providing assessments of the types of measures taken in accordance with the provisions of the Convention; and responding to questions that the COP may put to the body".[11] The meeting in 2019 was the 23rd held by the SBSTTA.

The meetings of the SBSTTA are essential moments in the life of the CBD. Organized every two years and alternated with the COPs, they allow for scientific and technical clarification of debates within the CBD, although numerous debates are politically tinged. The items addressed during these meetings guide the various negotiations underway, which are subject to vote at the COPs.

Given the fundamental importance of the natural sciences in this domain, the default nature regime here is deeply naturalist. Nevertheless, the conception of nature mobilized in this arena is more open than elsewhere. The CBD recognizes an intrinsic value to biodiversity, a value that is not entirely anthropocentric (Larrère & Larrère, 2015; Uggla, 2010; Ulloa, 2010) or instrumental. In addition, the scientific approach put forward is systemic in the sense that it tries to take account of the various factors that influence biodiversity conservation, whether biological, human, or economic. Given the influence of the environmental sciences, one might call it ecosystemic.

This event brings together a thousand participants, mainly the representatives of various countries (State Parties) who are negotiators, but also representatives from the private sector, the academic world, and various groups such as youth, women, and IPLCs, the latter of whom are observers. The participation of IPLCs, since 1996, is organized, thanks to a specific structure, the International Indigenous Forum on Biodiversity (IIFB) formed in 1996 during the third COP of the CBD (COP III, Buenos Aires, Argentina). The IIFB gathers representatives from indigenous governments, indigenous non-governmental organizations, and indigenous scholars and activists. They provide support in coordinating IPLC strategies during important environmental meetings such as the CBD in order to further the recognition and respect of indigenous rights and to influence debates at international or national levels.[12]

To express their positions, the IPLCs are organized into a specific group, following the model of regional groups, a caucus whose representation is undertaken by the IIFB. At the SBSTTA meetings, however, the IPLCs are treated like the other actors without decision-making powers: they cannot intervene in the negotiations reserved for "decision-makers", which basically means the State Parties. In the context of plenary negotiations, they have the option of sending comments written before the meeting, and they can also express their positions during a period of time dedicated to non-decision-makers; they only have two minutes available to them each time, for each item or document that is negotiated. Within this framework, therefore, it is important for IPLCs to be able to agree among themselves in order to then express themselves with a single voice in a clear and concise way. Moreover, outside of the plenary negotiations, other spaces are accessible to IPLCs, whether through the representations of the different countries, within specific regional groups, or in a more informal way in the corridors where access is more difficult for an observer. Other spaces allow IPLCs to express themselves more freely and at greater length. IPLCs may be invited to make a presentation – for the opening of the event, for example, or at the moment when negotiations on a particular topic begin – but this is quite infrequent, given the extent to which a scientific and technical perspective predominates in SBSTTA debates. In addition, during these meetings, IPLCs may organize or participate in side events, which are privileged spaces for presenting the results of an ongoing project or piece of research (the Potato Park that is the subject of Chapter 3 has often been presented at side events during the various meetings of the CBD or other large international meetings of this kind, and this has greatly contributed to its fame). These plenary presentations and presentations at side events may be organized or taken charge of by the IIFB, but presenters can also speak in their own name.

At the SBSTTA, IPLCs have little margin for maneuver to participate in negotiations, and their participation demands a considerable effort at representation and synthesis, which rests largely on the IIFB – except in the case of more restricted negotiations. It is at the opening of the event, or when beginning the discussion of a specific item, that IPLCs find more open and less constrained spaces to express themselves, and in such cases the IIFB may be mobilized, although this does not necessarily happen. During SBSTTA meetings, nonetheless, these spaces are less open to IPLCs than during the WG8j.

At these meetings where scientific and technical considerations are in the spotlight, the naturalist nature regime that predominates at the CBD is, unsurprisingly, omnipresent. At the opening of the SBSTTA – which follows the WG8j – one feels a change in mood. There are more people, and certain countries that were not represented beforehand now are. The room is enlarged for the occasion. One sees more suits, ties, and smart dresses. All of this gives the impression that the serious business is beginning.

The Working Group on Article 8j (WG8j)

The complete name of this group is the "Ad Hoc Open-ended Working Group on Article 8(j) and Related Provisions of the Convention on Biological Diversity". This group was established in 1998 by the fourth meeting of the Conference of the Parties (COP4) and in 2000, at the fifth meeting,

> the COP adopted a program of work to implement the commitments of article 8 (j) of the Convention and to enhance the role and involvement

of indigenous peoples and local communities in the achievement of the objectives of the Convention.[13]

Article 8 pertains to *in situ* conservation of biological diversity and paragraph (j) specifically recognizes the importance of the "knowledge, innovation and practices of indigenous and local communities" in such conservation.

The WG8j has a strengthened participation mechanism for IPLCs that makes the CBD a particularly important instrument for these actors. Unlike in the SBSTTA, here they have a status equivalent to that of the State Parties; in other words, they can take part in plenary debates and give opinions outside of the two-minute format reserved for observers at the end of debates. During these meetings, the IPLCs are organized into a caucus by the IIFB. Moreover, certain members of IPLCs act as "Friends of the Secretariat", which means that they participate in Secretariat meetings and act as co-chairs of these meetings, thus contributing to the organization of the event, over which they also co-preside.[14] These mechanisms allow IPLCs to express themselves in a freer and less constricted way – especially during plenary sessions – as well as during negotiations.

This working group meets every two years, just before the SBSTTA, pushing IPLC demands into the last debates. This organization also allows certain representatives of the IPLCs to attend the two meetings.

To give a rough idea of numbers, in 2019, around 600 people attended the meeting, most of whom were the representatives of the various State Parties, but NGOs, the private sector, academics, youth representatives, women representatives, and of course IPLCs were also present.

At these meetings, there are various spaces and moments, each of which offers different constraints and opportunities. The kind of discourse produced by IPLCs, as we shall see, depends on the context. First of all, there are the plenary sessions, which take place in the large auditorium where all participants can be accommodated. Thus, one can distinguish between speeches on the margins of negotiations (ceremonies and presentations) and interventions made in the context of negotiation (either written or oral, given that IPLCs are considered to be a State Party). In the first case, a diversity of viewpoints may be presented, but during negotiations it is important to come to an agreement. The caucus allows for the formulation of a shared position. Smaller discussion groups, formed at a regional scale or together with the State Parties, also exist in a more or less formal manner. At the WG8j, the IPLCs take great advantage of side events, either directly or in partnership with other institutions such as the Secretariat of the CBD, the International Union for the Conservation of Nature, or the Global Environment Facility (GEF). All actors can propose side events and the Secretariat determines the timetable and allocates rooms on the basis of requests. IPLCs use these spaces in particular to show off successful initiatives. As we will see, these different spaces for expression offered during the WG8j are not all taken up in the same way by IPLCs.

Since the creation of the WG8j, some extremely important texts have been adopted, which aim for the culture of IPLCs to be taken account of: some are binding, such as the 2010 Nagoya Protocol, and others are not, such as the optional Akwé: Kon guidelines in 2004,[15] the 2010 Declaration on Biocultural Diversity,[16] and more recently the Sharm el-Sheikh declaration on nature and culture of 2018[17] (see Figure 1.1, this volume, Chapter 1). Parks (2017, pp. 14–15) notes an upsurge in the debates surrounding the link between IPLCs and nature at the beginning of the 2000s, which has translated into the mobilization of terms such as "worldviews" and "cosmology", with a high point in 2007.

It is important to mention that in 2019 the mandate of the WG8j expired and the type of structure to replace it was under discussion.[18] The issue at stake was whether a permanent group would be formed, as IPLCs were demanding. This led to a certain tension all throughout the debates within the IPLC caucus, since the IPLCs rightly saw this as a critical issue.

The Global Thematic Dialogue for Indigenous Peoples and Local Communities

The Global Thematic Dialogue for Indigenous Peoples and Local Communities is a working group created in the context of the elaboration of the Post-2020 Strategic Plan. It was designed to ensure the full and effective participation of local and indigenous communities at all the levels relevant to the Plan, in line with Aichi target 18, which aims to recognize, respect, and integrate traditional knowledge, innovations, and practices of IPLCs relevant for conservation and sustainable use of biological resources, and integrate them in the implementation of the CBD.[19] 2019 saw the first meeting of this working group, whose operating life is indexed to the formulation of the Post-2020 Strategic Plan, which should have been adopted during COP15.[20] This meeting was organized by the Secretariat of the CBD – the organism responsible for the WG8j – with the support of the IIFB. In line with the mandate of the Global Thematic Dialogue, reflections made by participants in the Dialogue were transmitted to the working group dedicated to the Post-2020 Strategic Plan, which examines if and how they can be integrated into the document. The recommendations of this working group were also transmitted to the other institutions of the CBD, especially the WG8j and the SBSTTA. It is thus a group where the modalities for holding meetings – relatively fluid since it is essentially a meeting between IPLCs – lead to a more open space for expression. The official mandate is relatively precise, but the space is also used by the Secretariat of the CBD to strengthen the participation of IPLCs within the WG8j and SBSTTA, in a context where the participation mechanisms for IPLCs in the CBD are being renewed.

The First Global Thematic Dialogue was a closed meeting of 60 people, of whom 50 were representing IPLCs via the IIFB, in addition to the representatives of certain CBD State Parties (CBD, 2021). Choosing candidates is a complex exercise: they must be put forward by an indigenous organization or recognized by the CBD Secretariat, and also have a certain knowledge of CBD mechanisms (Friis, 2020, p. 41). During the Global Thematic Dialogue, the IIFB acted as the co-chair of the meeting, which means that they participated in the organization of the meeting. At the event, IPLCs can speak freely and interact with the other actors present. As a member of an academic institution, I was not able to participate in this meeting; instead, it was via the report (CBD, 2021), which was written up at a later date and sent out at the WG8j and SBSTTA, that I learnt about the debates that took place. The creation of this kind of space should also undoubtedly be understood in the context of the WG8j coming to its end: the Global Thematic Dialogue creates a specific space that allows IPLCs to debate the future.

As we can see, the three working groups under study are very different not only in their objectives, but also in the number of participants, and especially in terms of IPLC participation. IPLCs have access to different spaces to express themselves, which vary from one meeting to another (Table 4.1). We will now see how they use these spaces, mobilizing different nature regimes and deploying parallel strategies, sometimes conjointly.

TABLE 4.1. The various meetings under study and the ways in which IPLCs participate (Hall, original material for the book)

Meetings	*Date*	*Actors present*	*Participation of IPLCs*
First Global Thematic Dialogue for Indigenous Peoples and Local Communities	17–18 November 2019	Meeting organized behind closed doors by the CBD secretariat with the support of the IIFB; 60 participants of whom 50 were representatives of IPLCs	Internal debate within the IPLCs, mobilization of the IIFB
Working Group on Article 8j (WG8j) -11th meeting	20–22 November 2019	IPLCs present, with various State Parties and observers (business, NGOs, academics, youth, women) 600 people	IIFB co-chair of the event Caucus of the IPLCs organized by the IIFB IPLCs are considered as a State Party
Subsidiary Body on Scientific, Technical and Technological Advice (SBSTTA) – 23rd meeting	25–29 November 2019	State Parties and observers, including IPLCs Around a thousand participants	Caucus of the IPLCs organized by the IIFB IPLCs are not considered as State Party and do not co-chair the event

"Integrating" Nature and Culture, Putting Forward an Organic Nature Regime

Within negotiations, one of the first ways of articulating the concepts of nature and culture put forward by IPLCs depends upon what certain meeting participants called a "holistic" approach, "cosmovision", or "integration". While the first two terms are used by IPLCs, the third has been used by other actors, in this case the CBD Secretariat in charge of organizing the WG8j meeting, the IUCN representative, but also the Colombian and Ethiopian delegations. It also appears in the documents negotiated during meetings.[21] I have kept to this last term ("integration"), since this choice helps to underline the extent to which the nature regime we are discussing relates, true enough, to ontological concepts, but whose expression must also be understood in a given context. From a conceptual point of view, this nature regime is quite close to the organic nature regime identified by Escobar (1999, p. 7), characterized by the fact that "nature and society are not separated ontologically", a regime, in other words, where "the natural world is integral to the social world" (Escobar, 1999, p. 8).

WG8j

It was at the WG8j meeting that the integration of nature and culture was put forward in the most explicit and articulate manner. Certain declarations submitted by IPLCs to the Secretariat beforehand made reference to the matter. Yet above all, it was during presentations – the opening ceremony, presentations in plenary sessions before negotiations, and side events – that IPLCs defended this nature regime, although these presentations have no legal weight.

At the opening of the meeting, a range of viewpoints was presented. First of all, the opening ceremony itself consisted of a welcome from indigenous peoples whose ancestral territory has been where Montreal stands now, where the meeting was taking place, as has become the custom in Canada over recent years. Charlie Patton, from the Mohawk Nation (Quebec, Canada), welcomed all participants onto the land of his nation. The other official presentations marking the opening of the meeting, especially a short audiovisual document, emphasised the importance of IPLCs in biodiversity conservation the world over, and the uniqueness of the relationship that these peoples have with nature.

At the following session, on the contribution of IPLC knowledge to the Post-2020 Global Biodiversity Framework,[22] various speakers intervened, including Josefa Isabela Tauli, an indigenous representative from the Philippines (for the Global Youth Biodiversity Network). In her intervention, she highlighted the importance of the concept of *ili* for the Kankanaey. This term can be translated as "place of origin", and it has been made use of because it shows in an exemplary way that nature and culture are intimately linked for this indigenous people of the Philippines (see Adonis & Couch, 2018 who elaborate on this concept).

This nature regime was also mobilized and made explicit at two different side events, one organized directly by the IIFB, the other by the CBD Secretariat together with the IUCN. In both cases, the IIFB is a stakeholder, and it was two representatives from Latin America who intervened, namely Ramiro Batzin (Maya Kaqchikel Guatemala, co-chair for the IIFB) and Yolanda Terán (Kichwa, Ecuador).

These four interventions (from Charlie Patton, Josefa Isabela Tauli, Ramiro Batzin, and Yolanda Terán) differed in offering specific viewpoints on the way that a particular group conceptualizes its relationship to nature. Nonetheless, they overlapped on a key point: in their interventions, the various speakers argued that it does not make sense to oppose nature and culture. This fundamental point has been emphasized by various writers such as Escobar (1999) and Viveiros de Castro (2014).

Moreover, in three of these interventions, a feminine figure was utilized in order to clarify this integration of nature and culture: Mother Earth, Mother Nature, or Pachamama. Significantly, it was the representatives of indigenous peoples from the Americas – Mohawk, Kichwa, and Maya – who made these references. The emergence of Mother Earth seems to have been a North American phenomenon in the first instance. Gill (1991), who has retraced the emergence of this figure, shows that it was not familiar to all indigenous peoples from North America, and that it was through dialogue between first nations and academics that it gradually became established in the 1960s, then becoming indispensable in the 1970s and afterward. The rise of Mother Earth has gone together with the rise of environmental struggles that mobilize the image of American first nations: the figure of the ecological Indian, in the words of Krech (1999). Yet, it is not only a North American image. Since the 1960s, there has been an internationalization of indigenous and environmental struggles. Links have been built with the Philippines and Latin America, in particular. The link with Latin America, as we will see, has helped to extend the reference to Mother Earth.

In the Latin American region in the 1980s, and even more so in the 1990s, environmental issues were uniquely linked to social and cultural demands, as Escobar (1999), Ulloa (2010, p. 31 and thereafter), and Ramos (1994) have shown. The wide-scale adhesion of Latin American countries to Convention n° 169 of the International Labor Organization (ILO, 1989) and the ratification of multicultural constitutions at the beginning of the 1990s allowed indigenous populations to claim a new status both nationally and internationally. In Latin America especially, social demands were articulated around cultural differences that are now recognized by multicultural constitutions (Escobar, 1998, 1999; Ulloa, 2010, p. 31 and thereafter). The Colombian case study analyzed by Escobar (1998) and the Bolivian case study analyzed by Canessa (2006) show in an exemplary manner how environmental issues are closely associated with issues of identity. Culture has thus gradually acquired a political value, a process that Escobar (1997) and Povinelli (1995) in particular identify as "cultural politics".

Moreover, the international arena has become a significant one for such demands, as various writers have noted (de Sousa Santos et al., 2007; Escobar, 1999; Ulloa, 2010). This is especially the case in fields dedicated to environmental and biodiversity issues (Aubertin, 2005; Boisvert & Caron, 2002; Escobar, 1997), to the point where Ulloa (2010, pp. 216–217) speaks of an "ecological identity" for IPLCs. As de Sousa Santos and his co-authors write (de Sousa Santos et al., 2007, p. xli), biodiversity has become a "contentious area".

Cultural politics is deployed in two different ways with regard to the CBD. First, the central position given to the traditional knowledge of IPLCs in *in situ* conservation of genetic resources has given rise to a specific form of cultural politics. The naturalist vision of this knowledge tends to equate it to a kind of cultural baggage that has the unique characteristic of helping to maintain biodiversity. A kind of cultural politics has therefore been developed, in and by the processes of negotiation, around the valorization of this knowledge.

Yet, this gives rise to another form of cultural politics that is based upon the figure of Mother Earth, which over the years has attained the status of an "icon without cultural context" (Ulloa, 2010, pp. 199–200) and has established itself as a reference point in institutions involved in the worldwide governance of genetic resources (Escobar, 1999; Ulloa, 2010). The recognition of this figure, in its Andean form of Pachamama, within the constitutions of two Andean countries – Bolivia and Ecuador – in 2008 and 2009 has marked a turning point.[23] Poupeau (2011), Bretón Solo de Zaldívar (2013), and Landivar and Ramillien (2015) all highlight, as Gill (1991) did regarding the North American version, that the emergence of Pachamama results from a complex process that indeed involves indigenous peoples, but also numerous civil society actors,[24] as well as academics such as anthropologists. By recognizing Pachamama in their constitutions, the Andean countries have given a new legitimacy to the figure of Mother Earth, which aims to embody nature and helps to elucidate the supposedly reciprocal and harmonious relationship that the indigenous population of Andean countries maintain with their environment. This opens up a reflection on the legal status of nature, which may become a legal question (Hermitte, 2011) and allows us to envision novel solutions such as the attribution of legal personality to elements of nature. Demands formerly phrased in terms of cultural politics, therefore, are evolving. By disputing the very status of nature, we proceed to a politics of nature or, more generally, political ontology.[25]

This has led to two different outcomes within the CBD. On the one hand, the representatives of Andean indigenous peoples have made use of this figure – as the Kichwa representative of the IIFB did during a side event in 2019 (mentioned earlier). On the other hand, in official negotiations and debates, the Bolivian and Ecuadorian State Parties have repeated loud and clear the need to rethink the rights of nature after the ratification of their new constitutions (i.e. 2008). At different meetings of the WG8j, different countries have defended a similar position; in 2019, the Colombian and Mexican representatives were the most active.

Various writers, including Ulloa (2010, pp. 199–200), highlight how nebulous the figure of Mother Earth is. It has a spell-binding power over non-indigenous populations through its link to the figure of the noble savage or "ecological native" as clearly shown by Ulloa (2010, on the same topic, see also Fabricant & Postero, 2018; Hall, 2022; Igoe, 2010; Ramos, 1994). These images tend to reproduce the subaltern position of indigenous populations, as reflected by the frequent reference to a feminine figure (given gender-based asymmetries in power relations). Thus, while potentially emancipating, this figure and its feminine allegory contribute to reproducing a colonial heritage (Ulloa, 2010, p. 201).

Yet in my opinion, the mobilization of the Mother Earth figure, especially by reference to Pachamama, marks a turning point in the strategy of indigenous peoples. Having been legally recognized, Pachamama – and Mother Earth – has acquired a new status and has become a powerful discursive weapon that helps to underscore the need to dispute the naturalist (or capitalist) regime, and thus also the need to integrate nature and culture. The issue is no longer, as it was in the 1990s, one of demanding respect for cultural differences, but one of having ontological differences recognized. It is about moving from cultural politics to political ontology, which, in the particular context of the CBD, takes the form of a politics of nature. The reference to Mother Earth and her Andean personification might make us think of a "return" to an organic nature regime, but let us not deceive ourselves. For IPLCs, it is not about going back in time, but about negotiating contemporary and future issues. This paradigm shift entails the formalization of a hybrid nature regime in line with the current context.

As Parks (2017, pp. 16–17) shows by analyzing texts produced by the CBD, this vision is making headway within the CBD, where it is now in principle accepted – at least on a rhetorical level. In the report of the 2019 WG8j and SBSTTA meetings provided in the *Earth negotiation bulletin*,[26] one finds a tendency to present transformative change as expected and it is mentioned that "many [state parties and observers] supported the draft recommendations [of the Global Thematic Dialogue] [...] highlighting that nature and culture are deeply integrated" (2019, p. 13). Yet in the final recommendations relating specifically to this point,[27] the term integration only appears in the title of the document, and in the body of the text the discussion is carefully restricted to the "link" between nature and culture, with the integration of nature and culture largely left out.

Global Thematic Dialogue

The report of the first Global Thematic Dialogue also highlights the need to rethink the connection between nature and culture (CBD, 2019).[28] In this way, the importance of adopting a "nature-culture approach", which should help to establish a more "harmonious" relationship with nature, is underlined. Yet, contrary to my expectations, given that it is a group where IPLCs can speak freely, the promotion of this nature regime does not take center stage in demands made in the document. If the Latin American delegation defends an approach that explicitly challenges the dominant nature regime in demanding "respect and recogni[tion for] the sacred and holistic approach that indigenous peoples have to nature and biodiversity", this section seems to have been added in a rather contrived manner at the end of a table synthesizing the main messages that IPLCs wish to convey to other institutions. Not only is the content quite different, but one also notices that the passage is specifically raised by indigenous peoples (and not local communities) of the Latin American delegate. This is significant, given that not all IPLCs agree to the promotion of this nature policy, especially as far as local communities, and even non-Latin American indigenous peoples, are concerned.[29]

Alliances and Spaces for Expression

Some non-indigenous agents position themselves favorably regarding this type of nature regime. The Secretariat of the CBD, and especially the "people and biodiversity" unit that coordinates the WG8j (Friis, 2020, p. 13) actively contribute to creating spaces for IPLCs to express themselves at various meetings, not only at the WG8j. In 2019, the Secretariat – and this unit in particular – was particularly active in bringing the CBD closer to the IUCN and UNESCO, and this was supposed to result in the creation of a Nature-Culture Alliance (the modalities of which should have been detailed at the IUCN congress in September 2021 and which should have been be officially ratified during the COP 15 of the CBD planned in October 2021). This builds upon a dynamic that has existed for several decades through the LINKS program dedicated to Local and Indigenous Knowledge Systems; more recently, in 2019, the North American Declaration on biocultural rights has also positioned itself in continuity with this.[30] Some programs (such as LINKS, mentioned above) and units (people and diversity) have a mandate to push for the development of these issues. They thus draw on the strong standpoints taken by IPLCs. Nevertheless, in the discourses produced by these non-indigenous actors, however, one perceives the valorization of a nature regime relating to integration that is not fully complete. The texts produced by these various institutions remain dependent on the same concepts of nature and culture, which contributes to maintain a naturalist nature regime. This is largely because they have to take into account the institutional language adapted to the various institutions they represent in line with inter-institutional dynamics. In this context, the creation of specific spaces of expressions is in itself a way to contribute to the reflection on the nature regime.

IPLCs exercise, therefore, a "politics of nature" in spaces that are largely marginal to negotiations. This allows them to display a collective identity that rests mostly on the image of the "ecological native" and quite often also on the image of Mother Earth. This allows IPLCs to join forces under the same flag and to make themselves identifiable in a context where they are an overwhelming minority. Moreover, these two images have a power of attraction for non-indigenous actors that is not to be underestimated. This strategy thus depends on what Campbell and her co-authors (2014, p. 6) call a politics of performance: in other words, it is through performance that diplomacy operates. This resonates with the concept of recognition put forward by Povinelli (1997, p. 20) insofar as IPLCs perform a certain self-image that positions them with regard to other actors. Recourse to this nature regime thus allows IPLCs to join together under a unifying discourse, to reinforce their collective identity in the framework of negotiations, and to establish alliances despite the flaws inherent to this strategy. In the performance of this collective identity, the IIFB plays an essential role and this has the consequence of over-valorizing the indigenous element, to the disadvantage of the local communities.[31]

"Articulating" Nature and Culture: A More Pragmatic Vision

Over the course of the different meetings, the IPLCs do not always keep to a discourse that highlights the integration of nature and culture (i.e. a non-naturalist ontological regime) or plays on the image of the ecological native. As we will now see, in the spaces for expression linked more directly to negotiations – as can be seen in the recommendations for the SBSTTA and the WG8j, or the main messages for the Global Thematic Dialogue – the naturalist nature regime is generally not called into question. This should be considered in relation to the fact that most non-indigenous actors are deeply committed to a naturalist regime that separates nature and culture. In 2019, most passages of text that spoke too openly of "significant change", as mentioned above regarding the SBSTTA, were fiercely contested during negotiations, which led to a "bracketing" of this expression until the next COP planned in 2021 (consensus is necessary for a text to be definitively approved (Maljean-Dubois, 2021, pp. 111–112)). This nature regime is linked to economic interests, central in the global neoliberal governance of biodiversity. It is in order to underscore the importance of these interests that Escobar (1999) proposes the term capitalist nature regime, rather than naturalist nature regime.

Given the forces of resistance at work, and insofar as it is necessary to establish dialogue with the various actors present to be able to negotiate, the acceptance of this dominant nature regime, naturalist and capitalist, seems to be a pragmatic and realistic choice. This is even more the case, given that outside of the WG8j, the spaces available for IPLCs to express themselves are reduced and constrained (as a reminder: IPLCs can send a commentary beforehand; otherwise, they only have two minutes, at most, to speak at the close of official debates). In the IPLC caucus, too, this is a handicap; with such constraints, IPLCs necessarily have to speak with a single voice. A need for efficiency thus imposes itself on IPLCs. In the following section, I will focus on the ways that IPLCs deal with these various constraints.

WG8j

Even during the meetings of the WG8j, where we saw that IPLCs practice a politics of nature, it is noticeable that during negotiations in plenary sessions, participants' speeches do not make reference to this politics of nature. In their interventions, IPLCs make reference, instead, to the words used in the texts of the CBD, the way this institution works, and highlight the importance of other institutions such as UNESCO. Thus at the opening of the WG8j, the IIFB, in its declaration, indicated the necessity – in a context where the erosion of biodiversity is accelerating, and given the expectation that IPLCs will conserve biodiversity in an exemplary manner – of implementing the various texts previously voted for in order to allow IPLCs to carry out their conservation work.

When presenting the results of the Local Biodiversity Outlook 2 (Forest peoples program, 2020) to which she contributed, and trying to evaluate the accomplishment of Aichi Target 18, Ms. Joji Carino Nettleton (member of the IIFB) reiterated the message. In her intervention, she spoke of failure: the measures voted for are not put into practice, and there is a flagrant lack of political will. In this intervention, she also identified the dysfunctional mechanisms of the CBD: the delegations of the various countries must produce reports in which they document – among other topics – the situation of IPLCs and how their traditional knowledge is taken into account. This knowledge is not taken into account, however, and this turns the achievement of the goals previously fixed into an unrealistic hope (ISSD, 2019, p. 3). Here, therefore, we see clearly that the strategy adopted is anchored in a detailed knowledge of the CBD's texts and mechanisms.

In 2019, with the WG8j's mandate reaching expiry, a large number of the positions taken up by IPLCs and by the IIFB had to do with the future participation structure of IPLCs. The hope is that a permanent structure will be created. Thus, at the closing of the event, the IIFB "reminded delegates and participants that the full and effective participation of IPLCs is crucial for a strong post-2020 framework" (ISSD, 2019, p. 7).

SBSTTA

During the SBSTTA, where the participation mechanism for IPLCs is much more restricted, the positions upheld by the IPLCs are even more limited and must be even more efficient. IPLCs took advantage of their rare opportunities to speak to defend their most essential points. Over the course of the meeting, the IIFB thus consecutively insisted on the "important contributions of IPLCs to biodiversity conservation benefitting society as a whole"; on "the full and effective participation of IPLCs in the post-2020 process" (ISSD, 2019, p. 11); on the respect for recommendations given by the WG8j; on the crucial nature of "collaboration between all actors, including IPLCs" (ISSD, 2019, p. 14) who must have access to information on conservation issues and should be included in related actions. These formulae are taken up by other participants, to the extent that they are almost commonplaces. As is evident, in such a context, the predominant nature regime is not contested. In fact, it appears uncontestable and IPLCs do not take the risk of disputing it within the framework of the SBSTTA, at least not directly.[32]

Global Thematic Dialogue

It is in the report on the Global Thematic Dialogue, surprisingly, that acceptance of the naturalist nature regime is most clearly manifest. Despite the position of the Latin American group (placed at the end of the document and mentioned earlier in this chapter) and despite invitations to take seriously the issue of the relationship between nature and culture, the majority of the document drawn up for the Post-2020 Strategic Plan is in continuity with debates taking place within the CBD and the nature regime mobilized is one that enshrines the dichotomy between nature and culture (CBD, 2019).

The five main messages identified in the document (CBD, 2019) are revealing in this sense. IPLCs listed the following priorities: (1) use a human rights-based approach, (2) implement equitable governance, (3) be evidence-based (IPBES Global Assessment, Global biodiversity outlook – 5th edition – and Local biodiversity outlook – 2nd edition), (4) ensure coherence and synergy across the United Nations system, and (5) align the structure of the Post-2020 Strategic Plan with the Sustainable Development Goals. The way that these various objectives are then explained and developed is entirely compatible with the concepts and vocabulary predominant at the CBD and in the SBSTTA. The use of the adjective "harmonious" links to the leitmotiv of the CBD, "Live in harmony with nature", which became established after the Nagoya agreements in 2010; in other words, this is a watered-down version of the "harmony" associated with Mother Earth or Pachamama, a naturalist version whereby nature – easily identifiable – provides for human needs in a satisfactory way. Once again, this should be understood in relation to the fact that this text is part of a negotiation process: it is destined to send up the demands of IPLCs for the Post-2020 Biodiversity Framework to higher institutional levels, and beyond this for WG8j, the SBSTTA and the COP. Even though the "main messages" are in no way binding, the manner in which they are formulated responds to a pragmatic logic.

This pragmatism translates into numerous references to the rules of access and benefit-sharing (APA/access benefit-sharing) written into the Nagoya Protocol. The objective is clearly to ask that the redistribution mechanisms enshrined in the CBD be implemented effectively. On this point, recall that Joji Carino Nettleton at the WG8j highlighted a lack of political decisiveness in implementing the measures voted for in the context of the CBD, including the Nagoya Protocol. Moreover, the "full and effective participation" of IPLCs in negotiations is also demanded with insistence. This document has therefore been drawn up in view of the COP 15 to come.

As we have seen, in this document, a "nature-culture approach" is foregrounded, and the document requests that such an approach be a "cross-cutting element" in the context of the Post-2020 Biodiversity Framework and later negotiations. Yet, the reach of this argument is limited, since this approach is immediately associated with "cultural diversity", which identifies a cultural component considered separately from nature. It is then requested that "measures [be taken] to promote the respect and recognition for the value of traditional cultures to biodiversity"; the definition of that which relates to culture is therefore restricted to that which is relative to biodiversity.

This argumentation tends to envisage the conceptual differences regarding the relationship between nature and culture in cultural terms, as an issue of cultural politics. With regard to culture, the universality of nature – mononaturalism – is taken for granted. This is fiercely criticized by writers such as Viveiros de Castro (2014), Escobar (1997), and de la Cadena and Blaser (2018), who militate in favor of a multinaturalism that cross-cuts the cosmopolitical considerations of Stengers (2005) and Latour (2007). In other words, while IPLCs tend to forgo cultural politics in favor of nature politics, the international texts are still stuck in a naturalist nature regime, which IPLCs must take account of if they wish to position themselves favorably.

Yet, the document makes a call for "build[ing] understanding on the links between nature and culture". If we dig a bit deeper on this point, we see that the line of argument chosen rests not on advocacy for an organic nature regime, but on the deepening of scientific knowledge (a science-based approach). As such, there is a request to carry out studies that would help to better understand IPLCs' relationship to "nature" and, more specifically still, to develop indicators that would reflect "Nature-Culture values". This comes down to a politics of knowledge (Campbell et al., 2014, p. 5; de Sousa Santos et al., 2007). The valorization of IPLCs' knowledge is linked to a demand in terms of production, sharing, and access to scientific knowledge. Concretely, the establishment of "Community-based Monitoring and Information Systems (CBMIS)" is invoked, and these are presented as "indispensable tools for equitable and transparent environmental governance at all levels". Ferrari et al. (2015) show the potential of this approach, which in their study helps to strengthen IPLC participation and fine tune local-level strategies. By participating in the production of specific criteria adapted to their conceptions and realities, IPLCs contribute to influencing future policies: they produce knowledge with a political application, which allows the implementation (or otherwise) of broad objectives such as the Aichi Targets to be measured. Furthermore, IPLCs have recourse to what Campbell and her co-authors (2014) call a "politics of scale", which allows them to make use of effects of scale, being part of the production of information which can translate into actions.

This argumentation is very skilled: it builds upon the fact that it is nowadays morally impossible for others to stand against the recognition of IPLC rights, as Parks (2020) shows clearly. It is science that has been mobilized in order to push forward negotiations: it is in this sense, moreover, that we should understand the mention made, within the Global Thematic Dialogue report, of the Intergovernmental Science-Policy Platform on Biodiversity and Ecosystem Services (IPBES), an institution linked to the CBD, and the mention of the publications Global Biodiversity Outlook and Local Biodiversity Outlook (CBD, 2019).

The IPBES was modeled upon the Intergovernmental Panel on Climate Change (IPCC) in 2010. It is independent of the CBD, although closely linked to it, and situates itself at the interface between science and politics, as its name indicates (Le Prestre, 2017, p. 115; Vadrot, 2016). Yet, the mission of the IPBES is also to recognize and valorize the traditional knowledge of IPLCs and to fill a knowledge gap: in 2016, the IPBES positioned itself in favor of taking account of the cultural dimension of biodiversity,[33] setting itself the goal of better understanding other knowledge systems by adopting a scientific approach that would inform the decisions taken at the CBD. Moreover, in decision IPBES-7/1 (May 2019), the Plenary extended the mandate of the existing task force on indigenous knowledge in order for it to work on the implementation of objective 3 (b) "Enhanced recognition of and work with indigenous and local knowledge systems" of the rolling work program of IPBES up to 2030.[34]

In this sense, within the Global Thematic Dialogue report, one finds a great degree of reflection on the indicators that account for IPLC perspectives and strengthen their participation in gathering data. The importance of these indicators was highlighted at the WG8j meeting by Eleanor Sterling, Chief Conservation Scientist of the Center for Biodiversity and Conservation of the American Museum of Natural History, as she presented results in a plenary session and during a side event organized by the Secretariat and the IUCN.

In addition, this reflection on indicators helps to further "the human rights approach", also mentioned in the Global Thematic Dialogue report. In this document, a more integrated approach is taken, one which takes into account the various human components involved in the management and conservation of natural resources; components that are recognized by the other instruments developed regarding human rights, specifically those dedicated to the recognition of indigenous rights (UN Declaration on the Rights of Indigenous Peoples (UNDRIP) – 2007).

This multifactorial approach also resonates with the "holistic approach" developed within the FAO,[35] which aims to take account of the ecological, economic, and social dimensions of organic agricultural production. It is noteworthy that the term "holistic", which has become so common nowadays, is here used in the context of a naturalist nature regime, where its meaning differs from the meaning given to it by IPLCs who use it instead to dispute the naturalist nature regime.

During negotiations, although IPLCs leave out demands framed in terms of a non-naturalist nature regime, they develop nuanced strategies that respond to a complex (and dynamic) interplay of elements and that allow them to take advantage of texts favorable to them or to establish alliances with institutions. In

2019, IPLCs, together with the CBD Secretariat, favored a rapprochement with the IPBES and UNESCO. The conceptual tools developed by these institutions are often ambiguous; one notices the pervasive influence of naturalism, but also the way that certain openings are facilitated at certain moments,[36] openings which IPLCs seek to exploit. Accepting the terms of reference and using the tools available allow for the creation of equivalences or synergies that can have positive outcomes.[37]

This type of diplomacy builds not only upon a politics of nature, but also upon what Campbell and her co-authors (2014) identify as a politics of knowledge and a politics of scale. These negotiations aim to build upon science while also insisting on the need to take account of indigenous knowledge in the production and publicization of data. In this way, IPLCs build networks of allies according to the context and the occasion. This allows IPLCs to negotiate what Povinelli (1997, p. 20) calls identification, namely the concrete place attributed to a group in relation to others and the rights attributed to it.

Conclusion

Unsurprisingly, the nature regime that predominates within the CBD is naturalist (or capitalist). Nature and culture remain two distinct concepts and, at the meetings mentioned that took place in 2019, the questions asked concern the "link" that joins them. Although the importance of culture is by now widely recognized, its integration with nature remains difficult to conceptualize for most of the participants, especially from Western countries.

In analyzing the way that IPLCs mobilized themselves at meetings taking place in Montreal in 2019, it appears that disputing this nature regime is not an easy task. In this chapter, I have shown that IPLCs do not operate together as a bloc or in a strictly defined way in favor of an organic nature regime alternative to naturalism. In fact, the standpoint taken by these actors, especially through their representation in the IIFB, is much more nuanced and strategic. Their argumentation is adapted to the meeting, to the mode of participation that IPLCs are allowed, to the exact type of space in which they express themselves, to the concepts held by the other participants present, to the alliances that can be mobilized, and to specific issues.

I have identified two types of standpoint that build upon two different nature regimes: non-naturalist/organic and naturalist/capitalist. These two nature regimes correspond to two distinct strategies, one relating to a politics of nature, the other to a politics of culture. The first nature regime, non-naturalist, has to do with the valorization of a "holistic" worldview, which disputes the distinction between nature and culture and valorizes indigenous "cosmovisions". In 2019, the non-indigenous participants tended to call this approach "integration". This position is mostly defended by the delegations from the Americas, especially by the Latin American delegation, and more specifically still by the Andean countries. Pachamama – the Andean version of

Mother Earth – tends to personify this approach. This display of uniqueness comes at a price; the IPLCs bring along diverse concepts that tend to be assimilated to each other. It seems, therefore, as if all IPLCs attribute a central position to Mother Earth, which is not the case. Yet, this kind of politics has an essential symbolic reach by allowing the IPLCs to position themselves as a relatively homogenous group. This message is invoked in the spaces for expression situated at the margins of negotiations, spaces that allow IPLCs more significant room for maneuver – namely the presentations in plenary sessions and side events – but which are marginal with respect to the negotiations underway. Foregrounding ontologies alternative to naturalism thus depends on a political statement at the margins of negotiations: such statements aim to underscore the specific characteristics of IPLCs and to have the legitimacy of their claims recognized. This ontological diplomacy is linked to a process of symbolic recognition, which consists of "performing" the idea that IPLCs exist as a collective and that the nature regime proper to them is non-naturalist/organic. The identity mobilized in this way is close to the image of the ecological native, which allows for the establishment of certain strategic alliances, but has its limits. The strategy depends, therefore, upon a politics of nature.

In the context of negotiations, IPLCs position themselves differently. The nature regime upon which their standpoints is apparently aligned with the naturalist or capitalist type. The opposition between nature and culture is accepted, and it is on this basis that negotiations are carried out. In view of the importance that the texts give to traditional knowledge linked to genetic resources, a focus on culture is indispensable. It is regarding this point in particular that various strategic alliances may be worked out and that synergies may develop between participants, between institutions, and between legal texts. Thus, in 2019, the Secretariat of the CBD (in particular the "people and biodiversity" unit), together with the IUCN, foregrounded the importance of UNESCO and the creation of a Nature-Culture Alliance was announced. Even if the naturalist nature regime still predominates, a process of reflection on the link between nature and culture, led by various institutions and based upon an alliance with IPLCs, is flourishing. The various participants involved do not necessarily share the same viewpoints, but by working together they contribute to transforming the standpoints of their institutions. In this way, the holistic approach, understood in ecosystemic terms, tends now to recognize that the naturalist/capitalist nature regime should be challenged. This ambiguity allows the IPLCs who interest us here to practice a more discrete ontological diplomacy than beforehand. This diplomacy is pragmatic, it is carried out on moving ground, and it rests upon alliances that seem promising. At first sights, it rests on cultural politics, but it is more complex.

In this context, accepting the naturalist nature regime allows IPLCs, in my view, to "control" the deep misunderstandings that would arise from taking account of ontological differences, to take up the terminology proposed by Viveiros de Castro (i.e. controlled equivocation (2004)), and to place themselves in a framework shared by other participants.

This kind of diplomacy, then, also questions the naturalist nature regime, but in an indirect way. Through the scientific study of the link between nature and culture, whose cornerstone is currently the fine-tuning of indicators that integrate the viewpoint of IPLCs, this kind of diplomacy relates to a politics of knowledge. Furthermore, by enhancing the participation of IPLCs in projects carried out by the CBD, the position of IPLCs in their localities may be strengthened; at issue, therefore, is a politics of scale that allows IPLCs to strengthen their role as protectors of the environment.

Finally, these two types of ontological diplomacy are complementary and depend both, ultimately, upon a politics of nature. We can merely note that for IPLCs the issue is not one of imposing a precisely defined, rigid, or fixed nature regime that refers to a distant, idealized past. This diplomacy mobilizes various actors, allies, and detractors; and it is in these interactions that zones of rupture are constantly made and remade.

These two strategies are also complementary in jointly helping to develop the concepts prevalent in global arenas of biodiversity governance. In fact, despite incredible inertia, things seem to be moving forward bit by bit: the naturalist nature regime is (beginning to be) overturned, and a hybrid nature regime is in development. At the opening of the SBSTTA, Eduardo Brondizio (an anthropologist and co-chair of the IPBES) indicated that a "transformative change 'of our norms and values' is needed for a meaningful post-2020 framework" (ISSD, 2019, p. 8). The expression "transformative change" is now part of the IPBES' vocabulary, an assessment is being prepared on the topic,[38] and although the negotiated official definition remains relatively anchored in naturalism, Brondizio's remarks give a glimpse of the transformative potential held by the process of reflection underway, which tends toward disputing the predominant naturalist/capitalist nature regime. From a politics of culture, we move forward, with the burden entailed by already ratified texts, to a politics of nature. Let us not cry victory yet, however; at the SBSTTA, the use of the expression "significant change" was not formally agreed to, and the term stayed "between brackets". Let us hope that these are taken out at COP 15.

Notes

1 See the Zero draft from January 2021, https://www.cbd.int/article/2020-01-10-19-02-38. *All URLs retrieved on 1 September 2021.

2 Here, I am referring to the work of Stenger (2007) and Latour (2004) on cosmopolitics that invokes a recomposition of worlds, as well as that of Tsing (2005, p. 270) who speaks of "world making"; Haraway (Harvey & Haraway, 1995) also makes a plea for this. This proposal generates debate on the very possibility of taking account of such a diversity of perspectives (Blaser, 2016). Various authors insist on the fact that the outcome cannot be known in advance (Blaser, 2004; Cadena, 2012; Cadena & Blaser, 2018), which pushes Stengers (2007) to say that the pluriverse should be understood as "emergent" over time.

3 See https://www.cbd.int/tk/post2020.shtml.

4 See https://www.cbd.int/conferences/sbstta23-8j11/wg8j-11/documents.

5 See https://www.cbd.int/conferences/sbstta23-8j11/sbstta-23/documents.
6 See WG8j, SBSTTA, https://www.cbd.int/doc/c/c013/d7a5/f9b18a002b273903332ffdcf/wg8j-11-05-en.pdf.
7 For an explanation of the workings of the CBD more generally, see Le Prestre (2017).
8 I use the term negotiation in a broad sense in this chapter. I consider that the preparation of the texts that will be presented to the COP are part of a long and complex process of negotiation. During WG8j and SBSTTA, there are some formal negotiations around the texts themselves, which will be approved only during COPs. I consider that, although the texts produced by the WG8j or the SBSTTA are non-binding instruments, they are part of the elaboration of new norms. This choice allows us to grasp the kind of diplomacy IPLCs practice at the CBD.
9 Müller and Cloiseau (2015) give an account of the debates that took place over a single paragraph at the negotiations surrounding the Rio+20 Declaration.
10 On the SBSTTA, see Le Prestre (2017, pp. 103–108) and Maljean-Dubois (2021, pp. 113–114).
11 https://www.cbd.int/sbstta/.
12 https://iifb-indigenous.org/.
13 https://www.cbd.int/convention/wg8j.shtml, to see the program: COP CBD, Decision V/16, UNEP/CBD/COP/5/23 (22 June 2000): https://www.cbd.int/decision/cop/?id=7158 or COP CBD, Decision IV/9, UNEP/CBD/COP/4/27 (15 June 1998): https://www.cbd.int/decision/cop/?id=7132.
14 See WG8J, CBD/WG8J/11/7 (22 November 2019): https://www.cbd.int/doc/c/bd0a/077c/9af6c8783485bec92396af97/wg8j-11-07-en.pdf (section 15, p. 21).
15 COP CBD, Decision VII/16.F, UNEP/CBD/COP/7/21 (13 April 2004).
16 COP CBD, UNEP/CBD/COP/10/INF/3 (2 September 2010).
17 North American dialogue on biological and cultural diversity, 5–8 May 2019, Montreal, Canada; Center for Indigenous Conservation and Development (Joint program "Linking Biological and Cultural Diversity" between UNESCO and the SCBD).
18 As mentioned, by way of a reminder, the mandate of the WG8j was extended to 2020; currently at stake is the creation of a multi-year program of work on Article 8(j) in line with decision X/43 COP CBD, Decision X/43, UNEP/CBD/COP/10/27 (29 October 2010).
19 COP CBD, Decision X/2, UNEP/CBD/COP/10/27 (29 October 2010); also see COP CBD, Decision 14/17, CBD/COP/14/14 (20 March 2019), para. 13.
20 A relatively similar meeting was organized in 2017 in the context of the IPBES Global Assessment on Biodiversity and Ecosystem Services (2019) just before the meetings of the WG8j and SBSTTA.
21 See WG8J, Options for possible elements of work aimed at an integration of nature and culture in the post-2020 global biodiversity framework, CBD/WG8J/11/L.4 (22 November 2019).
22 WG8J, CBD/WG8J/11/L.1 (22 November 2019), p. 9.
23 A large literature exists on the topic. In one of his texts, Escobar (2010) gives a detailed comparison of the various Andean countries.
24 These constitutions were debated in constituent assemblies (*asambleas constituyentes*) made up of all sectors of the population (Landivar & Ramillien, 2015).
25 On this issue, see Chapter 3 of this book.
26 This publication from the International Institute for Sustainable Development (IISD) provides daily and final reports on the negotiations, which are largely distributed to meeting participants but also on line to a broader audience.
27 WG8j, Recommendation 11/3, CBD/WG8J/11/7 (22 November 2019).
28 https://www.cbd.int/doc/c/245c/aee3/33cabfb2c1daa9c539b3c5ed/post2020-ws-2019-12-02-en.pdf.
29 This is clear in interactions with the representatives of indigenous peoples who are not from Latin America, or with the representatives of local communities.

30 https://fr.ccunesco.ca/blogue/2019/11/declaration-nord-americaine-diversite-bioculturelle.
31 The place of the Local Communities in this caucus would be interesting to investigate, given the extent to which indigenous peoples predominate, as Friis (2020, p. 60 and thereafter) also notes.
32 The debates surrounding the nature of traditional knowledge, which is the cornerstone of the recognition of IPLC rights in the CBD and the Nagoya Protocol, help to rethink the opposition between nature and culture. Yet, this was not apparent at meetings in 2019, undoubtedly due to the context and due to the need to renew the institutions for IPLC participation within the CBD. On this point, see the discussion in the introduction to this book.
33 IPBES/4/19 (29 March 2016). Also see this volume, Chapter 1.
34 IPBES/7/10 (22 May 2019), Figure A.1, this book, Chapter 1.
35 http://www.fao.org/organicag/oa-portal/orca-research/research-fields/orca-research-models/orca-holistic-approach/fr/.
36 On this point, the conceptual framework of the IPBES, which is the subject of debate at the time of writing this chapter, is exemplary (Diaz & allii., 2015).
37 For example, speaking of traditional knowledge but adding a holistic interpretation to it and making reference to the UNDROP (United Nations Declaration on the Rights of Peasants and Other People Working in Rural Areas), which is involved both with cultural heritage and with the intellectual property of indigenous peoples, helps to create a considerably broader conceptual space (Girard, p.c.).
38 Annex II to decision IPBES-8/1, (June 2021).

References

Adonis, D. L., & Couch, J. (2018). Ili-based Community Organising: An Igorot Indigenous Peoples' Concept for Grassroots Collaboration. *Journal of Social Inclusion, 9*(1), 58. https://doi.org/10.36251/josi.131

Aubertin, C. (2005). La biodiversité: Une notion en quête de stabilité. In C. Aubertin (Ed.), *Représenter la nature? ONG et biodiversité* (pp. 99–122). IRD éditions.

Aubertin, C., & Filoche, G. (2011). The Nagoya Protocol on the Use of Genetic Resources: One Embodiment of an Endless Discussion. *Sustentabilidade Em Debate, 2.* https://doi.org/10.18472/SustDeb.v2n1.2011.3906

Aubertin, C., Pinton, F., & Boisvert, V. (2007). *Les marchés de la biodiversité.* IRD éditions.

Bavikatte, S. K. (2014). Stewarding The Earth: Rethinking Property and the Emergence of Biocultural Rights. In *Stewarding the Earth.* Oxford University Press. https://oxford.universitypressscholarship.com/view/10.1093/acprof:oso/9780198098669.001.0001/acprof-9780198098669

Bellier, I. (2015). L'avenir que veulent les peuples autochtones. Stratégies institutionnelles, mobilisations collectives et force de négociation, à Rio+20. In J. Foyer (Ed.), *Regard croisé sur Rio+20. La modernisation écologique à l'épreuve* (pp. 259–280). CNRS éditions.

Blaser, M. (2004). Life Projects: Indigenous Peoples' Agency and Developement. In M. Blaser, H. A. Feit, & G. Banks (Eds.), *In the Way of Development. Indigenous Peoples, Life Projects and Globalization* (pp. 26–44). ZED Book, International Development Research Centre.

Blaser, M. (2009). The Threat of the Yrmo: The Political Ontology of a Sustainable Hunting Program. *American Anthropologist, 111*(1), 10–20. https://doi.org/10.1111/j.1548-1433.2009.01073.x

Blaser, M. (2016). Is Another Cosmopolitics Possible? *Cultural Anthropology, 31*(4), 545–570. https://doi.org/10.14506/ca31.4.05

Boisvert, V., & Caron, A. (2002). The Convention on Biological Diversity: An Institutionalist Perspective of the Debates. *Journal of Economic Issues, 36*(1), 151–166. https://doi.org/10.1080/00213624.2002.11506447

Bonneuil Christophe. (2011). Des ressources génétiques à la biodiversité cultivée. *Revue d'anthropologie Des Connaissances, 52*(2), 206–233.

Bretón Solo de Zaldívar, V. (2013). Etnicidad, desarrollo y 'Buen Vivir': Reflexiones críticas en perspectiva histórica. *European Review of Latin American and Caribbean Studies Revista Europea de Estudios Latinoamericanos y del Caribe, 95*, 71–95.

Büscher, B., Sullivan, S., Neves, K., Igoe, J., & Brockington, D. (2012). Towards a Synthesized Critique of Neoliberal Biodiversity Conservation. *Capitalism Nature Socialism, 23*(2), 4–30. https://doi.org/10.1080/10455752.2012.674149

Cadena, M. de la. (2012). Indigenous Cosmopolitics in the Andes: Conceptual Reflections Beyond 'Politics'. *Cultural Anthropology, 25*(2), 334–370. https://doi.org/10.14506/

Cadena, M. de la, & Blaser, M. (Eds.). (2018). *A World of Many Worlds*. Duke University Press.

Callon, M. (2013). Sociologie de l'acteur réseau. In M. Akrich & B. Latour (Eds.), *Sociologie de la traduction: Textes fondateurs* (pp. 267–276). Presses des Mines. http://books.openedition.org/pressesmines/1201

Campbell, L. M., Corson, C., Gray, N. J., MacDonald, K. I., & Brosius, J. P. (2014). Studying Global Environmental Meetings to Understand Global Environmental Governance: Collaborative Event Ethnography at the Tenth Conference of the Parties to the Convention on Biological Diversity. *Global Environmental Politics, 14*(3), 1–20. https://doi.org/10.1162/GLEP_e_00236

Canessa, A. D. M. (2006). Todos somos indígenas: Towards a New Language of National Political Identity. *Bulletin of Latin American Research, 25*(2), 241–263.

CBD. (2019). *Report of the Global Thematic Dialogue for Indigenous Peoples and Local Communities on the Post-2020 Global Biodiversity Framework, Montreal, Canada, 17–18 November 2019*. CBD/POST2020/WS/2019/12/2 (18 November 2019).

CBD. (2021). *Report of the second Global Thematic Dialogue for Indigenous Peoples and Local Communities on the Post-2020 Global Biodiversity Framework, 1–3 December 2020*. CBD/POST2020/WS/2020/5/2 (15 January 2021).

de Sousa Santos, B., Arriscado Nunes, J., & Meneses, M. P. (2007). Opening up the Canon of Knowledge and Recognition of Difference. In B. de sousa Santos (Ed.), *Another Knowledge Is Possible: Beyond Northern Epistemologies* (pp. XIX–LXII). Verso.

Descola, P. (2005). *Par-delà nature et culture*. Gallimard.

Diaz, S., & allii. (2015). The IPBES Conceptual Framework—Connecting Nature and People. *Current Opinion in Environmental Sustainability, 15*, 1–16.

Dunbar, W. (2019, March 17). From Biocultural Diversity to a Nature-Culture Alliance. *The Nature of Cities*. https://www.thenatureofcities.com/2019/03/17/biocultural-diversity-nature-culture-alliance/

Escobar, A. (1997). Cultural Politics and Biological Diversity: State, Capital, and Social Movements in the Pacific Coast of Colombia. In L. Lowe, D. Lloyd, S. Fisch, & F. Jameson (Eds.), *The Politics of Culture in the Shadow of Capital* (pp. 202–226). Duke University Press.

Escobar, A. (1998). Whose Knowledge, Whose Nature? Biodiversity, Conservation, and the Political Ecology of Social Movements. *Journal of Political Ecology, 5*(1), 53–82.

Escobar, A. (1999). After Nature: Steps to an Antiessentialist Political Ecology. *Current Anthropology, 40*(1), 1–30. https://doi.org/10.1086/515799

Escobar, A. (2010). Latin America at a Crossroads. *Cultural Studies, 24*(1), 1–65. https://doi.org/10.1080/09502380903424208

Fabricant, N., & Postero, N. (2018). Performing Indigeneity in Bolivia: The Struggle Over the TIPNIS. *Anthropological Quarterly, 91*(3), 905–936. https://doi.org/10.1353/anq.2018.0044

Ferrari, M. F., de Jong, C., & Belohrad, V. (2015). Community-based Monitoring and Information Systems (CBMIS) in the Context of the Convention on Biological Diversity (CBD). *Biodiversity, 16*(2–3), 57–67.

Forest peoples program (Ed.). (2020). *Local Biodiversity Outlooks 2: The Contributions of Indigenous Peoples and Local Communities to the Implementation of the Strategic Plan for Biodiversity 2011–2020 and to Renewing Nature and Cultures. A Complement to the Fifth Edition of the Global Biodiversity Outlook.* Forest Peoples Program. https://www.cbd.int/gbo/gbo5/publication/lbo-2-en.pdf

Foyer, J. (2011). Les multiples politiques de la diversité bio-culturelle: Entre modernités alternatives et rhétorique instrumentale. In C. Gros (Ed.), *Le multiculturalisme au concret* (pp. 377–389). Presses de la Sorbonne Nouvelle.

Foyer, J. (2015). *Regards croisés sur Rio+20: La modernisation écologique à l'épreuve.* CNRS éditions.

Friis, J. (2020). *La participation des peuples autochtones et communautés locales dans les instances internationales: Le cas de la Convention sur la diversité biologique* (p. 103) [Rapport de stage EDD6999 Maitrise en Environnement et Développement durable Option: Enjeux sociaux et gouvernance]. Université de Montréal, Faculté des Arts et des Sciences.

Gill, S. D. (1991). *Mother Earth: An American Story.* University of Chicago Press.

Hall, I. (2022). Tying Down the Soul of a Potato in the Southern Peruvian Andes: Performance and Frictions (pp. 157–186). In F. Dussart & S. Poirier (Eds.), *Sensing Religion in Indigenous Context: Towards New Methods.* Press of Toronto.

Hannigan, J. (2014). *Environmental Sociology.* Routledge.

Harrop, S. R. (2011). Living in Harmony With Nature'? Outcomes of the 2010 Nagoya Conference of the Convention on Biological Diversity. *Journal of Environmental Law, 23*(1), 117–128.

Harvey, D., & Haraway, D. (1995). Nature, Politics, and Possibilities: A Debate and Discussion with David Harvey and Donna Haraway. *Environment and Planning D: Society and Space, 13*, 507–527. https://doi.org/10.1068/d130507

Hermitte, M.-A. (2009). Souveraineté, peuples autochtones: Le partage équitable des ressources et des connaissances. In F. Bellivier & C. Noiville (Eds.), *La Bioéquité. Batailles autour du partage du vivant* (pp. 115–135). Autrement.

Hermitte, M.-A. (2011). La nature, sujet de droit ? *Annales. Histoire, Sciences Sociales, 66*(1), 173–212.

Igoe, J. (2010). The Spectacle of Nature in the Global Economy of Appearances: Anthropological Engagements with the Spectacular Mediations of Transnational Conservation. *Critique of Anthropology, 30*(4), 375–397. https://doi.org/10.1177/0308275X10372468

ISSD. (2019). 11th Meeting of the Ad Hoc Open-ended Working Group on Article 8(j) and Related Provisions and 23rd Meeting of the Subsidiary Body on Scientific, Technical and Technological Advice of the Convention on Biological Diversity: 20–29 November 2019. *Earth Negotiations Bulletin, 9*(749), 1–20.

Krech, S. (1999). *The Ecological Indian: Myth and History.* W.W. Norton & Co.

Kurasawa, F. (2004). A Cosmopolitanism from Below: Alternative Globalization and the Creation of a Solidarity without Bounds. *European Journal of Sociology, 45*(2), 233–255. https://doi.org/10.1017/S0003975604001444

Laclau, E., & Mouffe, C. (1990). *Hegemony and Socialist Strategy: Towards a Radical Democratic Politics.* Verso.

Landivar, D., & Ramillien, E. (2015). Reconfigurations ontologiques dans les nouvelles constitutions politiques andines. *Tsantsa, 20*, 29–40.

Larrère, C., & Larrère, R. (2015). Diversité culturelle et environnement. *Sciences humaines*, 301–322.
Latour, B. (2004). *Politiques de la nature: Comment faire entrer les sciences en démocratie*. Découverte.
Latour, B. (2007). 2. Quel cosmos? Quelles cosmopolitiques?. In Jacques Lolive (Ed.), *L'émergence des cosmopolitiques* (pp. 69–84). La Découverte.
Le Prestre, P. G. (2017). *Governing Global Biodiversity the Evolution and Implementation of the Convention on Biological Diversity*. Routledge.
Maljean-Dubois, S. (2021). *Le droit international de la biodiversité*. Brill/Nijhoff.
Mignolo, W. (2013, October 20). *On Pluriversality*. http://convivialism.org/wp-content/uploads/2015/12/Mignolo-On-Pluriversality.docx
Mol, A. (1999). Ontological Politics. A Word and Some Questions. *Sociological Review*, *47*(1), 74–89.
Müller, B. (2014). Introduction. Seeds and the Ontic in Political Anthropology. *Focaal*, *2014*(69), 3–11. https://doi.org/10.3167/fcl.2014.690101
Müller, B., & Cloiseau, G. (2015). La fabrique d'un paragraphe: Les investissements agricoles responsables entre multilatéralisme et gouvernance privée. In J. Foyer (Ed.), *Regards croisés sur Rio+20: La modernisation écologique à l'épreuve* (pp. 89–116). CNRS éditions.
Parks, L. (2017). Spaces for Local Voices? A Discourse Analysis of the Decisions of the Convention on Biological Diversity. *SSRN Electronic Journal*. https://doi.org/10.2139/ssrn.3073052
Parks, L. (2020). *Benefit-sharing in Environmental Governance: Local Experiences of a Global Concept*. Routledge.
Parks, L., & Schröder, M. (2018). What We Talk about When We Talk about 'Local' Participation in International Biodiversity Law. The Changing Scope of Indigenous Peoples and Local Communities' Participation under the Convention on Biological Diversity. *PACO, 11*(3) 2018, 743–785. https://doi.org/10.1285/I20356609V11I3P743
Poupeau, F. (2011). L'eau de la Pachamama. *L'Homme. Revue française d'anthropologie, 198–199*, 247–276. https://doi.org/10.4000/lhomme.22781
Povinelli, E. A. (1995). Do Rocks Listen? The Cultural Politics of Apprehending Australian Aboriginal Labor. *American Anthropologist*, *97*(3), 505–518.
Povinelli, E. A. (1997). Reading Ruptures, Rupturing Readings: Mabo and the Cultural Politics of Activism. *Social Analysis: The International Journal of Social and Cultural Practice*, *41*(2), 20–28.
Ramos, A. (1994). The Hyperreal Indian—Alcida Rita Ramos, 1994. *Critique of Anthropology*, *14*(2), 153–171. https://doi.org/10.1177/0308275X9401400203
Reimerson, E. (2013). Between Nature and Culture: Exploring Space for Indigenous Agency in the Convention on Biological Diversity. *Environmental Politics*, *22*(6), 992–1009. https://doi.org/10.1080/09644016.2012.737255
Reyes-García, V., Fernández-Llamazares, Á., Aumeeruddy-Thomas, Y., Benyei, P., Bussmann, R. W., Diamond, S. K., García-del-Amo, D., Guadilla-Sáez, S., Hanazaki, N., Kosoy, N., Lavides, M., Luz, A. C., McElwee, P., Meretsky, V. J., Newberry, T., Molnár, Z., Ruiz-Mallén, I., Salpeteur, M., Wyndham, F. S., … Brondizio, E. S. (2021). Recognizing Indigenous Peoples' and Local Communities' Rights and Agency in the post-2020 Biodiversity Agenda. *Ambio*. https://doi.org/10.1007/s13280-021-01561-7
Stengers, I. (2005). The Cosmopolitical Proposal. In B. Latour & P. Weibel (Eds.), *Making Things Public* (pp. 994–1003). MIT Press. http://mnissen.psy.ku.dk/Undervisning/Stengers05.pdf
Stengers, I. (2007). La proposition cosmopolitique. In Jacques Lolive (Ed.), *L'émergence des cosmopolitiques* (pp. 45–68). La Découverte.
Suiseeya, K. R. M. (2014). Negotiating the Nagoya Protocol: Indigenous Demands for Justice. *Global Environmental Politics*, *14*(3), 102–124.

Thomas, F., & Boisvert, V. (Eds.). (2015). *Le pouvoir de la biodiversité: Néolibéralisation de la nature dans les pays émergents*. IRD éditions; Éditions Quae.

Tsing, A. L. (2005). *Friction: An Ethnography of Global Connection*. Princeton University Press.

Tsing, A. L. (2007). Indigenous Voice. In M. de la Cadena & O. Starn (Eds.), *Indigenous Experience Today* (pp. 33–68). Berg.

Uggla, Y. (2010). What Is this Thing called 'natural'? The Nature–Culture Divide in Climate Change and Biodiversity Policy. *Journal of Political Ecology*, *17*, 79–91.

Ulloa, A. (2010). *The Ecological Native. Indigenous Peoples' Movements and Eco-Governmentality in Colombia*. Routledge.

Vadrot, A. (2016). The Birth of a Science-Policy Interface for Biodiversity: The History of IPBES. In M. Hrabanski & D. Pesche (Eds.), *The Intergovernmental Platform on Biodiversity and Ecosystem Services (IPBES) Meeting the Challenges of Biodiversity Conservation and Governance* (pp. 41–76). Routledge.

Vadrot, A. B. M. (2014). *Politics of Knowledge and Global Biodiversity*. Routledge. http://search.ebscohost.com/login.aspx?direct=true&scope=site&db=nlebk&db=nlabk&AN=770442

Viveiros de Castro, E. (2004). Perspectival Anthropology and the Method of Controlled Equivocation. *Tipití: Journal of the Society for the Anthropology of Lowland South America*, *2*(1), 1.

Viveiros De Castro, E. (2014). Perspectivisme et multinaturalisme en Amérique indigène. *Journal des anthropologues. Association française des anthropologues*, *138–139*, 161–181. https://doi.org/10.4000/jda.4512

5

FROM OBSTRUCTION TO DECOLONIZATION?

Contested Sovereignty, the Seed Treaty, and Biocultural Rights in the U.S./Turtle Island and Beyond

Garrett Graddy-Lovelace[1]

Introduction

The concepts and legal precedents of biocultural rights and Biocultural Community Protocols carry transformational potential and noteworthy momentum in general, and within agrobiodiversity governance in particular. Yet, they face a number of logistical and ideological obstacles, which this edited volume explores. This chapter addresses the contested sovereignties at work and at odds in encoding cosmologically grounded biocultural protections into contemporary national and international governance systems still beholden to settler colonial scales of reference. In particular, this chapter focuses on the U.S., a nation-state emblematic for its colonial settler racial capitalism and (neo) liberalist orientation as well as for the robust diversity of indigenous, African diaspora, and local community biocultural heritages and agrarian resurgences that have long been placed therein. The chapter references biocultural rights invocations far beyond North America because these have the potential to inform and inspire broader political engagement with Biocultural Protocols in the U.S. context. The U.S. has long played an obstructionist role in international biodiversity negotiations, as the sole holdout to the Convention on Biological Diversity (CBD, which 196 other countries have ratified). The U.S. did recently ratify the International Treaty for Plant Genetic Resources for Food and Agriculture (Treaty), but what does this official affirmation mean? How do the ethics and premise of biocultural rights and community protocols circulate in the lands and waterways falling under the U.S. jurisdictions, and to what degree do they influence the U.S. involvement in the Treaty? How do the contradictions of representation (and lack thereof) in this U.S. case study shed light on the broader tensions of international liberalism and multilateralism and the related coloniality of intellectual property regimes?

DOI: 10.4324/9781003172642-6

To help answer these questions, I draw upon two realms of indigenous-led assertions of indigenous knowledge sovereignty: (1) the 2019 Atateken North American Regional Declaration on Biocultural Diversity and Recommended Actions (Atateken Declaration),[2] and the four Atateken Declaration-affiliated Policy Briefs, concerning Biocultural Indicators; Livelihood, Food Sovereignty, and Health; Information and Communication Technologies; and Extractive Contexts[3]; and (2) the Biocultural (BC) and Traditional Knowledge (TK) Labels initiative and underlying CARE Principles for indigenous data sovereignty and stewardship. Though largely based in the Kanien'kéha (Mohawk) regions of Canada and led by Canadian First Nations, the indigenous governance visions and recommendations of the Atateken Declaration pertain in the U.S. and beyond. The TK and BC Labels and CARE Principles, though oriented toward digital object metadata, chart a course for official recognition of indigenous knowledge and sovereignty.

Ironically, I concentrate this chapter on the U.S. so as to help dismantle the presumptions of American Exceptionalism, and I focus on the contradictions of this modern nation-state so as to help move scholarship, policymaking, and coalition-building beyond the nation-state as a dominant scale of reference. Also, of note, this is an ongoing project at a preliminary stage; this draft highlights a prominent example of Biocultural Protocols, but more exist. I look forward to continuing to learn from indigenous, African Diaspora, and local community initiatives in and beyond the U.S. that work to advance biocultural heritage rights, recovery, and sovereignty.

Biocultural Protections and Protocols on Turtle Island and Beyond

Long before the U.S. existed, indigenous communities on the continent were navigating how to protect biocultural knowledge and practices in the face of Spanish, French, British, Dutch, and other colonial forces. And of course, long before the term "biocultural" existed, indigenous movements launched brave and creative resistance to Iberian conquest and colonial settler invasions, occupations, and displacements – layers of resistance grounded in the explicit intersection of realms deemed biological and cultural. Grounded in multiple, parallel cosmovisions and creation stories, indigenous activists and scholars have invoked the term "Turtle Island" to refer to North America. Concurrently, movement leaders and scholar-activists refer to South America as "Abya Yala", drawing on the Kuna (indigenous Panamanian) geographic term. Zaragocin centers Abya Yala as a feminist "Indigenous decolonial imagined geography [...] a utopic territory [...] and counter-geography" (Naylor et al., 2018, p. 204). Within academic publications, Turtle Island figures more prominently in the work of Canadian-based scholars. Hunt and Steven trace how First Nations deploy digital counter-mapping for decolonial geographic imaginaries in Canada (2017); Fitznor draws upon her Cree/Nahayow background from Manitoba to think through indigenous-perspective education (2019); and Zellars traces the possibilities of Black commons and liberation in solidarity with indigenous territorial sovereignty "on Turtle Island" Canada (2021). Abya Yala and Turtle Island connect and overlap in the pre- and decolonial space of Aztlán, the Nahuatl word for Aztec/Mexica territories in Mesoamerica (Medina, 2019).

Within Turtle Island, a range of tribal and indigenous communities are navigating territorial sovereignty and recovery through place-based and plant-based mobilizations. These broadly fall under the conception of biocultural heritage and are thus biocultural political claims, even if they don't use the exact term. These movements work beyond the colonial nation-state scale of reference, even as they transgress actual borders. Audra Simpson chronicles how her Kahnawà:ke communities and their broader Mohawk nations straddle, transgress, and transcend the U.S.-Canadian border (Simpson, 2014). This is particularly relevant to international biocultural governance.

Because the Secretariat of the United Nation's CBD is located in Montreal, Quebec, Canada, which is the traditional ancestral land of the Kahnawà:ke (Mohawk) nation, key Kahnawà:ke cosmological principles have infused and informed the CBD's policies and paradigms as a prominent scale of reference in transnational and supranational biocultural governance:

1 The **Akwé: Kon Guidelines**, adopted in CBD COP-7,[4] guide "the conduct of cultural, environmental and social impact assessment regarding developments proposed to take place on, or which are likely to impact on, sacred sites and on lands and waters traditionally occupied or used by indigenous communities". Akwé: Kon means "everything in creation" in Kanien'kéha, the Mohawk language.
2 The Tkarihwaié:ri Code of Ethical Conduct to Ensure Respect for the Cultural and Intellectual Heritage of Indigenous and Local Communities Relevant to the Conservation and Sustainable Use of Biological Diversity, adopted by COP-10, guides models and codes of ethical research and "access to, use, and exchange and management of information concerning traditional knowledge", including prior informed consent, approval, and involvement in research and collection.[5] Tkarihwaié:ri means "the proper way" in Kanien'kéha.
3 The **Atateken North American Regional Declaration on Biocultural Diversity.**[6] Atateken means "brothers and sisters" in Kanien'kéha.

The CBD itself demonstrates the expanding and expansive potential of international forum. The CBD began in the late 1980s dialogues, opened for signature at the 1992 "Earth Summit" Rio Conference, and went into force the next year. The CBD's 2000 Cartagena Protocol on Biosafety (going into force in 2003) and 2010 Nagoya Protocol on Access to Genetic Resources and the Fair and Equitable Sharing of Benefits Arising from their Utilization (going into force in 2014) have foundational impacts on the International Treaty as well as on agrobiodiversity – and thus biocultural – governance at large. In the Global Biodiversity Outlook 5 report (issued in September 2020), the CBD chronicled dozens of biological indicators failing to reach the Aichi Biodiversity Targets. The report concludes

> [...] that opportunities for effective action in support of the Strategic Plan for Biodiversity 2011–2020 were missed due to insufficient involvement of women, indigenous peoples and local communities [...]. The new global framework can set stronger requirements for future action on biodiversity to include all of these considerations as foundational prerequisites.
>
> *(Secretariat of the Convention on Biological Diversity, 2020, p. 136)*

The press release for the 2nd edition of the Biodiversity Loutlooks (the sister publication to the Outlook, which presents the expertise and experience of indigenous peoples) went further to warn of "catastrophic loss of nature and biodiversity and increasing risk of pandemics" unless indigenous peoples and knowledges were heeded and honored. The summary reads:

- There is a way to protect the world's biodiversity by listening to indigenous peoples and local communities who have sustainably managed biodiversity for millennia
- The majority of the world's most biodiverse areas are found within the ancestral lands of indigenous peoples and local communities; securing their rights to land would secure biodiversity
- More than a quarter of the global land area is traditionally owned, managed, used, or occupied by indigenous peoples, and has been for millennia (Forest Peoples Programme et al., 2020, p. 1).

Indigenous peoples and local communities have worked hard for a generation in and through the CBD. Though the CBD retains the tensions and limits of liberal internationalism to some degree, it has remained a key forum for IPLCs to voice their claims and even actualize their worldviews. Importantly, it was within the context of the CBD that Biocultural Protocols were first officially recognized a decade ago.[7] Tellingly, the U.S. has remained a global outlier in refusing to ratify this important convention. Yet, all the while, perhaps because of the lack of U.S. presence, the CBD has continued to advance and apply biocultural frameworks of governance.

Over the past decade, indigenous and local communities around the world have begun articulating their place-based, plants-based, agroecologically grounded protections as "biocultural heritage" and territories, as outlined and enacted by the Potato Park in Andean Peru (Argumedo, 2008), and "biocultural rights", as formulated by the South African organization Natural Justice:

> The term 'biocultural rights' denotes a community's long established right, in accordance with its customary laws, to steward its lands, waters and resources. Such rights are being increasingly recognized in international

> environmental law. Biocultural rights are not simply claims to property, in the typical market sense of property being a universally commensurable, commodifiable and alienable resource; rather, as will be apparent from the discussion offered here, biocultural rights are collective rights of communities to carry out traditional steward- ship roles vis-à-vis Nature, as conceived of by indigenous ontologies.
>
> *(Bavikatte & Bennett, 2015, p. 7)*

The biocultural rights framework has helped mobilize indigenous groups to contest dominant intellectual property regimes, particularly in nations with high levels of biodiversity and with contested histories with their own indigenous peoples (India, Peru, and South Africa, for instance).

In the Turtle Island North American context, the legal framework for integrating biocultural territories or legal rights has only begun to make its way into national policies. Mexico has instituted various legislation for protection of indigenous maize territories, most notably the recent moratorium on genetically modified corn. Canada, meanwhile, has laid groundwork for such legal rights in the Truth & Reconciliation Commission and in its application at the intersection of genetic resources, justice, and reconciliation (see Oguamanam's 2019 edited volume). The Innu First Nation in Canada have successfully declared the River Magpie as a Living Being with Rights (Alliance Muteshekau-shipu, 2021). The U.S. however has largely failed to integrate biocultural rights, much less recognition of biocultural heritage into policies.

Nevertheless, diverse biocultural heritages abound as do the movements for their upkeep and protection. Indigenous, Black agrarian, and grassroots seed sovereignty initiatives that have arisen in the U.S. and "territories", and that align with the central concepts of BCP – farmers rights and self-determination – as well as its central principles: interconnectedness, pluralism, ecocentrism, and environmental stewardship.

As an exemplar of these movements, indigenous Hawaiian leaders, scholars, and activists have created the *Moku* system to manage biocultural resources "for abundance within social-ecological regions" in Hawaii (Winter et al., 2018, p. 1). This overlaps with efforts to restore "Aina Malo'o" on Hawaii island, with the goal of "expanding biocultural relationships" (Lincoln et al., 2018, p. 1). Such biocultural restoration, rooted in indigenous Hawaiian cosmological principles, has proven its efficacy in achieving dominant conservation goals – but has gone such much further to expand and indigenize the very notion of conservation (Winter et al., 2020).

Meanwhile, scholars and activists are documenting indigenous movements under the term "biocultural". A 2012 compilation *Sacred Species and Sites: Advances in Biocultural Conservation* featured a Diné-Navajo initiative to protect Dook' oo' sliid, or the "Abolone Shell Mountain", which is the westernmost of the four pillars of Dine cosmology/land. Known by its colonial name, the Holy San Francisco Peaks of northern Arizona, the mountain range is sacred to 13 different indigenous nations and cultural significant to 22 in the region: it controls the adulthood of peoples' lives, according to Jeneda Banally's overview of the movement (Pungetti & Cinquepalmi, 2012, p. 409). Yet, it is situated on National Forest Service land, not on reservations or tribal territories. The U.S. Forest Service has been working to manufacture snow utilizing wastewater from the peaks; the Diné community leadership has worked diligently to defend the sacred mountains by invoking their cosmological, biocultural significance. A team of conservation scientists focusing on pollinators co-authored a 2019 intervention for Nature Sustainability asserting the necessity of a biocultural perspective in analyses of root causes of biodiversity loss and ways forward for equitable and effective protections – globally, and thus in the U.S. as well (Hill et al., 2019). Other conservationists focusing on resilience indicators for the U.S. landscapes and species contend that biocultural approaches are key for "moving beyond the human-nature dichotomy" (Caillon et al., 2017).

Meanwhile, the world-renowned Biocultural Heritage Territory of the Potato Park in Andean Peru remains an ongoing influence, inspiring and informing initiatives across Abya Yala and Turtle Island. In one instance, Alejandro Argumedo and others in the Lares Barter Market Biocultural Heritage Territory collaborate with Appalachian-based seedkeeping coalitions (and Mayan ones in Yucatan, Mexico) in a transnational, indigenous-led project: "Agrobiodiversity Nourishes/ Agrobiodiversidad Nutre: Community-Based Research, Policy & Guidance for Agrarian Equity & Well-being", now a Special Feature at the journal Elementa: Science of the Anthropocene.

In short, there exists an abundance of initiatives percolating in a broader emergence of biocultural recognition and recovery, orientation and political commitment. Yet, they are often underrepresented, unsupported, and undervalued at the federal – and thus the international – level of policy.

Ratified

As Biocultural Community Protocols and indigenous-led initiatives for biocultural rights proliferate across Turtle Island, the U.S. government continues to play an obstructionist role as the sole non-party of the CBD. But, meanwhile, in 2016, the U.S. did ratify the International Treaty on Plant Genetic Resources for Food and Agriculture, 15 years after the Treaty was approved and opened for signatures and ratification. The U.S. government website announces and explains this move as one of securing easy access:

> Ready access to plant genetic resources – that is, seeds, bulbs, roots and other materials from which plants can be reproduced – is needed to meet the daunting challenges of feeding a growing global population, combating environmental degradation and constantly evolving pests and diseases, and facilitating continued productivity of agriculture in the United States and worldwide.
>
> *(U.S. Mission to the UN Agencies in Rome, 2017)*

The press release promptly situates the beneficiaries of ratification as "both the public-sector and commercial interests" (U.S. Mission to the UN Agencies in Rome, 2017). Indeed, the Acting Assistant Secretary of State's 2016 testimony to the Senate Foreign Relations Committee (which oversees International Treaty negotiations) argued that the U.S. needed "facilitated access" to the treasury of germplasm held in genebanks around the world to support agricultural industry needs. As a non-party, the U.S. institutions "have to engage in lengthy ad hoc negotiations of terms of access" to comply with the Nagoya Protocol for Access and Benefit Sharing, to receive prior informed consent from indigenous and local communities and stewards of such valuable germplasm. The numerous testimonies positioned these requirements as mere obstacles to the global flows of agri-food genetic resources. Moreover, the testimonies regularly couched the value of germplasm in terms of the Green Revolution, which, they erroneously argued, "saved hundreds of millions of lives"[8] (see Kumar et al., 2017; Patel, 2013; Perkins, 1997; Shiva, 1991; Stone, 2019). After lauding Borlaug again, the chairman of the American Seed Trade Association testified that "Public and private plant breeders once enjoyed much freer access to seeds for research and development".[9] This time, the Nagoya Protocol is called out by name as the enemy as "further threatening our ability to exchange germplasm globally. With ratification, the US would be able to resume its leadership position to enhance the functioning of the Treaty and greatly diminish the uncertainty created by the CBD and Nagoya".[10] Overall, the testimonies concurred that ratifying would "enable the United States effectively to guide the trajectory of the Treaty and its Material Transfer Agreement".[11] The arguments proved effective.

At the first Governing Body Session that the U.S. attended as a ratifying party, in November 2017, the U.S. representative Christine Dawson was elected as the next Chair of the Commission on Genetic Resources for Food and Agriculture: a two-year term. It is the State Department that represents the U.S., and Dawson was a longtime State Department official, serving as the director of Office of Conservation and Water in the Bureau of Oceans and International Environmental and Scientific Affairs. Her experience in such biological aspects of diplomacy as the CITES Treaty earned her the appointment, despite lack of experience or expertise in agriculture or food systems. Dawson chaired a Conference of the Parties session focused on longstanding axes of tension in the Treaty, such as expanding the Multilateral System of ABS (MLS) and the Standard Material Transfer Agreement (SMTA); funding the Benefit Sharing program; actualizing Farmers' Rights; and regulating access to Digital Sequence Information (DSI) of PGRFA. Dawson presented a compromise package on MLS which comprised a slightly revised SMTA with expanded Annex 1 list of MLS crops. It did not pass. She did agree to scoping research for DSI (iisd Reporting Services, 2019). In short: a largely status quo meeting. Even well-discussed solutions such as a ratio-based subscription system for the MLS to fund the Benefit Sharing fund were not formally considered. Industrial countries blocked the Farmers' Rights Ad

Hoc Technical Expert Group according to the International Food Sovereignty Planning Committee (iisd Reporting Services, 2019).

Though the highly anticipated focus of the meeting, DSI did not receive formal deliberation. Dawson kept discussions on DSI "informal for the first three days and waited until Wednesday to establish a contact group" which was limited to two Contracting Parties per region (Muzurakis, 2019). The 54-country continent of Africa had merely two representatives to engage with North America's two representatives. This allegedly equal system of representation ended up operationalizing a vestigially colonialist false equivalence of continents. The Commission on PGRFA established a new "work stream" on DSI in 2017, and initiated an "exploratory fact-finding scoping study" on DSI in 2018. At the November 2019 in-person session, under Dawson's lead, the Commission called for even further review (Commission on Genetic Resources for Food and Agriculture, n.d.): at the 2021 session, they would address "the innovation opportunities" and the "challenges related to capacity to access and make use" of DSI (ibid.). Meanwhile, however, the actual development, use, and circulation of the technology and sequencing of genetic data expand rapidly, even as the governance and collective deliberation of biocultural ethics and accountability of such data all stall and stagnate at the international level. Civil society and scholars working with *in situ* initiatives came to troubling conclusions: "With the ongoing omics developments, if not correctly addressed, the 'DSI issue' might threaten the stability of the Seed Treaty and possibly the entire ABS framework" (Aubry, 2019, p. 9).

As the introduction to this book contends, "BCPs could be an integral part of the domestic ABS legislation",[12] but the inspiring BCP-oriented initiatives in the U.S. have not made their way to the U.S. involvement in ITPGRFA implementation. Conversely, official U.S. leadership and participation in the Treaty process leave little room for biocultural rights, responsibilities, principles, and support – or even acknowledgment. What accounts for the disconnect between grassroots biocultural initiatives and official governance in international fora? In general terms, the *long durée* of coloniality. More specifically, the postcolonial legacy of the modern nation-state, and its inordinate power, even – especially – in (neo) liberal internationalism.

Liberalism's Limits: Multiple Critiques of Multilateralism and Multiculturalism

The international liberal order has temporal, spatial, and theoretical limits that manifest in the form of paradoxes. Temporally, international liberalism is recent, beginning after World Wars, in the throes of anti-colonial struggles to throw off the shackles of empire. In conjunction with the Bretton Woods international financial institutions, the United Nations (UN) began in 1945 "by 51 countries committed to maintaining peace and security, developing friendly relations among nations and promoting social progress, better living standards and human rights".[13] The UN's Food and Agricultural Organization (FAO) began the same year, with the U.S. designated as its first, temporary headquarters. Only recently have scholars chronicled the history of FAO (Pernet & Ribi Forclaz, 2019), its close relationship with the Green Revolution, and its developmentalist orientation (Ribi Forclaz, 2019); throughlines emerge however, such as the foundational discourse of modernizing the world's food and agricultural systems.

The temporality of modernity obscures how limited liberal internationalism has been. With progress as the central framework, the present dominates over the past – with the former deemed advanced, and the latter primitive. From this developmentalist perspective, the recentness of its reign confers its relevance. It overshadows other non-linear temporalities that value TK and center ancestral legacies. Paradoxically, the modernity-coloniality at the heart of international liberalism deploys the attribute of "traditional" as its necessary antithesis; it deploys a reductive vision of "the traditional" and positions it as pitied Other, from which to allegedly advance and progress.

This temporal paradox of international liberalism comprises a longstanding tension even in Biocultural Community Protocols: the colonial legacy of objectifying indigenous people and agricultures as subjects of the past, to be discovered, collected, and preserved. Studied by the colonial gaze, they come to constitute expertise itself. From this light, modernity is constructed in oppositional relation to traditional – with modernity meaning dynamic, adaptive, forward-oriented, while traditional is, and must stay by definition, static and historical. This imposes a linear temporality, of linear progress and development. Modernity/coloniality seeks "production" of knowledge and data, with epistemological assumptions of individualized innovation, and corresponding proprietary intellectual property regimes and enclosures.

Modernism hinges on modern nation-states, but goes further into epistemologies of "nature" as such. Activists and scholars have long chronicled the Western tendency to dichotomize nature and culture. Invoking the plurality of ontologies helps show how this Eurocentric materialist worldview is in fact one of many, and not the universal worldview it purports to be. Building on decolonial theory, Mario Blaser shows that modernist ontological assumptions impose reductive framework on science, culture, nature, and alleged expertise thereof. Phillipe Descola wrote that:

> it is not enough to show that the opposition between nature and culture is meaningless for non-modern societies, or that it emerges lately in the course of the history of the West; it must be integrated to a new analytical framework within which modern naturalism, far from constituting the template which allows to gauge cultures that are distant from ours in space and time, would be but one of the possible expressions of more general schemes regulating the objectification of self and non-self.
>
> *(Descola, 2009, p. 150)*

Blaser goes further to challenge even this as a universalizing statement. In fact, multiple worlds and worldings co-exist all the time, and require analytic – and, I would argue, governance – approaches that "delineate a picture of socio-material worlds as always-emergent heterogeneous assemblages of humans and more-than-humans" (Blaser, 2014, p. 50). Sarah Hunt, a Kwakwaka'wakw scholar, also lays out the power and risks of invoking ontology to counter coloniality; it "requires destabilizing how we come to know Indigeneity and what representational strategies are used in engaging with Indigenous ontologies, as differentiated from western ontologies of Indigeneity" (Hunt, 2014, p. 28). Destabilizing colonialist/modernist temporalities helps counter reductive multiculturalism. Yet, liberal internationalism retains its temporal limitations if the multilateralism deploys a reductive multiculturalism.

Liberal internationalism limits spatially as well. This also leads to paradoxes, wherein alleged universalism obscures and enables Eurocentric myopia. Moreover, in an attempt to transcend nationalism, modern international order reinscribes the nation-state as the dominant scale of reference. Indigenous and First Nations remain systematically erased, as do migrants, nomads, and the pan-African Diaspora the world over. In her 2020 book *Naming a Transnational Black Feminism: Writing in Darkness*, K. Melchor Quick Hall foregrounds Diasporic lineages of liberation transgressing and transcending state borders (Quick Hall, 2020). These alternate scales of reference relate directly to who feeds and nourishes whom.

Liberalism depends on the paradox of government power only insofar as this government power defers to globalized markets and private property. Neoliberalism furthers the paradox: it entrenches the power of the state only if the state submits to the reign of the allegedly free market, transnational corporate actors, and the "rights" of individualized real and intellectual property. The paradoxical dominance – and limits – of liberalism hinges directly on intellectual and physical property rights regimes. Accordingly, it is unsurprising that the tensions of the Treaty hinge on the fierce contestations around intellectual property of plant genetic resources and their digital manifestations as data. It is also foreseeable that international PGRFA governance stumbles over the fundamental paradox of liberalism: ideals of international peace, amidst realities of racial capitalism, ethno-nationalism, imperial legacies, and resurgent authoritarianism.

Political and social scientists – and civil society – have grappled with the tensions of how neoliberalism entrenches the nation-state as a dominant scale of reference even as transnational corporations have weakened it economically through political-economic processes of globalization. Many multinational companies do without the nation-state, as their increasing reliance on transnational law and free trade zones testify. Saskia Sassen demonstrates how "understanding the epochal transformation we call globalization must include studying these processes of denationalization" (Sassen, 2009, p. 14): "Today it is, then, the foundational features of multiple global, rather than national, systems that get partly structured inside nation-states" (ibid.).

Even liberal internationalism's rallying cries – of multilateralism and multiculturalism – belie its appropriative tendencies. The Treaty serves as an illustrative microcosm of these tensions: the MLS strives for cooperative governance even as its La Via Campesina and other civil society coalitions critique it for entrenching disproportionate Global North access to and benefit from the alleged global commons. Likewise, the Treaty celebrates the cultural and agricultural diversity of the world's farmers, fishers, pastoralists, gardeners, and food systems. Yet, civil society have long critiqued the Treaty for treating such (agri)cultural diversity as objects of value, rather than subjects needing a seat at the table.

The limits of liberal internationalism manifest acutely in the case of the U.S., which has historically imposed its agricultural surpluses on international markets, pressured "harmonization" of intellectual property regimes via bilateral trade agreements, and until very recently, obstructed international cooperation on environmental regulation. The U.S. has paradoxically had an outsized role in international fora, from UN to the WTO, even as it has refused to ratify key international instruments. Again, the Treaty historically demonstrated the U.S.'s heavy-handed isolationism. The recent ratification merely proved the point: that the U.S. plays an over-represented role in the Treaty's governance, even as the actual seedkeepers and agrobiodiversity practitioners of Turtle Island remain starkly underrepresented in the U.S. delegation to the Treaty. The burgeoning assertions of biocultural rights, principles, protocols, and even data labels, however, are emerging as important counterweights to the limits of liberal internationalism – in the U.S. and beyond.

Declaring Biocultural Diversity: Contested Seed Sovereignties

Indigenizing governance of plant genetic resources for food and agriculture entails moving into legal pluralism so as to center the diversity of indigenous customary governance. The South African legal advocacy organization Natural Justice contends that Biocultural Protocols need only be used "where the communities are confronted with an identified threat or where they have already been approached and there is a reasonable prospect that an agreement will be reached" (this book, Chapter 13). They can only be studied through the lens of a political ontology of IPLCs, and should be read as attempts at challenging Western worldviews and dominant forms of legal modernity (see Chapter 11 of this book by Anquet and Girard and Chapter 3 of this book by Hall). Accordingly, biocultural rights serve as a direct response to and defense against bio-coloniality. Biocultural Protocols deploy at the level of the law and politics, in an attempt to reclaim the legal terrain of policy, accountability, and binding legislation. The idea was expressed by Pierre du Plessis – expert at CRIAA SA-DC, Namibia, and one of Africa's lead negotiators of the Nagoya Protocol. Namibia was very active during the negotiations toward the Nagoya Protocol – du Plessis was advised by Natural Justice and Bavikatte (see IIED et al., 2012). The African Union Commission guidelines on ABS specifically recognizes and respects community level processes, empowers local resource managers, and channels benefits to local levels to create incentives.

In this way, biocultural rights assertions parallel indigenous-led manifestos, declarations, and portions of treaties that assert collective self-governance. A key recent example is the 2019 Atateken North American Regional Declaration on Biocultural Diversity and Recommended Actions (Atateken Declaration).

In May 2019, more than 100 participants, Indigenous Peoples of North America, together with partners and supporters, "united by their passion and concern for nature and culture", gathered on traditional lands of Kanien'keha:ka Nation (Mohawk of Canada) for the first North American Dialogue on Biocultural Diversity. Institutional affiliations ranged from the UN to governmental to First Nations tribal to scholarly consortiums to NGOs (Figure 5.1). The Declaration begins with land acknowledgment, allegiance to the Canadian Truth and Reconciliation process, and contextualization within other major policies for indigenous rights and justice, most notably the UN Declaration on the Rights of Indigenous People (UNDRIP).

In this way, the Atateken Declaration follows a long, undervalued, but potent lineage of grassroots, frontline, indigenous, and Black-led collective articulations of political demands firmly and explicitly rooted in cosmovisions and indigenous worldviews. Kabir Bavikatte and Daniel Robinson point to this phenomenon as the crucial "concept of (subaltern) 'cosmopolitan legality', which posits law as a site of struggle and implicates a grassroots movement that "seeks to expand the legal canon beyond individual rights and focuses on the importance of political mobilisation for the success of rights centered strategies" (Bavikatte & Robinson, 2011, p. 43: quoting de Sousa Santos & Rodríguez-Garavito, 2005). This comprises an "insurgent cosmopolitanism" (de Sousa Santos, 2006) grounded in agrarian justice and cosmological principles.

This work of mobilizing worldviews that have been systematically ravaged by colonialism, objectified by developmentalism, and appropriated by academia – and re-asserting their legitimacy and agency in international policy fora – is hard work. And yet, it has been key part of international negotiations for seed, food,

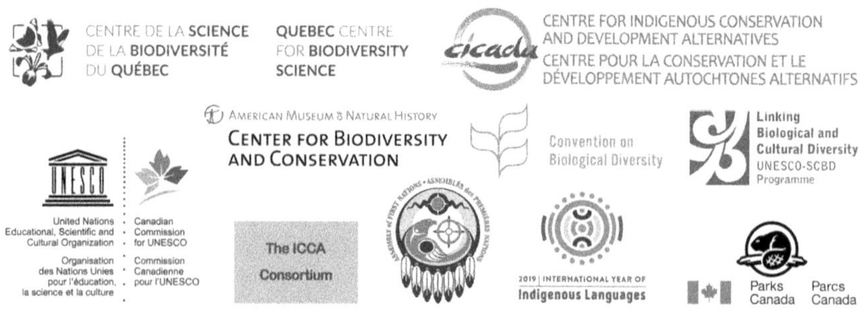

FIGURE 5.1 Atateken Declaration.
Source: CICADA, used with permission of CICADA.

land, and data governance for decades, with UNDRIP standing as a key victory. One of the many international agreements, treaties, UN declarations that resulted from frontline grassroots community-based organizing, coalition-building, and negotiation was the ITPGRFA itself – hence the centrality of Farmers' Rights.

Another key example is the CBD Nagoya Protocol itself. Maria Yolanda Terán Maigua documented her and others' experiences in a 2016 article "The Nagoya Protocol and Indigenous Peoples" in *International Indigenous Policy Journal*:

> I was part of several access and benefit sharing meetings and made note of their complicated processes. The Indigenous Peoples and local communities had to overcome confusion, fear, and disappointment in order to finally converge to support the urgent need for national and international instruments to defend our right to have our vision and input represented at meetings where key concepts, articles, and negotiations were discussed [...]. Many Indigenous Peoples felt that the Access and Benefit Sharing International Regime was written from a Western perspective.
>
> *(Teran, 2016, p. 9)*

Even the logistics of the meetings – the cost to travel and lodge on another continent, the session timed after 6pm, with no translation service or childcare – obstructed meaningful indigenous participation. Indigenous women and youth formed networks and coalitions to strategize and overcome barriers. Nevertheless, indigenous communities participated, according to F Lopez,

> with honesty, consistency, perseverance, and decision. We put our hearts and minds into each meeting and received the guidance and spiritual strength from our Elders, families, and communities. Our ceremonies, offerings, prayers, chants, reciprocal support, and tears helped us, the Indigenous women from Latin America and the Caribbean to continue calmly in these tiring, technical, and difficult dialogues under an umbrella of Western paradigms.
>
> *(ibid.: 11)*

For these coalitions that formed in and around Nagoya conferences, the Biocultural Community Protocols are new iterations of ancient knowledge and methods.

Indigenous agrarian justice leaders mobilized through and with La Via Campesina to craft and enact the UN International Declaration on the Rights of Peasants and Other People Working in Rural Areas (28 September 2018). This directly asserted sovereignty and rights over plant genetic resources for food and agriculture, and the knowledges therein (see Articles 19(a); 20(2); 23(2); 26(1); 26(3); and others). They have also helped design and populate the Civil Society Mechanism of the UN FAO's Committee on Food Security. They have helped craft and mobilize the Escazú Regional Agreement on Access to Information,

Public Participation and Justice in Environmental Matters in Latin America and Caribbean, which explicitly protects land and water defenders from criminalization and state-sanctioned or corporate-backed violence (Graddy-Lovelace 2021).

In this way, the Atateken Declaration continues the long tradition of contesting colonialism and coloniality at the level of the law, working to wrest the contested terrain of legality from its colonial and imperial origins toward emancipatory and reparative directions. It lays out sweeping array of Next Steps Forwards in terms of Actions at Local, Provincial, Indigenous Territorial, National, and International Levels – steps that in the aggregate would radically transform practice, policy, and landscapes from colonial settler, to decolonial, on to re-indigenized worldviews and biocultural principles and protocols writ large. The Atateken Declaration produced four addendum Policy Briefs to help activate the vision, the first for "Measuring Biocultural Diversity: Biocultural Indicators and the Nexus of Nature, Culture, and Well-Being" (Vaziri et al., 2020). It begins with the 1988 Declaration of Belem, which indigenous leaders launched at the first International Conference of Ethnobiology to demand scientists recompense indigenous communities for the knowledge they gain from them (Posey & Dutfield, 1996). The brief warns against "(Mis)appropriating and using Indigenous Knowledge" reductively, through "romanticization, decontextualization, asymmetrical power relations" and such interventions as Ecological Services, which reduce multi-dimensional value of ancient, place-based indigenous ecological knowledges and practices to "recreational" value (Vaziri et al., 2020, p. 1). The brief then lays out explicit steps for "Weaving knowledge systems for effective conservation and resource management". These necessitate biocultural restoration of indigenous languages concurrently with sacred landscapes: "healing the river as part of the process of healing itself from colonial trauma" (ibid: 4). The brief specifies that weaving together indigenous and "Western" sciences requires respecting indigenous institutions and authority. Socio-ecological well-being demands recovery of customary governance and affiliated epistemologies, and thus a transformation of policy and research at large toward indigenous knowledge, Elders, and frameworks.

The second Policy Brief moved to "Nurturing Biocultural Diversity: Livelihoods, Food Sovereignty, Health and Well-Being" (Sarrazin & Scott, 2020a) which requires "Control over and management of lands and waters". Restrictions on indigenous governance, management, and use of natural resources threaten indigenous livelihoods and food sovereignty. In many contexts, indigenous ownership and stewardship on their traditional territory are denied or are only partially recognized by the state, and indigenous institutions have limited participation, if any, in determining how their ancestral lands and waters are used and managed. One example of such issues is the criminalization of traditional harvesting in national or provincial parks located on ancestral lands and waters. This brief lays out how indigenous well-being and landscape-based food sovereignty diminishes precipitously with the violence of "Extractives and other 'development' pressures" (ibid.: 2). Climate change aggravates the injustices wrought by the historical trauma of colonial legacies: "Forced displacement and sedentarization on reservations as well as boarding/residential school systems are flagrant examples of states preventing the reproduction of Indigenous livelihoods, languages, and cultures, which are intrinsically linked" (ibid: 2). Public health depends upon revitalization of indigenous foodways, medicines, as well as livelihoods – the destruction of which has led to health crises. "Historical trauma is also one of the root causes of violence suffered disproportionately by Indigenous women, girls, and 2SLGBTQQIA people in Canada" (ibid: 2). Recovering indigenous health necessitates supporting "the repatriation and restoration of Indigenous languages, knowledge and related information, and artefacts" (ibid.: 4). Only with biocultural restorations and repatriations, can biological diversity itself recover from the current mass extinctions and erosions.

The third Policy Brief outlines the need for "Supporting Biocultural Diversity" through "Information and Communication Technologies" – which is centered as its subtitle (Sarrazin & Scott, 2020c). Indeed, digital knowledge on biodiversity has become crux to their analog reality. This brief dives in to how important and contested data on biodiversity is – from big datasets to apps, videos to GIS maps. It asserts a goal of "Data sovereignty and knowledge integrity" through "Culture and language preservation and revitalization" through such digital methods as intergenerational biocultural videos and community-led and controlled ecological mapping.

The fourth and final Policy Brief guides the "Safeguarding Biocultural Diversity: Territorial Defense in Extractive Contexts" (Sarrazin & Scott, 2020b). Moving through and beyond "Free, prior, informed consent", "Public Consultations", "Private Negotiations", this brief outlines the need for and the methods of supporting "Community protocols, rules, and management plans" and "Community-based assessment and monitoring" of extractive industries on indigenous lands, waterways, and people. Policies, the brief recommends, need to ensure the UNDRIP as the bare – but non-negotiable – minimum.

Biocultural and Traditional Knowledge Labels and CARE Principles

Amidst the impasse of (neo)liberal internationalism enabling collusion of nation-states and corporate capitalism, indigenous-led initiatives remain strategic and resourceful. The intractable case of intellectual property regimes illustrates the steep obstacles and the creative resistance. Biocultural rights and protocols stand as exemplary alternatives to appropriative tendencies of (neo)liberal internationalism.

Any analysis of biocultural initiatives in the U.S. would need to learn from TK and BC Labels and Notices. Begun in 2011 with an organization called Local Contexts, these Labels strive to counter coloniality of intellectual property rights with assertions of indigenous knowledge and indigenous data sovereignty. Though the TK and BC Labels began in Maori contexts, indigenous leaders and allies in the U.S./Turtle Island have been key to developing and advancing them. As Jane Anderson, a co-founder of Local Contexts, writes: "Despite its claims to universality, there exist knowledge systems that intellectual property law was never designed to incorporate or protect" (Anderson, 2015, p. 769). Following model of Creative Commons, Local Contexts worked with indigenous and tribal nations councils to develop a suite of TK and BC Labels to affix to indigenous digital cultural heritage objects that were either under copyright by museums or universities or were left-wide open in the public domain. As an extra legal instrument (meaning: beyond the authority of existing law), the "Labels function as an educative strategy to help users make more informed decisions about material that might generically be deemed 'public domain' but that still has Indigenous rules and obligations about access and use" (ibid: 777).

While developed much earlier, the Labels compliment and align with growing consortiums like the Global Indigenous Data Alliance and the International Indigenous Data Sovereignty Group, which together developed the CARE Principles for Indigenous Data Governance' (Collective benefit, Authority to control, Responsibility, and Ethics) in close consultation with indigenous peoples, scholars, non-profit organizations, and governments. They "complement the existing data-centric approach represented in the 'FAIR Guiding Principles for scientific data management and stewardship' [Findable, Accessible, Interoperable, Reusable]" (Carroll et al., 2020, p. 1). The CARE Principles emerged from earlier work by Te Mana Raraunga Maori Data Sovereignty Network, U.S. Indigenous Data Sovereignty Network, Maiam nayri Wingara Aboriginal and Torres Strait Islander Data Sovereignty Collective, and many other indigenous peoples, nations, and communities:

> The Notices are highly visible, machine-readable icons that signal the Indigenous provenance of genetic resources, and rights of Indigenous communities to define the future use of genetic resources and derived benefits. The Notices invite collaboration with Indigenous communities and create spaces within our research systems for them to define the provenance, protocols, and permissions associated with genetic resources using Labels.
>
> *(Liggins et al., 2021, p. 2477) (see Figure 5.2).*

These Labels and CARE Principles deliberately connect the global debates and contestations over genetic resources with those over their digital representations. In both realms, colonialist intellectual property regimes grow more proprietary, territorial, and subservient to commodification. The fact that Local Contexts has expanded from TK Labels to BC Labels shows how intricately linked these struggles for sovereignty and restoration remain.

The TK/BC Labels and CARE Principles speak directly to and from the realm of indigenous-led research, libraries, archives, and museums; they serve as a sort of meta-metadata. They unfold from indigenous worldviews, with layers of scholarly collaboration (indigenous scholars and non-indigenous scholars

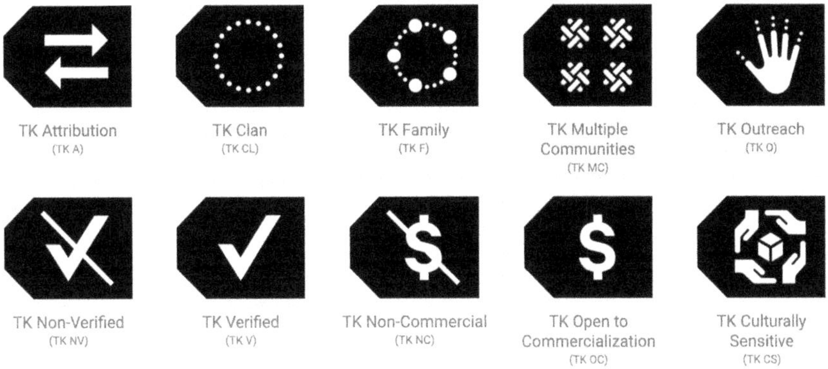

FIGURE 5.2 Some TK Labels.
Source: Local Contexts, used with permission of Local Contexts.[14]

following the lead of indigenous scholars), in much the same way as Biocultural Community Protocols. Many of the people who formulated the concepts of Biocultural Community Protocols negotiated and established them in close collaboration with researchers – or who were themselves community-based scholars, following the Tkarihwaieri Code of Ethical Conduct.[15] The TK Labels draw upon the CBD's Article 8(j), but deliberately remain at the educational – not the legal – level. This initiative seeks to guard indigenous materials from copyright enclosure but also by the open knowledge movement itself, which can reinscribe colonial appropriation.

Academia has a long history and wide geography of cultural appropriation and racialized/racist and gendered/misogynist epistemic hierarchy. Meanwhile, scholars working to counter the colonialist tendency toward extractive research methodologies have developed community research protocols that parallel Biocultural Community Protocols: from principles of partnership and terms of engagement on to memorandums of understanding. Following the primary principle of community-partnered research, these are developed in collaboration with and following the lead of frontline community representatives (Montenegro de Wit et al., 2021).

In the face of a centuries-long legal hegemony of copyright, the open knowledge movement arose and is expanding, particularly with mass digitization of cultural heritage objects held in museums, libraries, archives, conservatories, and collections. But merely opening access risks further cultural appropriation or reproducing racist harms. The TK and BC Labels work as an epistemic, not legal, tool. They emerged as public educational, public research, and community knowledge response to coloniality of intellectual property regimes. In this way, they parallel and actualize key principles of Biocultural Community Protocols regarding ownership of data in that they address crucial questions overlooked in colonialist epistemologies, such as who is an expert on whom, who gets to speak on behalf of whom, and who gets to take credit for the knowledge? At issue remains epistemic tensions between what constitutes intellectual property, data, consent, and innovation – the fundamental tensions of agrobiodiversity governance and policy.

The TK/BC Labels and Notices serve as a key product, but even more generatively, as a foundation for a process: a process of building relationships and collaboration toward decolonizing intellectual property relations into indigenous and Black Diasporic epistemic justice, sovereignty, and recovery. One of the key Labels declares "Non-Verified" to contest generations of appropriative and incorrect classifications. Other labels, such as "Outreach", work to lay groundwork for constructive community-building between institutions and indigenous nations. The TK/BC Labels project seeks to "decolonize attribution" and extricate from colonialist intellectual property regimes (Anderson & Christen, 2019). Going even further, following the lead of such scholars and activists as Miri (Margaret) Raven,[16] the goal is to move beyond using colonialism as a referent – even in post-colonial, anti-colonial, or decolonial efforts – but rather to indigenize intellectual property law and agrobiodiversity policy all together (Argumedo, 2012; Conway, 2009).

What would it take for the U.S. representation at the UN and the FAO, and the Treaty to move in the direction of Biocultural Community Protocols, TK and BC Labels and CARE Principles, and even the Atateken Declaration and Policy Briefs? What would it take for the U.S. to even ratify the CBD itself, after a generation of obstruction? These are the questions moving forward, for frontline community-based organizations, indigenous and African Diaspora-based seedkeeper practitioners, and civil society organizations. It is also a pressing question for policymakers and for scholars. Within academia, the discipline of international relations is acutely relevant and implicated. The fact that the U.S. representation operationally comes from the State Department risks entrenching the state-centric, Bretton Woods geopolitical imaginary, with its Eurocentric biases, colonial erasures, and developmentalist hierarchies of modern nation-states over indigenous, nomadic, diasporic, and migrant geographies.

Conclusion: Scales of Reference and Reckoning

This chapter began in recognition of the diverse worldviews and practices working to upkeep agricultural biodiversity in the U.S./Turtle Island and beyond, even in the face of steep legal, intellectual property, epistemic, financial, and material obstacles, much less, lack of support. Amidst this admirable topology of (agri)cultural survival grounded in biocultural principles are explicit attempts to protect and advance the work through Biocultural Community Protocols. This comprises a key research, advocacy, and policy engagement need: a compilation of all the initiatives within Turtle Island territories that have expressly endeavored Biocultural Community Protocols, or that want to. These initiatives, indigenous and African Diaspora-led, have already been informing and inspiring seed sovereignty and seedkeeping practices across the region. Even more expansively, such a compilation could learn from Biocultural Community Protocols across the Abya Yala continent (Las Américas) broadly.

The next research, advocacy, and policy need is an application of BC and TK Labels and Notices, along with CARE Principles, on data related to plant genetic resources for food and agriculture. This would expand knowledge and actualization of indigenous data sovereignty as it relates to digital and analog resources: PGRFA and the knowledge therein.

Finally, agrobiodiversity governance needs a compilation of indigenous and indigenous- and Black-led public declarations on the subject. Following the lead of the Atateken Declaration, what Policy Briefs would emerge from these cosmologically grounded political articulations.[17] Concurrently, the U.S. involvement in the ITPGRFA needs investigations regarding political economy and the role of big philanthropy and agribusiness and data industries in any U.S. national plan of action.

What are the diverse epistemologies at work – and at odds – in plant germplasm conservation governance in the U.S., and how do these dynamics impact international governance paradigms and resources? Just as hallmarks of biocultural jurisprudence and challenges to Western legal categories of privatized property and individualized innovation (as laid out in this book's introduction), Biocultural Community Protocols fundamentally challenge the U.S. colonial settler and imperial paradigms. Yet, they also speak to longstanding modes of resistance. The U.S. government has long played an obstructionist role in international forum related to agrobiodiversity, as the flagrant non-ratifier of the CBD. Even the U.S. government's ratification of the ITPGRFA carries the tensions and risks of increased IPR, entrenched agro-industry lobbyists, and indigenous erasures. Yet, possibilities abound for the U.S. leadership to listen to and learn from the Biocultural Community Protocols, indigenous-led biocultural initiatives, and the movements for indigenous data sovereignty, all unfolding and expanding across Turtle Island territories. Community-based action-research is needed to explore how the multiple perspectives and stakeholders of U.S.-based civil society, indigenous nations, farmer- and practitioner-of color organizations, and community-partnered researchers could make their way into CBD ratification as well as into the national focal point office at the U.S. State Department at the ITPGRFA meetings. This has the potential to inform, reform, and transform agrobiodiversity governance at large – to move from obstruction to decolonization and even indigenization.

Notes

1 *All URLs retrieved on 1 September 2021.
2 WG8J, First North American Dialogue on Biocultural Diversity, CBD/WG8J/11/INF/612 October 2019, Annex I.

3 https://www.iccaconsortium.org/index.php/2020/08/20/cicada-launches-four-policy-briefs-on-biocultural-diversity/
4 COP CBD, VII/16. Article 8(j) and related provisions, UNEP/CBD/COP/DEC/VII/16 (13 April 2004), F. Annex.
5 COP CBD, Decision X/42. The Tkarihwaié:ri Code of Ethical Conduct to Ensure Respect for the Cultural and Intellectual Heritage of Indigenous and Local Communities, UNEP/CBD/COP/10/27 (20 January 2011),
6 See above note (2).
7 See, this book, Chapter 1.
8 Statement of Hon. Judith G. Garber, Acting Assistant Secretary, Bureau of Oceans and International Environmental and Scientific Affairs, U.S. Department of State, Washington, DC, Thursday, May 19, 2016 U.S. SENATE, COMMITTEE ON FOREIGN RELATIONS, Washington, DC, S. Hrg. 114–324.
9 Statement of John Schoenecker, Director, Intellectual Property, Hm.Clause, On Behalf of American Seed Trade Association, Davis, Ca, Washington, DC, Thursday, May 19, 2016 U.S. SENATE, COMMITTEE ON FOREIGN RELATIONS, Washington, DC, S. Hrg. 114–324.
10 Additional Material Submitted for the Record. Responses to Questions Submitted to Judith G. Garber by Senatorcorker, May 19, 2016 U.S. SENATE, COMMITTEE ON FOREIGN RELATIONS, Washington, DC, S. Hrg. 114–324.
11 Statement of Hon. Judith G. Garber, Acting Assistant Secretary, Bureau of Oceans and International Environmental and Scientific Affairs, U.S. Department of State, Washington, DC, Thursday, May 19, 2016 U.S. SENATE, COMMITTEE ON FOREIGN RELATIONS, Washington, DC, S. Hrg. 114–324.
12 See, this book, Chapter 1.
13 https://www.un.org/un70/en/content/history/index.html
14 https://localcontexts.org/labels/traditional-knowledge-labels/
15 UNEP/CBD/COP/DEC/X/42, 29 October 2010, para. 25.
16 Raven shared these informative insights and analysis in the Online Panel Discussions: Biocultural Rights and Community Protocols, 5 November 2020–6 November 2020, Bioculturalis research project: https://bioculturalis.univ-grenoble-alpes.fr/news/online-panel-discussions-biocultural-rights-and-community-protocols
17 See Veronica Limeberry's forthcoming dissertation "Fast Violence, Slow resistance: Territoriality, Land Rights and Collective Identity for Agrobiodiversity Governance" (American University School of International Service).

References

Alliance Muteshekau-shipu. (2021). *For the First Time, a River Is Granted Official Rights and Legal Personhood in Canada*. CISION. https://www.newswire.ca/news-releases/for-the-first-time-a-river-is-granted-official-rights-and-legal-personhood-in-canada-848414747.html

Anderson, J., & Christen, K. (2019). Decolonizing Attribution: Traditions of Exclusion. *Journal of Radical Librarianship*, *5*, 113–152.

Anderson, J. E. (2015). Indigenous Knowledge and Intellectual Property Rights. In J. D. Wright (Ed.), *International Encyclopedia of the Social & Behavioral Sciences* (2nd ed., pp. 769–778). Elsevier. https://doi.org/10.1016/B978-0-08-097086-8.64078-3

Argumedo, A. (2008). The Potato Park, Peru: Conserving Agrobiodiversity in an Andean Indigenous Biocultural Heritage Area. In T. Amend, J. Brown, A. Kothari, A. Phillips, & S. Stolton (Eds.), *Protected Landscapes and Agrobiodiversity Values* (Vol. 1, pp. 45–58). Kasparek Verlag.

Argumedo, A. (2012). Decolonising Action-Research: The Potato Park Biocultural Protocol for Benefit-sharing. In K. Swiderska, A. Milligan, K. Kohli, H. Jonas, H. Shrumm, W. Hiemstra, & M. J. Oliva (Eds.), *PLA 65—Biodiversity and Culture: Exploring Community Protocols, Rights and Consent* (pp. 91–100). IIED.

Aubry, S. (2019). The Future of Digital Sequence Information for Plant Genetic Resources for Food and Agriculture. *Frontiers in Plant Science, 10*, 1046. https://doi.org/10.3389/fpls.2019.01046

Bavikatte, K. S., & Bennett, T. (2015). Community Stewardship: The Foundation of Biocultural Rights. *Journal of Human Rights and the Environment, 6*(1), 7–29. https://doi.org/10.4337/jhre.2015.01.01

Bavikatte, K. S., & Robinson, D. F. (2011). Towards a Peoples History of the Law: Biocultural Jurisprudence and the Nagoya Protocol on Access and Benefit Sharing. *Law, Environment and Development Journal (LEAD Journal), 7*(1), 35–51.

Blaser, M. (2014). Ontology and Indigeneity: On the Political Ontology of Heterogeneous Assemblages. *Cultural Geographies, 21*(1), 49–58. https://doi.org/10.1177/1474474012462534

Caillon, S., Cullman, G., Verschuuren, B., & Sterling, E. J. (2017). Moving Beyond the Human–Nature Dichotomy through Biocultural Approaches: Including Ecological Well-being in Resilience Indicators. *Ecology and Society, 22*(4), art27. https://doi.org/10.5751/ES-09746-220427

Carroll, S. R., Garba, I., Figueroa-Rodríguez, O. L., Holbrook, J., Lovett, R., Materechera, S., Parsons, M., Raseroka, K., Rodriguez-Lonebear, D., Rowe, R., Sara, R., Walker, J. D., Anderson, J., & Hudson, M. (2020). The CARE Principles for Indigenous Data Governance. *Data Science Journal, 19*, 43. https://doi.org/10.5334/dsj-2020-043

Commission on Genetic Resources for Food and Agriculture. (n.d.). *Digital Sequence Information*. FAO. Retrieved 15 July 2021, from http://www.fao.org/cgrfa/topics/digital-sequence-information/en/

Conway, D. M. (2009). Indigenizing Intellectual Property Law: Customary Law, Legal Pluralism, and the Protection of Indigenous Peoples' Rights, Identity, and Resources. *Texas Weleyan Law Review, 15*(2), 207–256.

de Sousa Santos, B. (2006). Globalizations. *Theory, Culture & Society, 23*(2–3), 393–399. https://doi.org/10.1177/026327640602300268

de Sousa Santos, B., & Rodríguez-Garavito, C. A. (2005). Law, Politics, and the Subaltern in Counter-Hegemonic Globalization. In B. de Sousa Santos & C. A. Rodríguez-Garavito (Eds.), *Law and Globalization from Below* (1st ed., pp. 1–26). Cambridge University Press. https://doi.org/10.1017/CBO9780511494093.001

Descola, P. (2009). Human Natures. *Social Anthropology, 17*(2), 145–157. https://doi.org/10.1111/j.1469-8676.2009.00063.x

Fitznor, L. (2019). Indigenous Education: Affirming Indigenous Knowledges and Languages from a Turtle Island Indigenous Scholar's Perspective: Pikiskēwinan (Let Us Voice). In C. Cocq & K. Sullivan (Eds.), *Perspectives on Indigenous Writing and Literacies* (pp. 29–66). Brill. https://doi.org/10.1163/9789004298507

Forest Peoples Programme, IIFB, IWBN, & CBD. (2020). *Media Kit: Ways forward for Biodiversity and People. Press Briefing for 2nd Edition of Local Biodiversity Outlooks.* Forest Peoples Programme; IIFB; IWBN; CBD. https://www.cbd.int/doc/press/2020/pr-2020-09-23-lbo2-en.pdf

Graddy-Lovelace, G. (2021). Leveraging Law & Life: Criminalization of Agrarian Movements & the Escazú Agreement. In J. Shapiro and J. A. McNeish (Eds.), *Our Hyper-Extractive Age: Expressions of Violence and Resistance* (pp. 94–113). Routledge.

Hill, R., Nates-Parra, G., Quezada-Euán, J. J. G., Buchori, D., LeBuhn, G., Maués, M. M., Pert, P. L., Kwapong, P. K., Saeed, S., Breslow, S. J., Carneiro da Cunha, M., Dicks, L. V., Galetto, L., Gikungu, M., Howlett, B. G., Imperatriz-Fonseca, V. L., Lyver, P. O., Martín-López, B., Oteros-Rozas, E., … Roué, M. (2019). Biocultural Approaches to Pollinator Conservation. *Nature Sustainability*, *2*(3), 214–222. https://doi.org/10.1038/s41893-019-0244-z

Hunt, D., & Stevenson, S. A. (2017). Decolonizing Geographies of Power: Indigenous Digital Counter-Mapping Practices on Turtle Island. *Settler Colonial Studies*, 7(3), 372–392. https://doi.org/10.1080/2201473X.2016.1186311

Hunt, S. (2014). Ontologies of Indigeneity: The Politics of Embodying a Concept. *Cultural Geographies*, *21*(1), 27–32. https://doi.org/10.1177/1474474013500226

IIED, Asociacion ANDES, & Natural Justice. (2012). *Using Biocultural Community Protocols to Implement MEAs and UNDRIP at the Local Level for Sustainable Development* (Workshop of the ISE Global Coalition for Biocultural Diversity). 13th ISE Congress. https://pubs.iied.org/pdfs/G03388.pdf

iisd Reporting Services. (2019). Summary of the Eighth Session of the Governing Body of the International Treaty on Plant Genetic Resources for Food and Agriculture: 11–16 November 2019. *Earth Negotiations Bulletin*, *9*(740), 1–15.

Kumar, P., Lorek, T., Olsson, T. C., Sackley, N., Schmalzer, S., & Soto Laveaga, G. (2017). Roundtable: New Narratives of the Green Revolution. *Agricultural History*, *91*(3), 397. https://doi.org/10.3098/ah.2017.091.3.397

Liggins, L., Hudson, M., & Anderson, J. (2021). Creating Space for Indigenous Perspectives on Access and Benefit-Sharing: Encouraging Researcher Use of the Local Contexts Notices. *MEC Molecular Ecology*, *30*(11), 2477–2482.

Lincoln, N. K., Rossen, J., Vitousek, P., Kahoonei, J., Shapiro, D., Kalawe, K., Pai, M., Marshall, K., & Meheula, K. (2018). Restoration of 'Āina Malo'o on Hawai'i Island: Expanding Biocultural Relationships. *Sustainability*, *10*(11), 3985. https://doi.org/10.3390/su10113985

Medina, N. (2019). Indigenous Decolonial Movements in Abya Yala, Aztlán, and Turtle Island: A Comparison. In R. Barreto & R. Sirvent (Eds.), *Decolonial Christianities* (pp. 147–163). Springer International Publishing. https://doi.org/10.1007/978-3-030-24166-7_8

Montenegro de Wit, M., Shattuck, A., Iles, A., Graddy-Lovelace, G., Roman-Alcalá, A., & Chappell, M. (2021). Operating Principles for Collective Scholar-Activism: Early Insights from the Agroecology Research-Action Collective. *Journal of Agriculture, Food Systems, and Community Development*, *10*(2), 1–19. https://doi.org/10.5304/jafscd.2021.102.022

Muzurakis, M. (2019, November 25). How Much Biodiversity will be Available for Future Generations? IPC Participation at the Governing Body of the International Seeds Treaty. *International Planning Committee for Food Sovereignty (IPC)*. https://www.foodsovereignty.org/gb8-itpgrfa/

Naylor, L., Daigle, M., Zaragocin, S., Ramírez, M. M., & Gilmartin, M. (2018). Interventions: Bringing the Decolonial to Political Geography. *Political Geography*, *66*, 199–209. https://doi.org/10.1016/j.polgeo.2017.11.002

Oguamanam, C. (Ed.). (2019). *Genetic Resources, Justice, and Reconciliation: Canada and Global Access and Benefit Sharing*. Cambridge University Press.

Patel, R. (2013). The Long Green Revolution. *Journal of Peasant Studies*, *40*(1), 1–63. https://doi.org/10.1080/03066150.2012.719224

Perkins, J. H. (1997). *Geopolitics and the Green Revolution: Wheat, Genes, and the Cold War*. Oxford University Press.

Pernet, C. A., & Ribi Forclaz, A. (2019). Revisiting the Food and Agriculture Organization (FAO): International Histories of Agriculture, Nutrition, and Development. *The International History Review, 41*(2), 345–350. https://doi.org/10.1080/07075332.2018.1460386

Posey, D. A., & Dutfield, G. (1996). *Beyond Intellectual Property: Toward Traditional Resource Rights for Indigenous Peoples and Local Communities*. International Development Research Centre.

Pungetti, G., & Cinquepalmi, F. (2012). Sacred Sites, Sacred Landscapes and Biocultural Diversity: Applying the Principles. In G. Pungetti, G. Oviedo, & D. Hooke (Eds.), *Sacred Species and Sites* (pp. 407–441). Cambridge University Press. https://doi.org/10.1017/CBO9781139030717.036

Quick Hall, K. M. (2020). *Naming a Transnational Black Feminist Framework: Writing in Darkness*. Routledge.

Ribi Forclaz, A. (2019). From Reconstruction to Development: The Early Years of the Food and Agriculture Organization (FAO) and the Conceptualization of Rural Welfare, 1945–1955. *The International History Review, 41*(2), 351–371. https://doi.org/10.1080/07075332.2018.1478873

Sarrazin, M.-L., & Scott, C. (2020a). Livelihoods, Food Sovereignty, and Well-Being. *Policy Brief 2: Nurturing Biocultural Diversity*. https://cicada.world/files/pb/CICADA_PolicyBrief2_Livelihoods.pdf

Sarrazin, M.-L., & Scott, C. (2020b). Safeguarding Biocultural Diversity: Territorial Defense in Extractive Contexts. *Policy Brief 4: Safeguarding Biocultural Diversity*. Supporting Biocultural Diversity. Information and Communication Technologies. https://cicada.world/files/pb/CICADA_PolicyBrief4_Extractives.pdf

Sarrazin, M.-L., & Scott, C. (2020c). Supporting Biocultural Diversity. Information and Communication Technologies. *Policy Brief 3: Supporting Biocultural Diversity*. https://cicada.world/files/pb/CICADA_PolicyBrief3_Technologies.pdf

Sassen, S. (2009). *Territory, Authority, Rights: From Medieval to Global Assemblages*. Princeton University Press.

Secretariat of the Convention on Biological Diversity. (2020). *Global Biodiversity Outlook 5*. UNEP, CDB. https://edepot.wur.nl/531233

Shiva, V. (1991). *The Violence of the Green Revolution: Third World Agriculture, Ecology, and Politics*. Zed Books, Third World Network.

Simpson, A. (2014). *Mohawk Interruptus: Political Life Across the Borders of Settler States*. Duke University Press.

Stone, G. D. (2019). Commentary: New Histories of the Indian Green Revolution. *The Geographical Journal, 185*(2), 243–250. https://doi.org/10.1111/geoj.12297

Teran, M. Y. (2016). The Nagoya Protocol and Indigenous Peoples. *International Indigenous Policy Journal*, 7(2). https://doi.org/10.18584/iipj.2016.7.2.6

U.S. Mission to the UN Agencies in Rome. (2017, April 14). *The United States Joins the International Treaty on Plant Genetic Resources for Food and Agriculture*. U.S. Mission to the UN Agencies in Rome. https://usunrome.usmission.gov/united-states-joins-international-treaty-plant-genetic-resources-food-agriculture/

Vaziri, L., Sarrazin, M.-L., Sterling, E. J., & Pascua, P. (2020). Measuring Biocultural Diversity: Biocultural Indicators and the Nexus of Nature, Culture, and Well-Being. *Policy Brief 1: Measuring Biocultural Diversity*. https://cicada.world/files/pb/CICADA-CBC_PolicyBrief1_Biocultural-indicators.pdf

Winter, K. B., Beamer, K., Vaughan, M. B., Friedlander, A. M., Kido, M. H., Whitehead, A. N., Akutagawa, M. K. H., Kurashima, N., Lucas, M. P., & Nyberg, B. (2018). The Moku System: Managing Biocultural Resources for Abundance

within Social-Ecological Regions in Hawai'i. *Sustainability*, *10*(10), 3554. https://doi.org/10.3390/su10103554
Winter, K. B., Ticktin, T., & Quazi, S. A. (2020). Biocultural Restoration in Hawai'i also Achieves Core Conservation Goals. *Ecology and Society*, *25*(1), art26. https://doi.org/10.5751/ES-11388-250126
Zellars, R. (2021). Dreams of a Black Commons on Turtle Island. *Studies in Social Justice*, *14*(2), 454–473. https://doi.org/10.26522/ssj.v14i2.2243

6

THE LEGAL FRAMEWORK BEHIND BIOCULTURAL RIGHTS

An Analysis of Their Pros and Cons for Indigenous Peoples and for Local Communities

Giulia Sajeva[1]

Scope and Objective

When indigenous people or a local community decides to engage in the creation of, for instance, a Biocultural Community Protocol, it enters the realm of international, regional, and national laws with the precise intention of becoming an actor more conscious of its rights, and better able to vindicate their respect and implementation. However, choosing which path to follow in terms of rights vindication is not always easy nor straightforward. Currently, there are an increasing number of international and national laws, policies, court cases, declarations, and guidelines concerning indigenous peoples, local communities, and the protection of the environment, which provide rights regarding access to land, benefit-sharing, traditional knowledge, carbon emissions, protected areas, and much more. These rights are essentially fragmented (ILC Conclusions, 2006; Payandeh, 2015): addressed by different bodies, found in diverse sources (Jonas, 2020, p. 4; Jonas et al., 2014), and differently treated by different courts (Abrusci, 2017). Consequently, indigenous peoples and local communities must engage with a plethora of legal sources to obtain protection for interconnected aspects of their lives which are all part of the same biocultural landscape.

The idea of biocultural rights, which was developed predominantly by Sanjay Kabir Bavikatte (2014; Bavikatte & Bennet, 2015), strives to look at and deal with the overall biocultural landscape of indigenous peoples and local communities, conflating many of the different rights they need to promote their self-government and the conservation of their cultural identities. This chapter explores the pros and cons of claiming biocultural rights, but it does so separately for indigenous peoples and for local communities (Sajeva, 2018). They appear, in fact, as different subjects in international law, whose positions are somehow similar but sufficiently diverse to require distinct considerations.

DOI: 10.4324/9781003172642-7

Protecting Indigenous Peoples and Welcoming Local Communities

> [T]he debate over who is 'indigenous' should not side-track the important task of valuing local communities – whether or not they are indigenous. The important task is to rekindle and enhance the spiritual and cultural values that cultures have used effectively to conserve biodiversity.
>
> *(Posey, 1999, p. 4)*

Darrel Posey was right. The spiritual and cultural values, practices, and knowledge of local communities and indigenous peoples, and their capacity to help us through the current environmental crisis, should not be shadowed by tribulations concerning what a local community is and who is indigenous. However, indigenous peoples have indigenous rights, while local communities do not.

Indigenous rights are neither fully respected nor completely implemented (UN Department of Economic and Social Affairs, 2019), but are recognized by international human rights law and entrenched, at least on paper, in the Constitutions, statutory laws, and court decisions of many countries and regional organizations. On the contrary, the term "local community" is still a vague and murky concept, with uncertain contours, status, and recognition. Its legal subjectivity is tentatively and timidly emerging in national and international law debates, and its rights are, currently, just a shadow of indigenous peoples' rights. So far, *non-indigenous* local communities are welcomed into the conversation about collective rights, although usually only when joining conversations about the protection of the environment (Jonas, 2020; Jonas & Godio, 2020). The larger step to welcome them into the realm of human rights – outside of considerations concerning the environment – still remains to be taken. This consideration, as we will see below, is paramount for the analysis of biocultural rights' pros and cons.

Indigenous Peoples and Their Rights

The definition of indigenous peoples is not yet a settled dispute, as the very need and appropriateness of having a definition are still widely questioned.[2] The 2007 UN Declaration on the Rights of Indigenous Peoples (UNDRIP) – a benchmark for the acceptance of the status of indigenous peoples as subjects of international law – stresses the importance of peoples' *self-identification* as *indigenous* and lists the main characteristics that most indigenous peoples are likely to have.[3] However, linkages to pre-colonial times practices, beliefs, and institutions remain the most essential feature to qualify as indigenous and act as the main justification (foundation, *raison d'être*) for indigenous rights (see also Jonas, 2020, p. 21).

Indigenous peoples are recognized as holders of indigenous rights, which are, mostly, collective human rights. The more important one is the right to (internal)[4] self-determination,[5] which is accompanied by many specific collective cultural, property, and social rights whose protection was acknowledged as essential to preserving their existence as people, and their identity and well-being. The rights recognized by UNDRIP and by the International Labour Organization Indigenous and Tribal People Convention No. 169[6] include the right to the lands, territories, and natural resources they have traditionally owned, occupied, or used; the right to maintain distinctive spiritual traditions, customs, and land tenure systems; the right to conservation and protection of the environment of their lands and territories; and the right to free, prior, and informed consent (including the right to refuse consent) in all cases where their lands or natural resources are affected by external agents; the right to non-discrimination and cultural integrity,[7] based on the recognition of the paramount importance of language, and cultural, social, and religious practices for the preservation of people's identity; the right to welfare, including the right to maintain traditional health practices; and the right to development[8] that encompasses the right to determine priorities and strategies for their lands, territories, and resources.

Matters are instead more complicated and less developed for local communities.

Local Communities and Their Rights

Local communities may be defined as those small-scale communities that "do not fit the strict test of indigeneity but nevertheless" (Bessa, 2015, p. 332) identify themselves as a community – either traditional or relatively new – thanks to shared spatial, sociocultural, and economic elements. They often hold a collective relationship with the local environment and natural resources – which may be concentrated on permanent places or nomadic routes and habits. They also share dynamic and evolving social and cultural elements, such as a common history, traditional practices, institutions, language, values, and life plans that bind them together and distinguish them from other groups of society (Sajeva et al., 2019, p. 12). Their internal economic organization reflects their common interest in, and special collective relationship with, the local environment and resources, most often containing elements of common-pool resource management. Importantly, the community has, within its defined boundaries, institutional processes, and leadership perceived as legitimate by a majority of members, who can exercise sufficiently effective control and compliance of rules and conflict-resolution. Moreover, local community members are likely to share a common political identity that enables them to exercise and claim for the recognition of collective rights and responsibilities.

This list of features mirrors those employed to describe indigenous peoples, of which *indigeneity* is the key differentiating feature.[9] As stated above, the distinction between local communities and indigenous peoples is not simply an identity issue, but also a legal one.[10] Local communities are, like all human beings, holders of the fundamental rights recognized by the Universal Declaration of Human Rights and the UN Conventions and Covenants on Human Rights.[11] However, these documents predominantly recognize individual rights,[12] which may not be sufficient to safeguard the survival, identity, and flourishing of local communities as groups (Golay, 2013, 2015) and to react to their specific vulnerabilities and struggles. These peculiarities require collective rights to be fully protected, as it was widely recognized for indigenous peoples.

Currently, when identifying specific references to local communities' collective rights in international law, it is necessary to look at documents strongly related to the protection of the environment. Unsurprisingly, it was the UN Special Rapporteur on Human Rights and the Environment, calling for special attention to the role of non-indigenous local communities for environmental protection (Knox, 2017, sec. 59), who raised awareness of their special vulnerability to environmentally harmful actions (Knox, 2017, sec. 52) and lack of effective legal protection (Knox, 2017, 2018). He asserted that States owe them obligations that are "not always identical to those owed to indigenous peoples" (Knox, 2017, p. 71, 2018, sec. 48) but that entail the protection of "the special relationship of people with the territory that they have traditionally occupied when their subsistence and culture is closely linked to that territory" (Knox, 2017, sec. 56). Protecting their rights, he continued, "is not just required by human rights law; it is also often the best or only way to ensure the protection of biodiversity" (Knox, 2017, sec. 59).

In particular, the Special Rapporteur referred to the first acknowledgment of local communities' collective rights in international law, due to the Convention on Biological Diversity (CBD), which has made significant contributions to the recognition of the rights of local communities (and of indigenous peoples) (Morgera, 2014, p. 984).[13] Article 8(j) requires member states to respect, preserve, and maintain the knowledge, innovations, and practices of local communities (and of indigenous peoples) that have preserved lifestyles relevant for the conservation and sustainable use of biodiversity. States are further required to promote the wider application of such sustainable practices and associated traditional knowledge, and ensuring communities and peoples of origin are consulted and involved, and that any benefits that arise from this utilization are fairly and equitably shared with them. The 2010 Nagoya Protocol on Access to Genetic Resources and the Fair and Equitable Sharing of Benefits Arising from their Utilization further requires (art. 5, 6, 7) States to ensure that genetic resources and traditional knowledge of local communities are accessed with the prior informed consent of the communities; that benefits arising from their use are equally shared with them; and that customary laws and community protocols of local communities are taken in consideration.[14]

The CBD was soon followed by the adoption of the UN Convention on Combating Desertification, in 1994, which simply cited local communities as subjects of special consideration that should be included in development programs (Ziegler et al., 2008). Equally soft are the provisions of the 2016 Paris Agreement, which acknowledges that actions to address climate change should respect, promote, and consider the rights of local communities (and indigenous peoples) and that adaptation measures "should be based on and guided by the best available science and, as appropriate, traditional knowledge of indigenous peoples and local knowledge systems" (art. 7).

One of the difficulties of listing the rights of local communities lies in the fact that they are heterogeneous.[15] They may indeed be "small-scale farmers, artisanal fishing communities, island, and mountain communities" (Bessa, 2015, p. 332), as well as communities of mainstream religions living in European countries that have protected sacred natural sites (Frascaroli et al., 2016; Verschuuren et al., 2010), or may even be newly born local communities of peoples that gathered with the precise aim to live sustainably (e.g. ecovillages; see Ergas, 2010, p. 34), or communities that have migrated to peri-urban or urban areas, which maintain a special relationship with the environment, and groups of citizens claiming the right to care for urban green areas, without necessarily "liv[ing] geographically close to natural environment[s] hold[ing] spiritual, social, and cultural value" (Cocks, 2006, p. 194).

Consequently, very often, national and international laws and policies concern local communities and recognize them as holders of collective rights, without using this general term, because they exclusively address more specific types of local communities such as farmers, peasants, and tenure holders; small-scale fishing communities, pastoralists, traditional hunting, herding, or nomadic peoples.

Farmers, for example, are recognized as holders of farmers' rights under the 2001 FAO International Treaty on Plant Genetic Resources for Food and Agriculture (ITPGRFA) that requires States to give them recognition and implementation through national law. ITPGRFA, established to facilitate the exchange of seeds and germplasm to develop new plant varieties for food and agriculture, recognizes the contribution that "local and indigenous communities and farmers of all regions of the world" have made "for the conservation and development of plant genetic resources" (art. 9). It calls States, through discretional national measures (Frison, 2018, p. 91), to promote the conservation of traditional knowledge and practices of farmers, share benefits arising from the use of farmers' plant genetic resources, and involve them in relevant decision-making procedures. These rights of farmers stand in explicit contrast with "plant breeders' rights and patents" – professional selectors that breed, discover, or develop new plant varieties – but their contours are still quite unclear (Lawson, 2015). Indeed, ITPGRFA does not seem to impose specific binding obligations on states to protect farmers' rights, but simply to suggest that they adopt policies, fund projects, and administrative measures to promote the realization of farmers' rights (Haugen, 2020; Lawson, 2015).

The remaining imbalance of rights between breeders and farmers – that *de facto* lack formalized recognition of "the right to save, use, exchange and sell farm-saved seeds" (Frison, 2018, p. 90) – led to a strong politicization of agricultural negotiations and, thanks to the work of international movements such as *La Via Campesina*,[16] to the adoption of a new instrument: the UN Declaration on the Rights of Peasants and Other Peoples living in Rural Areas (UNDROP; see Claeys & Edelman, 2020). UNDROP is addressed to all peasants (Edelman & Carwil, 2011), *the people of the land*, without any necessary reference to them being the owners or tenants (Paoloni & Vezzani, 2019, p. 11),[17] as long as production is related directly with the land (including costs, waters, sea, etc.) and is on a small-scale (subsistence or little more). The Declaration reaffirms the special relationship with the environment that peasants and other people working in rural areas have and recognizes their contribution to conserving and improving biodiversity. In order to promote and protect this special interaction, it contains a long list of very diverse rights. Grounded on the importance of non-discrimination, these rights partially reaffirm human rights (Paoloni & Vezzani, 2019, p. 24) already included in international human rights law,[18] or are still under debate (such as the right to a safe, clean, and healthy environment and right to water; see Boyd, 2019; Knox, 2017), expanding (or narrowing?[19]) them onto local communities. A second category of rights partially overlaps with indigenous peoples' rights, focusing on issues mostly related to traditional practices and legal systems, and collective rights related to access and use of natural resources. Furthermore, UNDROP recognizes more specific collective rights, whose origins can be traced through the struggle for adequate recognition of farmers' rights, and that may be considered as new rights in international human rights law (Golay, 2015, p. 24). Among these, there is the right to food produced and consumed sustainably and equitably respecting their cultures; the right to food sovereignty as a right to determine their food and agriculture systems; the right to engage in traditional ways of farming, fishing, livestock rearing, and forestry; the freedom to determine prices and markets for agricultural production and access to markets; the right to the protection of traditional knowledge, innovation, and practices, including traditional agrarian, pastoral, forestry, fisheries, livestock, and agroecological systems relevant to the conservation and sustainable use of biological diversity; the right to save, use, exchange, and sell their farm-saved seed or propagating material; the rights to rely either on their own seeds or on other locally available seeds of their choice, and to decide on the crops and species that they wish to grow; the support to seed systems, and promotion of the use of peasant seeds and agrobiodiversity; the right to maintain, control, protect, and develop seeds and their traditional knowledge; and the right to biological diversity.

The above list of rights was approved with the favorable vote of 122 states[20] at the UN General Assembly; however, its legal force is limited to that of a simple recommendation with no legally binding force.[21]

UNDROP aligns with other non-binding instruments that may be relevant to understand the rights (and at least their current process of evolution) of local communities: the FAO 2012 *Voluntary Guidelines on the Responsible Governance of Tenure of Land, Fisheries and Forests in the Context of National Food Security* directed toward very diverse local communities sharing special relationships with lands and natural resources and contributing to their conservation and sustainable use, and followed by the 2016 *Governing Tenure Rights to Commons. A guide to support the implementation of the Voluntary Guidelines on the Responsible Governance of Tenure of Land, Fisheries and Forests in the Context of National Food Security* aimed at protecting tenure rights to commons and community-based governance structures; the 2013 *Voluntary Guidelines for Securing Sustainable Small-scale Fisheries in the Context of Food Security and Poverty Eradication*, addressed at small-scale fishing communities; and the 2005 *Voluntary Guidelines to support the progressive realization of the right to adequate food in the context of national food security* meant for "local and indigenous communities and farmers". These instruments were adopted by the FAO Committee on World Food Security after years of negotiations and deliberation that involved governments, civil society organizations, and representatives of the private sector and the of philanthropic organizations. The Guidelines in fact, even though they are a *voluntary* instrument, might be said to represent the position of the international community "as to how global food security should be addressed" (De Shutter & Rajagopal 2020, p. 215) and might be able to provide frameworks that States may use when developing relevant strategies, policies, laws, programs, and activities.

Biocultural Rights: A More Comprehensive Approach

Chaos

This chaotic list of collective rights of local communities creates a complex scenario of entitlements concerning the environment that can easily confuse any local community. Such complexity often concerns indigenous peoples as well, and is due partially to the nature of international law, whose lack of centralized hierarchy makes it an ever-evolving corpus, pulled and twisted by heterogeneous interests, needs, and institutions (Payandeh, 2015); and partially to a lack of consensus on the definitions of local communities and indigenous people. Indeed, the aforementioned rights also depend on several different factors (Jonas, 2020, p. 2), including their ways of life, the type of territory where they reside, and the institutional settings (protected area, recognized/non-recognized territories and areas conserved by indigenous peoples and local communities[22] – community-owned land, private land) they live within. Each of these factors intersects with the others, making the description of the collective rights of indigenous peoples and local communities an exceedingly difficult enterprise.

To further complicate matters, it shall be reminded that the history of environmental protection and conservation movements bears the original sin of a cumbersome colonial past (Adams & Milligan, 2003; Dowie, 2009; MacKenzie, 1988), whose influence on indigenous peoples and local communities is not yet completely over (Brechin et al., 2002; Wilshusen et al., 2002). Disregarding the stewardship-based relationship they have kept with nature, the actions aimed at nature's conservation have often resulted in the creation of only nature areas that have led (and still lead) to the eviction of indigenous peoples and local communities from their ancestral lands (Dowie, 2009; Tauli-Corpuz, 2016), and have hampered their access to natural resources on the ground of assumptions about the incompatibility of people's lives and conservation activities.[23]

Order

Rights-based approaches to the conservation of the environment – approaches that build on the respect of the rights of indigenous peoples and local communities and that include them as active stakeholders and rightsholders in the management and governance of protected areas and other environmental conservation actions and projects – have, however, found their way, thanks to the increased concomitant understanding of the advantageous inclusion of communities and peoples in conservation activities, and the increased recognition of indigenous peoples and local communities' rights. Through the amalgamation of these streams,[24] Bavikatte suggested that a new basket of rights – biocultural rights – may be emerging in international law: a *de iure condendo* construct whose (yet) non-explicit recognition might be interpreted from international environmental and human rights hard and soft law and policies (Bavikatte, 2014, pp. 2, 21; Bavikatte & Bennett, 2015)[25] and have evolved – in legal literature and court cases[26] – to become an important construct for interpreting the rights of indigenous peoples and local communities that relate to the environment.

Biocultural rights[27] are defined as the basket of collective rights that indigenous peoples and local communities need to maintain their role as stewards of the environment (Bavikatte, 2014, p. 16). They "initially appeared [...] as 'farmers' rights', 'livestock keepers' rights' and rights to traditional knowledge" and "sought to assert" group rights "not only based on ethnicity or religion or minority status but primary on a history of stewardship of ecosystems" (Bavikatte, 2014, p. 28). Indeed, "despite the seeming differences" between these rights "they all have the same pith and substance – that is, they seek to secure the stewardship role of communities over their cultures, lands, and waters" (Bavikatte, 2014, p. 29). Stemming from these considerations, biocultural rights can be said to build on a double foundation, i.e. two heterogeneous protected interests[28] – *raisons d'être* – the protection of the environment, and the protection of the lifestyles, practices, beliefs, and cosmovision of indigenous peoples and local communities. The idea of biocultural rights gravitates around the understanding that the protection of the ways of life and stewardship role of indigenous peoples and local communities can be beneficial for the conservation of biodiversity and ecosystems. As the Executive Secretary of the CBD recently stated: "If we are to achieve by 2050 the Convention's vision of 'Living in harmony with nature', it is critical that we harness the power of the collective and local actions of the world's indigenous peoples and local communities" (Mrema, 2020).

These rights were "developed as a people-led alternative to state-led technocratic solutions to the environmental crisis" (Bavikatte, 2014, p. 18), and include all "the rights of communities to fulfil their role as trustees of their cultures, lands, waters, and resources" (ibid., p. 21). The rights found within the biocultural basket will therefore differ from one community or people to another: the heterogeneity of their ways of life, practices, knowledge, and worldviews implies a complex and flexible set of rights needed to maintain them (for instance, farming, fishing, or nomadic communities will have different needs). These ever-changing rights (depending on each community needs) may be clustered under four, very general, themes (see Sajeva, 2018, pp. 105–109):

- rights to land, waters, and natural resources: including the right to access and use of traditional lands and waters; special access to sacred natural sites; access to and use of biotic and abiotic resources present in the land; protection from external threats to the environment;
- rights to self-government: including the right to internal self-determination and to regulate its internal matters through the use of its legal institutions and rules;
- rights to cultural identity: including the rights necessary to safeguard the integrity of values, worldviews, practices, and knowledge;
- associated procedural rights: such as the right to access to justice, the right to free, prior, and informed consent, and the right to the application of a precautionary approach.

Biocultural rights place themselves in the Anthropocene debate as powerful tools that provide answers to both human rights and environmental issues. Moreover, it is precisely because of their environmental genesis and focus that they bring to life a basket of rights which conflates many of the different elements of the biocultural landscapes of indigenous peoples and local communities. They may be used by indigenous peoples and local communities to put forward a single overarching claim for the protection of their needs and interests vis-à-vis their lands, waters, and natural resources, without having to reference many different treaties, conventions, declarations, and guidelines. Moreover, when dealing with projects and policies addressed at the conservation of the environment, biocultural rights may be a (political and legal) trump to be used against the threat of being evicted from traditional lands or impeded to pursue traditional practices and governance of natural resources. As many developing countries are facing strong international pressure to conserve their ecosystems, while having to deal with a constant lack of economic resources, biocultural rights may be an instrument to seek a balance between conservation and human rights, the interests of the environment, and the interests of peoples and communities. Indeed, biocultural rights simultaneously imply the possibility to protect the environment and the cultural diversity and self-governance of a community or people, precisely because they are rights to environmental stewardship.

Beware

Nevertheless, the environmental core of biocultural rights reveals to be both their power and limit: they are environmentally conditioned rights. Their double foundation may be interpreted as meaning that, with both indigenous peoples and local communities on the one side and the environment on the other, all should hold the rights found in the biocultural rights basket: both "subjects" are thus protected through the realization of biocultural rights. The protection of nature's interest[29] (to thrive, to continue to be, to evolve) grounds biocultural rights and justifies (in legal terms) their recognition. Indeed, as Bavikatte (2014, pp. 142 *et seq.*) underlines

> [t]he demand for biocultural rights does not take as its point of departure the inherent right of a group or community to flourish, but rather [...] the ethic of stewardship: it is the ethic of stewardship and not the group *per se* that justifies the right.

Biocultural rights stem from the interpretation of environmentally relevant documents whose ultimate goal is the protection of the intrinsic value of biodiversity, and indigenous peoples and local communities emerge as subjects whose rights are to be protected *because* they have preserved ways of life relevant for biodiversity,[30] not merely because they are holders of intrinsic value *as local communities or indigenous peoples.* Therefore, biocultural rights may only be claimed by sustainable indigenous peoples and local communities, binding them to exercise the rights in their basket in ways that do not harm the environment, but instead promote its protection.

These implications may not raise any particular concern if we abandon ourselves to the illusion of the noble savage myth (Ellingson, 2001), according to which indigenous peoples and local communities will always remain pacific friends of the Earth, regardless of external and internal changes; regardless of the abundance of the species they traditionally hunt, fish, or domesticate; regardless of the number of people that makes up a community; and regardless of their changing desires, aspirations, and needs.

We may take it that most indigenous peoples and local communities are sustainable and desire to remain as such, but we cannot assume that each implementation of their right to self-governance and cultural identity leads to sustainable outcomes and promotes the conservation of the environment. We shall refrain from entrapping them into the simplistic and mistaken duality picturing them either as "intrinsically attuned to nature" noble savages or as fallen angels once in contact with mainstream society (Berkes, 2001, p. 116).

Therefore, recognizing that indigenous peoples and local communities may or may not live life fully compatible with the conservation of the environment, and given that both foundations and interests-holders – indigenous peoples and local communities; and the environment – have equal standing, biocultural rights not only hand their holders a set of positive legal positions but also a set of duties concerning the protection of the environment: as holders of biocultural rights, indigenous peoples and local communities are required to act as stewards of their lands and natural resources, and their rights to self-governance and cultural identity are to be exercised in ways that are not detrimental to the conservation of the environment: each foundation acts as a limit to the other.[31]

Limiting a set of indigenous peoples and local communities' human rights to environmental considerations raises many concerns. Is it fair? Or is it an instrument to shift burdens and responsibilities, once more, to the most vulnerable and less environment-detrimental peoples of the world? People or community remains, of course, free to renounce the duties biocultural rights come along with, but this will be at the cost of also renouncing the rights the biocultural rights basket contains.

To fully comprehend the implications of biocultural rights, regard needs to be given to the distinction, set out above, between indigenous peoples and local communities. While indigenous peoples hold indigenous rights – that are not linked to environmental considerations (be sustainable, be relevant for the conservation of biodiversity, etc.) – local communities' rights are still strongly conditional on their contribution to environmental considerations. This difference places them in dissimilar positions vis-à-vis biocultural rights claims.

Biocultural rights, as a framework, as an idea, as a right *de jure condendo*, are a card that could be used by local communities to acquire new rights in – or obtain stronger protection of those already recognized by – international and national laws. Local communities may conflate their different needs, interests, and desires into biocultural rights, accepting their environmental conditionality as an old "condition". One that is already present in the other collective rights, they are being recognized by international law.

Positive notes may be there for indigenous peoples as well. It is still often the case that states do not recognize indigenous peoples' rights or deny indigenous status to any indigenous people residing in its territory for political reasons, linked to the fear of self-determination (most often unnecessarily, as indigenous peoples claim internal self-determination). Compared with indigenous peoples' rights, biocultural rights' lack of emphasis on political issues makes them a more neutral tool vis-à-vis government still imbued with colonial hatreds, afraid of secessionist claims, and determined to deny the existence of (certain) indigenous peoples in their territories.[32] On such occasions, biocultural rights could be strategically employed as a tool to require at least some – environmentally conditional – protection. However, and importantly, biocultural rights for indigenous peoples must remain a second-best route to be used only when and if indigenous rights are, for one reason or another, impaired, ignored, or threatened by environmental considerations (such as the creation of an important protected area). Indigenous peoples' rights already include all the rights found in the biocultural rights basket and are not strictly conditional or limited to the protection of the environment: they are the most advantageous route to be used by indigenous peoples.

Besides the strategic considerations that indigenous peoples and local communities need to make when deciding whether to claim biocultural rights or not, it must be acknowledged that the concept of biocultural rights proposes a new way to combine human and environmental considerations. They incorporate both human's and nature's needs into one, powerful, legal construct: rights and their foundations. In this way, they suggest a possible route to move in the direction of human rights that not only embed the planetary boundaries of the Earth but that also strive to guide human behavior toward sustainable paths. However, their focus on indigenous peoples may limit any potential for development. It is probably better to disentangle them from indigenous peoples, whose rights are still widely threatened, so often disregarded, and vulnerable to political shifts and economic interests. A possible way forward would be to develop biocultural rights in the direction of rights to be granted in return for environmental stewardship duties of local communities, whether they may be farmers, urban dwellers, ecovillages, or fishermen; whether they may be in the so-called South of the world or its North, and whether they are already holders of other collective rights or not.

Notes

1 This project has received funding from the European Union's Horizon 2020 research and innovation program under the Marie Skłodowska-Curie grant agreement No 841546. Email: giulia.sajeva@strath.ac.uk. *All URLs retrieved on 1 September 2021.

2 There are many proposed definitions. The most commonly used is the one proposed in the *Study of the Problem of Discrimination Against Indigenous Populations*, commissioned by the United Nations in 1986, which underlines as the first and foremost of characteristics necessary to identify indigenous peoples the "historical continuity with pre-invasion and pre-colonial societies that developed on their territories" (UN Doc. E/CN.4/Sub.2/1986/7 and Add. 1–4, para. 379).

3 The peculiar features of indigenous cultures, institutions, and traditions have also been central in the description of indigenous peoples in the ILO Convention 169. For a closer look at these features, see Sajeva et al. (2019, p. 10).

4 Indigenous peoples have traditionally been claiming *internal* self-determination, which is aimed at asserting identities, preserving languages, cultures, and practices, and maintaining traditional governance structures (Anaya, 1996, p. 111) rather than at obtaining political independence from the State where they reside.

5 The use of the plural of the term people in the UNDRIP is a very important achievement because *peoples* are subjects of international law, and are subjects whose right to self-determination is recognized by the UN Charter (Art. 1.3) and by the two 1966 International Covenant on Civil and Political Rights and on Economic, Social and Cultural Rights (Art. 1.1).

6 From now on, ILO 169.

7 Recognized also by the Universal Declaration on Human Rights (1948), the 1966 Covenants, the International Convention on the Elimination of All Forms of Racial Discrimination (1966), the UN Declaration on the Rights of Persons Belonging to National or Ethnic, Religious and Linguistic Minorities (1992).

8 Recognized also by the UN Declaration on the Right to Development (1986).

9 In 2011, the CBD Working Group on Article 8(j) and related provisions of the Convention on Biological Diversity proposed a list of features a local community could possess "reflecting its own unique cultural, ecological and social circumstances" (p. 12), with particular focus on characteristics linked to traditional ecological knowledge, and environmental sustainability. These features very much resemble those usually employed to describe indigenous people, with the sole (important) exclusion of the existence of links with pre-colonial societies (Convention on Biological Diversity 2011).

10 It is to be noted that two decisions of the Inter-American Court on Human Rights went the opposite direction. In the 2005 Moiwana Village v. Suriname case (IACtHR, *Case of the Moiwana Community v. Suriname*, Series C No. 124 (15 June 2005)), and the 2007 Saramaka People v. Suriname case (IACtHR, *Case of the Saramaka People v Suriname*, Series C No 172 (28 November 2007)), the Court treated two local communities composed of descendants of Africans forcibly taken to South America during the 17th-century colonization as holders of the same rights of indigenous peoples. The Court employed the same rationale used for the Mayagna (Sumo) Awas Tingni Community v. Nicaragua case that dealt with indigenous peoples (IACtHR, *Nicaragua, Case of the Mayagna (Sumo) Awas Tingni Community v. Nicaragua*, Series C No. 79 (31 August 2001)). However, besides similar special cases, local communities cannot yet be considered holders of indigenous peoples' rights.

11 Among which, the 1966 Covenants on Civil and Political Rights, and on Economic, Social and Cultural Rights; the 1966 Convention on the Elimination of All Forms of Racial Discrimination; the 1979 Convention on the Elimination of All Forms of Discrimination Against Women.

12 An exception may be found in Art. 27 of the International Covenant on Civil and Political Rights that provides for both individual and collective rights: "persons belonging to minorities shall not be denied the right, *in community* with the other members of their group, to enjoy their own culture, to profess and practise their own religion, or to use their own language" (emphasis added).

13 See, for example, the CBD Strategic Plan for 2011–2020 (UNEP/CBD/COP/DEC/X/2 29 October 2010), target 14 and 18; Decisions XII/12 on the implementation of the program of work on Article 8(j) and related provisions and mechanisms to promote the effective participation of indigenous and local communities (UNEP/CBD/COP/DEC/XII/12 13 October 2014); Decision VII/28 on Protected Areas and art. 8(j) (UNEP/CBD/COP/DEC/VII/28 13 April 2004); the 2004 Akwé: Kon Voluntary Guidelines (UNEP/CBD/COP/DEC/VII/16 13 April 2004) and the 2006 Voluntary Guidelines on Biodiversity-inclusive Impact Assessments (UNEP/CBD/COP/DEC/VIII/28 15 June 2006) (also see Morgera, 2016; 2020).

14 On the interpretation of these stipulations, see Chapter 1.

15 A further differentiating feature of local communities is its "traditional" character (see Sajeva et al., 2019, p. 15).

16 See Larking (2019); and the official website at https://viacampesina.org/en/. On the unique process that led to the adoption of UNDROP through the "almost" direct representation of agrarian movements, see Claeys (2019).

17 On the difference between land ownership and land tenure, see Sajeva et al. (2019), "Land Tenure".

18 Mostly within the Covenants on Civil and Political Rights and Economic, Social and Cultural Rights, but also in the UN Convention on the Elimination of All Forms of Discrimination Against Women (1979) (see Golay, 2015, pp. 20 *et seq.*).

19 See Paoloni and Vezzani (2019, p. 24), for a comment on whether UNDROP may question the universality of human rights by "granting" to a specific category (peasants and local communities) rights which had already been recognized to all human beings. Interestingly, during the negotiations of UNDROP, the EU acknowledged that peasants "make an important contribution to the protection of the environment" but expressed concern about recognizing new human rights to local communities because they "would undermine the universality of human rights" (EU General Statement at the 4th Session of the Open-Ended Intergovernmental Working Group on UNDROP, 18–19 May 2017). This statement illuminates a concern about those who do not belong to local communities who would be excluded from these new rights. On the evolution on UNDROP as a claim for non-exclusion of existing human rights, to a claim for the recognition of new rights, see Edelman & James (2011).

20 54 states abstained and 8 voted against the adoption.

21 On the possibility of applying the supremacy of human rights norms over norms regulating different matters in the context of UNDROP and intellectual property rights and other rights protecting seeds of plant for food and agriculture, see Golay and Bessa (2019, p. 25) and Golay (2019).

22 ICCAs – a term that is not an acronym – are "territories and areas conserved by indigenous peoples and local communities", also called "territories of life" (Sajeva, 2019).

23 See for example the famous proposal of Wilson (2016) to turn *half the Earth* into a strictly protected area. For a critique on this proposal see Kothari (2020), and for a much-needed plea for a different approach to conservation see Fischer et al. (2017).

24 In his book, Bavikatte (2015) mentions the confluence of three streams: the post-development movement; the commons movement, and movement for the recognition of the rights of indigenous peoples and local communities (ibid., p. 16).

25 Biocultural rights may or may not move to the next stage of a fully recognized human right in international law, however, they deserve attention to understand their potential implications for indigenous peoples and local communities.

26 In 2016, in Opinion T-622, the Colombian Supreme Court made reference to Bavikatte's work and acknowledged the biocultural rights of the peoples and communities living alongside the Atrato River (see Castillo Galvis et al., 2019). On this case, also see, this book, Chapters 1 and 11.

27 Interest in the idea of biocultural rights has largely increased in the last few years, particularly because of the Colombia Supreme Court's first explicit reference to them in the Atrato case. For literature making reference to biocultural rights, see Macpherson et al. (2020); Carrillo Yap (2021); Gimenez (2020); Millaleo Hernández (2020); and Gilbert (2018).

28 The foundations of a right are here understood as the very reasons/interests/needs that a right was grounded upon or was recognized for (whether these are moral, legal, or political in nature).

29 Whether it is appropriate or not to treat nature as a holder of rights is a complex issue that is not relevant for the purpose of this paper. For the treatment of its main contours, see Stone (1972); Corrigan and Oksanen (2021); Studley and Bleisch (2018); and Kauffman and Martin (2019).

30 Quite explicatory is Article 8(j) of the Convention on Biological Diversity:

> Each contracting Party shall, as far as possible and as appropriate: [...] Subject to national legislation, respect, preserve and maintain knowledge, innovations, and practices of indigenous and local communities embodying traditional lifestyles relevant for the conservation and sustainable use of biological diversity;

as well as Article 10(c): "Each Contracting Party shall, as far as possible and as appropriate: [...] Protect and encourage customary use of biological resources in accordance with traditional cultural practices that are compatible with conservation or sustainable use requirements".

31 Of course, the rights in the biocultural rights basket are also limited – as all other human rights – by other human rights and by interests and values that are considered as particularly important for the general interest. For an analysis of the difference between classic external limits and biocultural rights internal limits, see Sajeva (2019).

32 See Sajeva (2018, Chapter 5).

References

Abrusci, E. (2017). Judicial Fragmentation on Indigenous Property Rights: Causes, Consequences and Solutions. *International Journal of Human Rights, 21*(5), 550–564.

Adams, W. M., & Milligan, M. (Eds.) (2003). *Decolonizing Nature. Strategies for Conservation in a Post-Colonial Era*. Earthscan.

Anaya, J. (1996). *Indigenous Peoples in International Law.* Oxford University Press.
Bavikatte, S. K. (2014). *Stewarding the Earth. Rethinking Property and the Emergence of Biocultural Rights.* Oxford University Press.
Bavikatte, S. K., & Bennett, T. (2015). Community Stewardship: The Foundation of Biocultural Rights. *Journal of Human Rights and the Environment, 6*(1), 7–29
Berkes, F. (2001). Religious Traditions and Biodiversity. In S. A. Levin (Ed.), *Encyclopedia of Biodiversity* (Vol. 5, pp. 109–120). Elsevier.
Bessa, A. (2015). Traditional Local Communities: What Lessons Can Be Learnt at the International Level from the Experiences of Brazil and Scotland? *Review of European, Comparative & International Environmental Law, 24*(3), 330–340.
Boyd, D. R. (2019). Right to a Healthy Environment: Good Practices. *Report of the Special Rapporteur on the Issue of Human Rights Obligations Relating to the Enjoyment of a Safe, Clean, Healthy and Sustainable Environment.* UN Doc A/HRC/43/53, 2019.
Brechin, S. R., Wilshusen, P. R., Fortwangler, C. L., & West, P. C. (2002). Beyond the Square Wheel: Toward a More Comprehensive Understanding of Biodiversity Conservation as Social and Political Process. *Society and Natural Resources, 15*(1), 41–64.
Carrillo Yap, S. L. (2021). The Role of the UNESCO Convention for the Safeguarding of the Intangible Cultural Heritage (ICH Convention) in the Protection of Traditional Forest-Related Knowledge (TFRK) of Amazonian Indigenous Peoples. *International Journal of Human Rights, 25*(5), 853–869.
Castillo Galvis, S. H., D'Janon Donado, M. L., & Ramírez Nárdiz, A. (2019). El control de convencionalidad y el diálogo judicial frente al medio ambiente como sujeto de protección y reparación. *Revista Mexicana de Derecho Constitucionales, 41*, 397–428.
Claeys, P. (2019). The Rise of New Rights for Peasants. From Reliance on NGO Intermediaries to Direct Representation. *Transnational Legal Theory, 9*(3–4), 386–399.
Claeys, P., & Edelman, M. (2020). The United Nations Declaration on the Rights of Peasants and Other People Working in Rural Areas. *Journal of Peasant Studies, 47*(1), 1–68.
Cocks, M. (2006). Biocultural Diversity: Moving Beyond the Realm of "Indigenous" and "Local" People. *Human Ecology, 34*(2), 185–200.
Convention on Biological Diversity. (2011). Report of the Expert Group Meeting of Local Community Representatives Within the Context of Article 8j and Related Provisions of the Convention on Biological Diversity. UNEP/CBD/WG8J/7/8/Add.1.
Corrigan, D., & Oksanen, M. (Eds.) (2021). *Rights of Nature. A Re-examination.* Routledge.
De Shutter, O., & Rajagopal, B. (2020). Conclusion. The Revival of the "commons" and the Redefinition of Property Rights. In O. De Shutter & B. Rajagopal (Eds.), *Property Rights from Below. Commodification of Land and the Counter-Movement* (pp. 203–231). Routledge.
Dowie, M. (2009). *Conservation Refugees. The Hundred-Year Conflict Between Global Conservation and Native Peoples.* The MIT Press.
Edelman, M., & Carwil, J. (2011). Peasants' Rights and the UN System: Quixotic Struggle? Or Emancipatory Idea Whose Time Has Come? *Journal of Peasant Studies, 38*(1), 81–108.
Ellingson, T. J. (2001). *The Myth of the Noble Savage.* University of California Press.
Ergas, C. (2010). A Model of Sustainable Living: Collective Identity in an Urban Ecovillage. *Organization & Environment, 23*(1), 32–54.
Fischer, J., Meacham, M., & Queiroz, C. (2017). A Plea for Multifunctional Landscapes. *Frontiers in Ecology and the Environment, 15*(2), 59.
Frascaroli, F., Bhagwat, S., Guarino, R., Chiarucci, A., & Schmid, B. (2016). Shrines in Central Italy Conserve Plant Diversity and Large Trees. *Ambio, 45*(4), 468–479.
Frison, C. (2018). *Redesigning the Global Seed Commons. Law and Policy for Agrobiodiversity and Food Security.* Routledge.

Gilbert, J. (2018). *Natural Resources and Human Rights*. Oxford University Press.
Gimenez, T. V. (2020). De la justicia climàtica a la justicia ecològica: los derechos de la naturaleza. *Revista Catalana de Dret Ambiental, 11*(2), 1–42.
Golay, C. (2013). *Legal Reflection on the Rights of Peasants and Other People Working in Rural Areas*. Background Paper Prepared for the First Session of the Working Group on the Rights of Peasants and Other People Working in Rural Areas (15–19 July 2013).
Golay, C. (2015). *Negotiation of a United Nations Declaration on the Rights of Peasants and Other People Working in Rural Areas*. Research Brief. Geneva Academy of International Humanitarian Law and Human Rights.
Golay, C. (2019). *The Implementation of the United Nations Declaration on the Rights of Peasants and Other People Working in Rural Areas*. Research Brief. The Geneva Academy of International Humanitarian Law and Human Rights.
Golay, C., & Bessa, A. (2019). *The Right to Seeds in Europe*. Research Brief. The Geneva Academy of International Humanitarian Law and Human Rights.
Haugen, H. M. (2020). The UN Declaration on Peasants' Rights (UNDROP): Is Article 19 on Seed Rights Adequately Balancing Intellectual Property Rights and the Right to Food? *Journal of World Intellectual Property, 23*(3–4), 288–309.
ILC Conclusions (2006). ILC, Report of the Study Group, Fragmentation of International Law: Difficulties Arising from the Diversification and Expansion of International Law: Conclusions (A/CN.4/L.702) (18 July 2006).
Jonas, H. (2020). *The Living Convention. Volume II: A Methodology for Counter-mapping and Recentering International Law*. (3rd ed.). Natural Justice.
Jonas, H., & Godio, M. J. (Eds.). (2020). *The Living Convention. Volume I: An Accessible compendium of the International Rights of Indigenous Peoples, Local Communities and Peasants*. (3rd ed.). Natural Justice.
Jonas, H., Jonas, H., & Makagon, E. J. (2014). Introducing the Living Convention and the Landscape Approach to Legal Empowerment. In W. Littlechild & E. Stamatopoulou (Eds.), *Indigenous Peoples' Access to Justice, Including Truth and Reconciliation Processes* (pp. 390–406). Columbia University, Institute for the Study of Human Rights.
Kauffman, C. M., & Martin, P. L. (2019). How Courts are Developing River Rights Jurisprudence: Comparing Guardianship in New Zealand, Colombia, and India. *Vermont Journal of Environmental Law, 20*, 260–289.
Knox, J. H. (2017). *Report of the Special Rapporteur on the Issue of Human Rights Obligations Relating to the Enjoyment of a Safe, Clean, Healthy and Sustainable Environment*. A/HRC/34/49.
Knox, J. H. (2018). *Report of the Special Rapporteur on the Issue of Human Rights Obligations Relating to the Enjoyment of a Safe, Clean, Healthy and Sustainable Environment*. A/HRC/37/59.
Kothari, A. (2020, January 13). Colonial Conservation in New Avatars Nature Conservation Approaches Will Work Only if We Move Towards Life-Affirming Ways of Being and Doing. *Economy & Politics*. https://wsimag.com/economy-and-politics/64574-colonial-conservation-in-new-avatars.
Lawson, C. (2015). Implementing Farmers' Rights: Finding Meaning and Purpose for the International Treaty on Plant Genetic Resources for Food and Agriculture Commitments? *European Intellectual Property Review, 37*(7), 442–454.
MacKenzie, J. M. (1988). *The Empire of Nature: Hunting, Conservation and British Imperialism*. Manchester University Press.
Macpherson, E., Ventura, J. T., & Ospina, F. C. (2020). Constitutional Law, Ecosystems, and Indigenous Peoples in Colombia: Biocultural Rights and Legal Subjects. *Transnational Environmental Law*, 9(3), 521–540.

Millaleo Hernández, S. (2020). Guarda de la Naturaleza: Conocimientos Ecológicos Tradicionales de los Pueblos Indígenas y Estrategias de Protección. *Cadernos de Dereito Actual, 13*, 202–230.

Morgera, E. (2014). Against All Odds: The Contribution of the Convention on Biological Diversity to International Human Rights Law. In D. Alland, V. Chetail, O. de Frouville, & J. E. Viñuales (Eds.), *Unity and Diversity of International Law* (pp. 983–995). Martinus Nijhoff Publishers.

Morgera, E. (2016). The Need for an International Legal Concept of Fair and Equitable Benefit-Sharing. *The European Journal of International Law,* 27(2), 353–383.

Morgera, E. (2020). *Biodiversity as a Human Right and its Implications for the EU's External Action*. European Parliament. Directorate-General for External Policies of the Union.

Mrema, E. M. (2020). *Statement by Ms. Elizabeth Maruma Mrema, Executive Secretary of the CBD on the occasion of the opening of the Second Global Thematic Dialogue for Indigenous Peoples and Local Communities on the Post-2020 Global Biodiversity Framework* (1 December 2020). https://www.cbd.int/doc/speech/2020/sp-2020-12-01-iplc-en.pdf.

Paoloni, L., & Vezzani, S. (2019). La Dichiarazione ONU sui diritti dei contadini e delle altre persone che lavorano nelle aree rurali: prime riflessioni. *Federalismi, 1*, 1–33.

Payandeh, M. (2015). Fragmentation within International Human Rights Law. In M. T. Andenas & E. Bjorge (Eds.), *A Farewell to Fragmentation: Reassertion and Convergence in International Law* (pp. 297–319). Cambridge University Press.

Posey, D. A. (Ed.). (1999). *Cultural and Spiritual Values of Biodiversity. A Complementary Contribution to the Global Biodiversity Assessment*. Intermediate Technology Publication.

Sajeva, G. (2018). *When Rights Embrace Responsibilities. Biocultural Rights and the Conservation of Environment*. Oxford University Press.

Sajeva, G. (2019). Inside-out. Internal and External Limits to Rights: Does it Matter?. *Diritto & Questioni Pubbliche, 19*(2), 201–223.

Sajeva, G., Borrini-Feyerabend, G., & Niederberberger, T. (2019). *Meanings and More... Policy Brief of the ICCA Consortium*. Issue no. 7.

Stone, C. (1972). Should Trees Have Standing? Toward Legal Rights for Natural Objects. *South California Law Review, 45*, 450–501.

Studley, J., & Bleisch, W. V. (2018). Juristic Personhood for Sacred Natural Sites: a Potential Means for Protecting Nature. *Parks Journal, 24*(1), 81–96.

Tauli-Corpuz, V. (2016). *Report of the Special Rapporteur of the Human Rights Council on the Rights of Indigenous Peoples*. General Assembly (A/71/229).

UN Department of Economic and Social Affairs. (2019). *State of the World's Indigenous Peoples. Implementing the United Nations Declaration on the Rights of Indigenous Peoples. Volume IV*. United Nations Publication (Vol. ST/ESA/371).

Verschuuren, B., Wild, R., McNeely, J., & Oviedo, G. (Eds.) (2010). *Sacred Natural Sites. Conserving Nature & Culture*. Earthscan.

Wilshusen, P. R., Brechin, S. R., Fortwangler, C. L., & West, P. C. (2002). Reinventing a Square Wheel: Critique of a Resurgent "Protection Paradigm" in International Biodiversity Conservation. *Society & Natural Resources, 15*(1), 17–40.

Wilson, E. O. (2016). *Half-Earth. Our Planet's Fight for Life*. Liveright Publishign Corporation.

Ziegler, J., de Kalbermatten, G., Ortega, L., Golay, C., Mahon, C., Way, S.-A., & Büschi, M. (2008). *Human Rights and Desertification. Exploring the Complementarity of International Human Rights Law and the United Nations Convention to Combat Desertification*. Desertification, land degradation and drought issue paper No. 1. Secretariat of the United Nations Convention to Combat Desertification.

PART 2

Biocultural Community Protocols, Access and Benefit-Sharing, and Beyond

7

COMMUNITY PROTOCOLS AS TOOLS FOR COLLECTIVE ACTION BEYOND LEGAL PLURALISM – THE CASE OF TRACKS IN THE SALT

Pía Marchegiani and Louisa Parks[1]

Introduction

This chapter explores the Kachi Yupi community protocol as a case study that provides a good basis for a discussion of these tools within a broad frame of local activism. In this case, a community protocol was used to underpin local community action for the stewardship of the earth, and this reveals how a community protocol can be linked to other forms of collective action in a longer timeframe. It also allows us to reflect on when and why community protocols might lead to legal change, foster dialogue with various institutional actors, or lead to other forms of collective action. The Kachi Yupi community protocol was produced by Kolla and Atacama communities in the Salinas Grandes and Laguna de Guayatayoc area of northwest Argentina. For these communities, drafting, publishing and following up on this community protocol constituted one form of collective action within a wider campaign to protect their lands, waters and livelihoods. The chapter draws on legal scholarship and seeks to place community protocols in the perspective of literature on collective action, drawing in particular on political and legal opportunity structure approaches to explain community action choices of different types.

We explore the case as follows. First, in the remainder of this introduction, we describe the political and legal contexts that preceded and framed the decision to draft the community protocol. We then give the reader an overview of early community actions as well as the process of drafting the community protocol, its aims and content. The discussion of the protocol then begins by considering the effects and impacts it had following its publication, before a reflection on why it did not become a tool for legal pluralism, but did become a key basis for other, including more contentious, forms of action. The latter reflection provides ideas about what conditions are needed for legal change to follow from community

DOI: 10.4324/9781003172642-9

protocols, as well as how protocols can contribute to activism of other types, which may also help communities to achieve their aims.

The Kachi Yupi/Tracks in the Salt community protocol was drafted by 33 communities from the Salinas Grandes and Laguna de Guayatayoc area: a shared watershed that straddles the two provinces of Salta and Jujuy in the northeast of Argentina. It was published in December 2015. To better interpret the protocol, it is useful to first place it against the backdrop of the political and legal contexts surrounding lithium mining and environmental regulation in Argentina. Lithium mining has been underway since 1997 in the country, but mining increased markedly from around 2010. The reasons for this are rooted in the increase in global demand and interest in the mineral that have accompanied the global push to move away from a fossil fuel-based global economy. Lithium, indeed, is the key component of li-ion batteries. These are used in key modern technologies like laptop computers and mobile devices and, crucially, in electric cars. Lithium thus plays a key role in current visions of the path away from fossil fuels. It is a metal that is technically challenging to extract, however, and one of the most important accessible sources is the "lithium triangle", a cluster of salt flats located high in the Andes where the borders of Argentina, Bolivia and Chile meet. Argentina is the only of these three countries in which lithium extraction can be freely licenced and has received much attention from international investors, seen, in turn, as opportunities for development by local authorities. Together, these reasons explain Argentina's push to grow its lithium sector and why the country became the fastest growing provider of lithium in 2016, when its global share in the market grew from 11% to 16%. Both before and after this increase, the government in Argentina actively pursued lithium mining projects and sought to boost the sector, creating public companies for battery production, and axing federal taxes on mineral exports, for example (see Marchegiani et al., 2018 for a full discussion).

The push for the "white gold" of lithium translated into the granting of exploratory and mining permits in the Argentinian area of the triangle, including the Salinas Grandes and Guayatayoc area discussed here. These permits raised questions of how local communities could effectively make their voices and concerns heard in procedures surrounding the granting of eventual permits. Thus, the legal situation surrounding mining permits and environmental impacts is another necessary piece of the puzzle for a more complete interpretation of the local community actions undertaken in this vein.

Argentina is a federal state and as such has distributed power between the central state and its autonomous provinces. The federal state retains all powers not specifically delegated in the provinces. Moreover, with the 1994 Constitutional reform, the country conferred supremacy to international human rights within its domestic legal framework. As such, Argentina has ratified international instruments relevant to the inclusion of local points of view in mining permit decisions. It ratified ILO Convention 169,[2] which recognizes indigenous and tribal peoples' right to participate in the use, management and conservation of natural resources pertaining to their lands, in 2000, and this has helped to underpin the case law of the Inter-American Court of Human Rights filed by communities in Salinas Grandes[3] (Marchegiani et al., 2020). In 2007, it signed the United Nations Declaration on the Rights of Indigenous Peoples (UNDRIP), which also contains legal provisions on consultation and consent.

In the same 1994 Constitutional reform, Argentina introduced two other changes with significant impacts on questions of local community involvement and mining decisions. First, the Constitutional reform introduced the right to a healthy environment,[4] which, in turn, opened the way for minimum standards for environmental protection to be set. Following the federal organization of the State, the Constitution gives power to the central State to set minimum standards of environmental protection for all citizens in the country. Provinces may regulate beyond these, increasing protection standards, but may not limit or regulate to standards lower than them.

Second, the reform recognized the ethnic and cultural pre-existence of indigenous peoples and community rights over lands traditionally occupied by them.[5] This opened the way for provincial governments to transfer communal land rights to indigenous communities. Although these transfers have not transpired in full for a host of reasons, including a lack of information about indigenous communities, a lack of State capacity, budget and political will, the reform remains important for the case in hand. This is because, as noted by former UN Special Rapporteur on the rights of indigenous peoples, James Anaya, the Argentinean Mining Code "requires the permission of the land 'owners' to explore for minerals" (Anaya, 2012, para. 45).[6] Although the Mining Code does not contain any other requirement for consultation, this gap is arguably filled by the existing environmental framework which calls for participation (Marchegiani et al., 2020).

The 2002 General Environmental Protection Law sets these minimum standards. This law introduced environmental policy principles to be mainstreamed across all policy areas, and a slew of tools for environmental management, including the environmental impact assessment (EIA) processes applied in the case of lithium mining in the Salinas Grandes and Laguna de Guayatayoc area discussed here. The new EIA tools were accompanied by minimum standards to be followed in their implementation in the provinces. EIAs are obligatory, and a report on their outcomes must be published before any activity with significant impacts on the environment or on local populations' quality of life (or both) begins.[7] The process itself must comprise a participatory phase allowing citizens to debate the proposed activity and its implications. Other minimum standards concern access to information. EIAs must include a statement from the proponent explaining the activity and its environmental impacts, and they must include a report identifying impacts and mitigation measures. Information must be provided in a timely fashion for the participatory phase. Finally, a public authority must make a decision about the proposed activity.

These minimum standards are then fleshed out in more detail at the provincial level. Here, however, we find lags in the implementation of both national and international laws. The province of Jujuy, the power in question for this chapter, named its Environmental Management Unit as the relevant decision-making body for mining activities and set up a procedure for an EIA (with provincial decree 5722-2010). The Unit comprises members from a range of provincial agencies (such as human rights, environmental management and industry) and stakeholders (such as geologists and indigenous communities). A specific procedure was decided for lithium mining in 2011, when this resource was declared a strategic natural resource for the province in line with federal government policy. An additional layer of review by an expert committee was to apply for proposed lithium mining projects. However, this additional review was abrogated in 2019.

This complex legal patchwork is also affected by significant implementation gaps. Local indigenous communities in the Salinas Grandes and Laguna de Guayatayoc have contested the correct application of EIA procedures, as access to information was not granted when different companies sought permits to explore in the area, and a consultation process in line with existing regulation was not devised and/or implemented. The government of Jujuy later claimed that as there was no specific provincial law to implement consultation as required in Convention 169 ILO and other international instruments, there were no clear provisions about how to consult communities to be followed or disregarded.

The next section discusses the actions taken to address these perceived shortcomings, with particular emphasis on the community protocol the indigenous communities drafted with a view to outlining rules for consultation.

Action on Lithium Mining – from Courts to the Community Protocol

Around 2010, mining companies began exploration activities for lithium in the Salinas Grandes and Laguna de Guayatayoc area, and the wheels of EIA processes were to be set in motion. However, as mentioned, this was done without providing enough information to communities and without opening proper consultation processes. As exploration activities began, community members began to organize with the aim of obtaining information on these activities. At an early stage, they formed a Roundtable of the Indigenous Peoples of Salinas Grandes and Laguna de Guayatayoc (the Roundtable) comprising representatives from 33 communities in the area (Comunidades indígenas de las Salinas Grandes y Laguna de Guayatayoc de Jujuy y Salta, 2015; Parks, 2020). The Roundtable was created as a main forum for community members to share information and organize collective action. Lithium was a prominent issue, but the Roundtable was created to discuss any matter that could pose a threat to the defence of their territory.

One of the main concerns expressed by community members about lithium mining centred on the possible water use and its effects. The communities had not received information about water use and impacts on local supplies at the exploration stage. Many of these communities depend on water for their livelihoods as well as normal use – for their artisanal salt extraction activities, but also for their small-scale agriculture and livestock. Fresh water is a resource in short supply in the area, which is one of the most arid on the planet. A central claim was thus for information about how much water would be used in mining, and about risks that saltwater could be introduced into fresh water sources.

From the beginning, then, the actions of the communities did not focus necessarily against mining, but rather against the failure to consult and inform them as rights-holders on their lands. The communities' main demand centred on their recognition as rights-holders. Although communities in the area hold legal status as indigenous communities, and as such have the right to the land in which they live, community land rights titles have not officially been granted and hence, a struggle for recognition was undertaken.

The recognition of communities as such, and especially the right to be consulted as rights-holders over their land, formed the basis of their first legal action. With the assistance of community lawyers based in the province, the communities filed complaints before Argentina's Supreme Court of Justice in 2010 (Ferradás Abalo et al., 2016). The case was rejected in a sentence that argued that there was no "case" due to the lack of factual evidence and enough proof provided by the communities on the existence of mining permits on their land that would enable the development of an FPIC process. The government of Jujuy had denied in the audience called by the Supreme Court on the 28th of March 2012 the existence of any permit granted by its mining authorities, and suggested that any company intervention in the area was not officially authorized.[8]

When the case was rejected, the communities filed another complaint, this time before the Inter-American Commission on Human Rights. In late 2011, the group also made presentations to the UN Economic Social and Cultural Rights Committee and to the UN Special Rapporteur Anaya, who visited the country at this time (Ibid).

After the rejection of the national case, and based on the fact that the government of Jujuy claimed there was no specific provincial or national regulation that would specify how consultation should be implemented, and how it should be done in the context of EIA process, which was true, the communities decided to proactively devise a way forward that would call for a substantial dialogue in a way that would support their legal claims.

This, combined with the experiences of other local communities in the area and exchanges with other external organizations, underpinned the decision to draft a community protocol. The communities were aware of similar processes that had unfolded at other salt flats, including the relatively nearby Olaroz flat. There, a small group of concerned community members had tried to express their concerns via EIA processes, but were frustrated with how these had been carried out – more as information meetings rather than opportunities for dialogue, but without the provision of accessible information before local meetings, which were also accompanied with various promises tied to community consent (for a full discussion, see Marchegiani et al., 2020).

In this context, the Roundtable decided to take steps to provide specific instructions on how to conduct a consultation process. The choice to address the gap in provincial law around specific provisions was clearly linked to the rejection of the case they had brought at the national level. Yet, the communities also wanted to move beyond this more bureaucratic view, and felt that consultations should respect their culture. Working once more with locally based lawyers as well as a national NGO, the Foundation for the Environment and Natural Resources (FARN) based in Buenos Aires, the communities began to consider a community protocol as a possible way forward that would allow them to frame a specific consultation process, linking it with their rights and culture and underpinning their claims with reference to different levels of law, including international sources.

As for the decision to engage in drafting a community protocol, this was not taken quickly, nor was it without controversy. Some representatives felt that outlining a process for engaging with external actors could be construed as an indication that communities would end up giving their consent – that consent would become the foregone conclusion of any dialogue in the eyes of authorities and those wishing to access their lands and resources. Nevertheless, the decision was made to begin the process. The intention for the protocol was an emerging strategy in the context of a lack of implementation of consultation rights as discussed. Communities decided to address the consultation conundrum by laying foundations for dialogue: by providing information about the communities and their worldview, underlining their knowledge of their rights and outlining a clear and detailed process – including the point that consent was by no means a foregone conclusion of consultation – that all external actors should respect and follow.

These decisions about the content of the protocol were developed in participatory ways. Before drafting the protocol, a series of workshops were carried out in each of the communities with legal experts who first explained the legal provisions on consultations to every community member interested, then collected their initial ideas about what a proper consultation process that respected their views would look like. After these initial workshops, a small group of around 15 community members was selected to drive the drafting process for the protocol on a consensus decision-making basis involving all of the communities. The group divided into teams formed to work on specific parts of the text, but the group as a whole answered to the Roundtable, and the process was also discussed with broader sections of the community on regular occasions. The process of drafting the protocol took nearly two years, from early 2014 to its publication in late 2015, and also included meetings between the small drafting group, local lawyers, FARN and other external organizations, general meetings and gatherings in the Roundtable space, training sessions on community protocols (led by the South African-based NGO Natural Justice), thematic workshops, workshops on the text of the draft protocol and the final consensus-based approval process.

Kachi Yupi in Detail

The resulting protocol is divided into an introduction and three distinct chapters (Comunidades indígenas de las Salinas Grandes y Laguna de Guayatayoc de Jujuy y Salta, 2015). The first chapter places consultation and participation processes in the context of local history. It describes Argentina as a State run by descendants of colonial power-holders. Those in power are described as responsible for the continued trend of exploitation of indigenous peoples through taxes and other means, and to understand this, the history of struggle of the communities must be recalled. The document thus describes the history of community struggles against colonial powers, including battles, the forced migration that occurred with the construction of a railway line in the area, and mining activities that obliged many to abandon traditional livelihoods. This history of oppression underlies the fundamental aim of the 33 communities to be recognized and allowed to enjoy their rights, including through proper consultation and FPIC. In this vein, as mentioned, the communities are recognized as indigenous and as pre-dating the existing State. However, the implementation of their rights deriving from this, ranging from the allocation of their communal land titles to the right to participate in decision-making, is still at an early stage. The historical context serves to substantiate the need for the State to ensure a heightened level of protection in the context of an FPIC process. In addition, the protocol calls upon the State to ensure both the transparency of the consultation process and genuine participation by indigenous communities, as well as necessary support to them.

The second chapter then recalls the rights recognized to indigenous peoples, placing these firmly in the context of the specific historical experience of the local communities. In other words, it ties national and international laws that alone appear technical, abstract and oblivious to local reality, to the communities' history. In particular, this chapter serves the purpose of articulating the communities' views of an ideal consultation and participation process clearly and in such a way as to assert their rights as recognized in Argentina's Constitution as well as international human rights treaties, declarations and conventions, particularly ILO Convention 169 and the UNDRIP. The third chapter then outlines the procedure of consultation and FPIC agreed by the communities. Importantly, this chapter describes this procedure in a comparison with the natural cycle of salt formation – once again, the more abstract procedure is firmly tied to the land, history and culture of the communities. A key point in the procedure described is that salt is not an economic or nutritional good alone, but a living thing. Thus, the protocol underlines that consultation is (or should be) a living, dynamic process. It also makes it clear that the basis for any consent is the compatibility of a project with *Buen Vivir*, which is "the process of full communal life on our land. It is being one and the same with the communities from its very roots. To achieve *Buen Vivir* means knowing how to live and thus how to live with others" (Comunidades indígenas de las Salinas Grandes y Laguna de

Guayatayoc de Jujuy y Salta, 2015). Consultation processes should respect this full view of community, and be based on unbiased information, good faith and genuine consultation where external actors are ready to listen and change. The processes should also respect communities' wishes as to their timing and, as such, leave sufficient time whenever needed for communities to discuss and define their views and positions without external interference or pressure.

Overall, the document defines a clear process for consultation in the context of the communities' culture, describing the history of the communities from their own point of view, their expectations regarding consultation and consent on the basis of international, national and provincial laws, and advice on how such procedures should unfold. Beyond consultation, the protocol is part of a wider struggle for the recognition of indigenous rights, expressed in the text through the retelling of a history of oppression (Flores, 2017). The name of the protocol, Kachi Yupi (Tracks in the Salt), is significant on all these fronts:

> Why did we think of tracks in the salt? Because this document is rooted in the essence of our identity, in the heritage of our grandparents, in the tracks left by their struggle for our territory, in the marks left by their feet, in the signs left by history, in the remains of their teachings and wisdom; in the deep and lasting impression of their culture.
>
> At the same time, a track represents a path to follow, a guide for the passage of people and animals, a furrow that we must follow. This document then, hopes to serve as a track, as a community conduit to channel our rights to participation, consultation and Free, Prior and Informed Consent and thus continue the legacy of our ancestors of the defence of the land and territories with which we are intimately connected.
>
> *(Comunidades indígenas de las Salinas Grandes y Laguna de Guayatayoc de Jujuy y Salta, 2015, p. 6)*

The path outlined in the Kachi Yupi community protocol can thus be thought of as stretching not only outwards, describing modes of respectful communication, but also from the past to the future, and from the local to the global. The change inherent to tracks drawn in salt that leave deep marks, as well as the space to carve new ways, also speaks to the intention of the communities.

After the Protocol – From Dialogue to Contention

The Kachi Yupi community protocol's publication and launch in December 2015 was timed to coincide with elections in Argentina, in a bid to attract attention from authorities. The elections were won by the opposition, and heralded a new governing coalition under the name Cambiemos at the federal level, led by new President Mauricio Macri's liberal *Propuesta Republicana* party (PRO). An ally from the same coalition (*Unión Cívica Radical* – UCR) won elections and took up government in Jujuy province. This provincial government had run a campaign that included attention to indigenous matters, including the community protocol. As it had promised before the elections, the new provincial government discussed the community protocol. In addition, permits stopped moving forward in the Salinas Grandes and Laguna de Guayatayoc area and lithium mining practically came to a halt, though it should be noted that this coincided not only with the community protocol but with a slackening in the pace of investments. A dialogue between the community members and the provincial government began, with various meetings and discussions in the following years, focusing for the most part on a draft provincial decree that would legally recognize the protocol, and make adherence to its terms compulsory. During the same period, the protocol was recognized publicly at the national level by the National Ombudsman. The National Ombudsman's office had intervened in an attempt to act as a neutral broker between communities and different external actors in a variety of cases in the country in previous years, and had taken an interest in the community protocol and the issues surrounding EIAs and consultation and consent before its public launch. When the protocol was published, the office praised it as a useful source for guiding consultation procedures, and recommended its application to all government authorities (both at the provincial and national levels) where any decision that would affect communities in the Salinas Grandes area was to be made.

Despite all these positive signs of an opening towards the claims of the communities for fairer and better specified rules about consultation procedures, it is worth recalling that economic reforms aimed at benefiting the extractives sector were brought in following Macri's election. Further moves in this direction were also introduced in 2016, when export duties were cut for the sector, and the geographical reach of mining projects extended. As the dialogue between the provincial government and the communities continued, the communities thus also engaged in other strategies. They sought to gather independent expertise about the potential impacts on water sources from the proposed mining projects, and met with other communities in the lithium triangle to exchange information. This is in line with the aims of the protocol, which underlines the need for independent expert information, given that in Argentina, and indeed many other countries, those applying for permits to mine also commission and supply expert information since local authorities lack the necessary resources. Understandably, this raises suspicions among community members about the independent nature of the information where it is produced by experts hired by mining companies (Marchegiani et al., 2020; Parks, 2020).

Concerns in the communities about the genuineness of the dialogue with the provincial government were also growing, and came to a culmination at the end of 2017, when new permits were granted ending the de facto pause that had commenced after the elections. In early 2018, with no legal decree forthcoming, the accumulation of authorities' actions in favour of mining at both local and federal levels and the recommencement of exploration work on their lands by mining companies, the communities began to question whether it made sense to continue their action through dialogue. As the year unfolded, a decision to change their approach was also driven by the complete lack of willingness among extractives companies now active on their lands to pay any attention to or respect any of the terms of the community protocol. For example, the protocol includes the condition that all communities in the area be consulted together, since the projects take place on a watershed that forms a single, connected ecosystem. Yet, companies have instead approached communities individually, providing information and promises in a manner similar to the aforementioned Olaroz case (Ibid). This meant that although communities had formed their own process of organizing internally, the Roundtable was not recognized by external actors, who continued to see individual community representatives as the correct contact point for dialogue. This demonstrates the communities' claims about being ignored by authorities.

In early 2019, the communities discovered another new bidding process involving studies to exploit lithium on their land about which they had not been consulted in any way. At this point, the group decided to change their course of action.[9] They mounted a peaceful demonstration, blocking the main highway that runs through their lands and distributing pamphlets to inform those passing through their lands and from other communities about their predicament. They sought a meeting with the governor of the province, and were invited to meet with him at his offices in the provincial capital of San Salvador de Jujuy, some 130 kilometres away. The community members, having engaged in dialogue and meetings for years at this point, refused the meeting, inviting the governor to visit their demonstration site instead. Rather than travelling to the communities' land, the governor now ordered the police to clear the mobilization site. Following this turn of events, community members held an emergency meeting and decided to change their position. While the community protocol and their previous work had been about proper consultation rather than against mining per se, they now decided that simply asking for consultation was no longer enough, and that a flat no and blanket opposition to mining was the better strategic choice. They thus stopped calling for consultation rights and the respect of the process outlined in Kachi Yupi, and adopted a "no to lithium" position. This could be interpreted as a new way of framing the cause after lack of progress in their original strategy.

The effects and impacts of the Kachi Yupi community protocol thus went well beyond the question of legal pluralism and can be read in a number of community and authority actions and reactions. In the short term, the protocol led to dialogue, opening an opportunity to discuss the protocol and future possible ways of engaging in dialogues on subjects beyond that of lithium too. The recognition conferred by the national ombudsman and the provincial government suggested that bridges were being built for a broader political cooperation agenda, constructing trust between the parties. This opportunity did not transpire however. The dialogue came to be viewed as fruitless in light of concrete moves to recommence and boost mining projects without any regard to consultation processes as described by communities themselves in the protocol, which led them to lose any trust that had been built. They thus moved into a more contentious stage of mobilization. At the time of writing, mobilization has become more difficult with the problems posed by the global pandemic. Communities remain firm in their new position against lithium.

Conclusions: Reflecting on Community Protocols and Their Role in Local Community Collective Action

One of the purposes of this volume is to reflect on how community protocols can lead to more legal pluralism. In this chapter, we have focused on different collective actions by the communities of Salinas Grandes and Laguna de Guayatayoc: complaints brought before national and subsequently international courts, the processes around a community protocol and other collective action to acquire information, build alliances and finally protest. These different modes of collective action, we argued, could be understood as strategic reactions to contexts. In the latest moves to protest, the communities have reacted, among other things, to a failed dialogue with the provincial government and the complete lack of respect for the community protocol among extractives companies. In a perspective of a formal reading of legal pluralism linking "[...] the local and the international legal levels, according to standards set out in customary, national and international law [...]" (Morgera et al., 2014, p. 157), the case was not successful. If we consider a more elastic perspective, however, the case does demonstrate some important aspects. First, the community protocol itself is a clear expression of legal pluralism in its elaboration of a process for consultation and consent that links national and international laws to a procedure rooted in the communities' worldview. Even if not formally recognized as expected (i.e. for the dialogue involving decisions affecting Salinas Grandes´ communities), the protocol remains a tool for legal pluralism, and has been introduced and acknowledged by different authorities with different scopes. It was recognized by the National Ombudsman as well as the provincial government initially. Moreover, it was recently mentioned as a precedent by the Instituto Nacional de Asuntos Indígenas (INAI) the National authority for indigenous matters, in a resolution that created – within the structure of the mentioned authority – a specific area to strengthen consultation rights.[10]

Reflecting on the case as a tool for collective action in a broader view can inform an understanding as to why it stopped short of a more formal recognition of legal pluralism. This can provide some ideas about what conditions might be needed for formal legal pluralism to come about, as well as revealing in more detail how the community protocol fed into logical decisions about which other types of collective action to pursue in light of their understanding of the legal and political context they found themselves within. To guide this reflection, we draw on a framework commonly used in the political sociological literature on collective action and social movements. Given the inherent political nature of introducing legal pluralism in its more and less formal dimensions, some of these approaches may prove useful to legal and socio-legal scholars, and we therefore use some space to give an overview of them here.

The first useful point taken from the literature on collective action is Tilly's concept of the action repertoire or, in the case of social movements, the repertoire of contention (della Porta and Diani, 2005; McAdam et al., 2001). This term is used to recognize that collective action takes many forms, and that those forms are linked to time, place and other considerations. There is thus a sort of menu of actions available to collective actors – like the local communities that drafted the Kachi Yupi community protocol discussed here – depending on what is feasible and deemed appropriate in light of circumstances. In this view, we can understand the decision to draft a community protocol as one available action in the repertoire available to these local communities. At the same time, the concept of the action repertoire also underlines that all collective action choices must be interpreted within specific contexts (Tarrow, 1998). To better reflect on the collective action choices of different types in the case described in this chapter, social movement studies supply another useful approach – political process or political opportunity (ibid). Although this approach began with attention to political contexts (a consequence of the clearly political aims and engagement of many social movements that have attracted scholarly attention), it was later integrated with attention to legal (and discursive) contexts too. The political opportunity approach is based on the observation that collective action choices are logical, and that to understand this logic, proper attention needs to be paid to context. It was first developed in scholarship on social movements to understand why and how social movements mobilize, and then used in the study of influence and outcomes (Meyer, 2004). It generally concerns opportunities and threats stemming from structural conditions determining how "open" or "closed" a polity is to different types of collective action, as well as describing various aspects that help researchers to identify more time-dependent or dynamic factors in political contexts that similarly facilitate or hinder certain actions, and thus ultimately shape outcomes (see, e.g., Giorgi, 2018).

The approach has expanded over time in response to critiques that other types of contextual factors are equally important to explain collective action. One such expansion is the legal opportunity approach, which aims to account more specifically for action in courts. Existing work focuses on the structural features of legal stock (the body of law applicable in a particular context), rules on legal standing (access to courts) and rules about legal costs (Vanhala, 2018). Community protocols would require further specification of the legal opportunity structure to accommodate the aim of achieving, or at least demonstrating the possibility of, legal pluralism. The approach as it is currently used focuses mainly on the decision to litigate, whereas community protocols challenge legal structures on the basis of customary and international laws. Essentially, they can be thought of as tools that seek to push the boundaries of legal systems and demonstrate where they can be more "convivial" (Bavikatte et al., 2015) and overcome the contradictions and difficulties where international law meets regional, national and customary laws (Bavikatte, 2014; Jonas et al., 2010; Tobin, 2013).

Bringing these observations together, thinking of collective action choices as many and varied, and logically chosen in the light of specific political and legal contexts, can help us to interpret the Kachi Yupi case from the initial complaints brought by the communities onwards, as well as point to what these choices suggest for community protocols elsewhere. The initial choice to bring complaints before courts can be interpreted using the legal opportunity approach. After the legal recognition of communities' cultural pre-existence and rights in the Constitutional reform of 1994, communities had for the first time in a long while a clear source to support their various and broader previous demands in their quest for recognition. Moreover, the legal element was central not only for the mobilization of demands and the organization of the communities but as a basis for any legal complaint. The lack of implementation of environmental standards, particularly where indigenous consultation and FPIC rights were concerned, in the context of lithium mining presented itself as a concrete opportunity for the communities to gain acknowledgement of their claims from a more progressive court compared to their conservative provincial counterparts. The decision to bring a case complaining about a lack of proper consultation to the Supreme Court of Justice thus appears all the more logical as well as a strategic move to draw attention to broader claims for recognition and rights.

In turn, the decision to draft a community protocol in such a way as to help move implementation forward[11] appears equally logical following the judgements that underlined the lack of provincial government rules as the reason for not implementing consultations. In addition, a lull in the immediacy of threats from mining exploration at the same time afforded the communities the possibility to pursue this lengthier type of action available in the repertoire. The process of drafting the community protocol following consensus-based decision-making can also be understood as an outcome of collective action. The in-depth discussions and debates about the protocol helped strengthen communities' ties to one another in preparation for a time when pressure from outside actors for their consent would be more present.

The outcomes of the community protocol itself in terms of legal pluralism can perhaps be more fully explained with reference to the changing political opportunities described. As mentioned, while the protocol did not lead to formal recognition and legal pluralism, its content, its recognition by the National Ombudsman, and the initial recognition given by the provincial government, or the recent acknowledgement in INAI's resolution, does underline some informal legal pluralism. To unpack these outcomes, the literature on political opportunity draws particular attention to elections as moments where collective action can extract promises for change from actors vying for power. Having a new elected government take office, with a campaign that integrated indigenous demands, could be argued as an opportunity for a new way forward in the relationships between communities and local authorities – the decision to launch the community protocol publicly during the peak of the election campaign was linked to this possibility, even if remote. In the event, the new government did at least appear to make concessions after taking office, by opening a dialogue on the protocol's recognition. This can be understood as an outcome of a more informal type of legal pluralism. The changes in circumstances as time wore on suggest that this concession was not as meaningful as hoped for in that it did not lead to the promised step to formal legal pluralism in the form of a provincial decree on the protocol. A stronger commitment was made to lithium as a strategic resource in line with the policies pursued by the new federal administration, and a lack of understanding about how to use the community protocol became clear (the provincial government wanted to make it applicable for all communities in the province and not only to communities in Salinas Grandes area). These changes also help to explain why these initial concessions failed to culminate in a decree. The provincial government was biased in favour of lithium exploitation and extraction, and this prevented them from seeing the opportunity to pursue that aim alongside communities in a path of cooperation.

This showed that the local authorities were not ready to engage in multi-cultural dialogues and advance to make clearer commitments to listen to and understand demands from communities at the local level. The unwillingness to keep the conversation going also speaks to the lack of understanding of the nature and scope of the protocol. Communities in Salinas Grandes and Laguna de Guayatayoc area did not seek to devise a tool to be applied to each and every community in the province of Jujuy; rather they were stating conditions for dialogues with them (33 communities) and thus, proactively paving a way for a multicultural dialogue with authorities and companies that wanted to engage with them. In a deeper perspective, the efforts made to "translate" *Buen Vivir* to external actors (de Sousa Santos, 2014) seem to have come up against a discursive barrier here. To the extent that dominant discourses based on a nature/culture divide, where the planet is essentially understood as separate from human communities, shape governance the world over (Uggla, 2010), communicating worldviews that do not separate the planet from human communities is challenging (Vermeylen, 2017). In this vein, some have called for a move beyond multiculturalism towards multinaturalism (Viveiros de Castro, 2004).

This failure to advance the dialogue around the decree on one side, while moving forward with decisions that would advance explorations for mining in the area, on the other, then ate away at what little trust had developed until then between communities and provincial government. The communities had, with their protocol, taken constructive steps to open a dialogue and build a space for understanding with provincial authorities and extractive companies who failed to recognize the opportunity this presented for advancing with projects in mutually beneficial ways. Ultimately, this failure to engage increases the costs for the provincial government, and potentially for extractive companies should protests persist. Instead of using this opportunity for dialogue with communities mobilized around a controversial issue, the government failed to listen, and the communities moved to more contentious action choices. This choice included a move in the communities' position from one of asking for dialogue to one of rejecting mining outright, which also signals a move away from the steps taken towards legal pluralism in the sense of shutting down channels of communication.

The choice of some of the communities to follow a different and more contentious path of action also makes logical sense with regard to other aspects of available political and legal opportunities. In terms of legal opportunities, the communities had now exhausted the path of litigation. They had then taken the much debated decision to accept the frame of "free, prior and informed consent" and sought to inject that framework with their own interpretation, based on a particular worldview. They later invested significant time and effort in a dialogue with the provincial government. Their political opportunities were proving empty however–as discussed, though dialogue continued, the provincial government continued to make choices that favoured extractives industries. At the same time, extractive companies were ignoring the protocol too, following their established and contested habits where consultation and consent processes were concerned. Where opportunities for dialogue and more conventional engagement with authorities close down, opportunities for more contentious actions open up. Opting for contentious forms of action when other paths are closed down is generally seen as clear and logical in the literature. The reasoning is that where advocacy types of actions are not, or are no longer, plausible, choosing a more disruptive path of action allows a group without formal power to apply pressure on authorities by attracting the support of public opinion, often via media coverage of the action in question. Contentious action also serves the purpose of directly disseminating information about the group's position (in this case by distributing pamphlets) and recruiting new supporters to the cause, with numbers being another important factor in wooing public opinion and pressuring authorities (Tarrow, 1998).

Another way of understanding this move to contentious action is to reflect on the resources created by the communities in drafting their protocol. It could be argued that the process of drafting the community protocol gave momentum for the communities to bolster their internal organization, strengthen their internal understanding, and become a more unified collective actor able to undertake a peaceful road block relatively quickly and easily. Considering the geographical spread of the single communities and the various challenges in meeting and communicating lends this view some credence, as does the fact that after the police intervention the communities were able to convene an emergency meeting and decide a change in their fundamental position. In that sense, community protocols can also be seen as more than collective action tools in a repertoire. They can also be seen as important opportunities for the development or reconstitution of communities as collective actors through the formation of different networks. In this case, the Roundtable can be understood as a mode of organizing the communities into such a network, allowing the different communities to come together to form a single collective actor that pursued a range of different modes of collective action around the lithium mining issue.

In a broader view, these reflections on how the Kachi Yupi community protocol unfolded and took its place in a longer chain of collective action suggest some ideas. First, it suggests some scope conditions that might be needed for community protocols to effectively translate into formal legal pluralism. These are mainly political. The case suggests that a community protocol is pitted against pre-existing and dominant discourses of commitment to an agenda that clashes with its aims, legal pluralism will be harder to achieve. The commitment to lithium, and its construction as a key part of moving away from fossil-fuelled economies, forms such a dominant discourse in this case, and dominant discourses have been argued to be particularly difficult to challenge where economically valuable resources are concerned (Nelson, 2010). It also suggests that both subnational and national politics need to be aligned in favour of legal pluralism, or at least one of these levels needs to be so. The only pressure from the national level in this case came from the independent office of the National Ombudsman, while the provincial and federal governments were politically aligned and saw no need to engage in cooperation with the communities. When the provincial government engaged in dialogue, it could not (or would not) commit to it in a real sense. These political scope conditions would seem helpful, if not crucial, for the success of legal pluralism (in addition to a conducive legal context of course). Second, the case is helpful for understanding community protocols both as tools in action repertoires and as bases for building on those repertoires. Community protocols may not be the only actions that communities undertake. By considering them as one action in a longer-term view, their drawbacks and any failures to achieve their aims appear in perspective. But they are also very peculiar types of collective action: the processes that are undertaken to draft protocols can strengthen communities' action repertoires by bolstering their standing as collective actors. In other words, community protocols can help build communities that are better placed to act together in a wider range of ways and with a better understanding of the legal and political contexts they are in, and of the claims they hold most dear.

Notes

1 The authors thank community members for the fruitful exchanges that drive the reflections of this contribution. *All URLs retrieved on 1 September 2021.

2 Convention Concerning Indigenous and Tribal Peoples in Independent Countries (ILO No. 169) (1989).

3 Communities in the area have been recognized by Argentincan law as indigenous communities and hold legal status as such.

4 National Constitution of Argentina, rev. 1994 (Constitución de la Nacion Argentina, *Boletín Oficial* [BO], Jan, 3, 1995), Art. 41:

> All inhabitants are entitled to the right to a healthy and balanced environment fit for human development in order that productive activities shall meet present needs without endangering those of future generations; and shall have the duty to preserve it [...].

English translation from http://www.biblioteca.jus.gov.ar/Argentina-Constitution.pdf, accessed 18 June 2021.

5 National Constitution of Argentina, Art. 75, point 17: "To recognize the ethnic and cultural pre-existence of indigenous peoples of Argentina". English translation from http://www.biblioteca.jus.gov.ar/Argentina-Constitution.pdf, accessed 18 June 2021.
6 While the indigenous conception of land ownership involves soil and subsoil, the Mining Code distinguishes between the ownership of minerals generally found in the subsoil, which belong to the State that then grants concession contracts to private actors, and the ownership of the soil, which follows the private conception of land ownership. The Salinas Grandes communities' land ownership involves both soil and subsoil.
7 Artt. 11–13 of the 2002 General Environmental Protection Law ("*Ley General del Ambiente*" Ley N.° 25.675, B.O. del 28/11/2002).
8 Corte Suprema de Justicia de la Nación, *Comunidad Aborigen de Santuario Tres Pozos y otros c/ Provincia de Jujuy y otros s/ amparo*, Expte. C.1196.XLVI, 18/12/2012. To see the text of the Supreme Court Decision, access:https://sjconsulta.csjn.gov.ar/sjconsulta/documentos/verUnicoDocumentoLink.html?idAnalisis=698081&cache=1621430473022
9 Over time, the initial group of communities that drafted the community protocol took different paths, and the decision to undertake protest concerned a smaller group concentrated in the province of Jujuy. As we do not have detailed data about these changes, we do not advance any discussion of this here, but simply remark that the literature on collective action has long noted the different paths that groups take, into different types of activism or indeed the exit from activism, over time (Tarrow, 1998).
10 For more details, see Instituto Nacional de Asuntos Índigenas, Resolución 30/2021 from the 5th of April 2021. Available at https://www.boletinoficial.gob.ar/detalleAviso/primera/243738/20210429
11 Since international human rights standards are operative (non-programmatical) in Argentina's legal system, they do not need specific legislation in order to be applied. Legal scholars have thus found those laws designed to help implementation in a context where indigenous rights are not well understood.

References

Anaya, J. (2012a). *Report of the Special Rapporteur on the Rights of Indigenous Peoples, James Anaya. Addendum. The Situation of Indigenous Peoples in Argentina* (A/HRC/21/47/Add.2). United Nations General Assembly, Human Rights Council, Twenty-fourth session, Agenda item 3, 2 September 2013.

Anaya, J. (2012b). *Report of the Special Rapporteur on the Rights of Indigenous Peoples, Progress Report on Extractive Industries.* United Nations General Assembly, Human Rights Council, Twenty-fourth session, Agenda item 3, 1 July 2013.

Bavikatte, K. (2014). *Stewarding the Earth. Rethinking Property and the Emergence of Biocultural Rights.* New Delhi, India: Oxford University Press.

Bavikatte, K., Robinson, D. F., and Oliva, M. J. (2015). Biocultural Community Protocols: Dialogues on the space within. *Ik: Other Ways Knowing*, 1(2), 1–31.

Comunidades indígenas de las Salinas Grandes y Laguna de Guayatayoc de Jujuy y Salta (2015). *Kachi Yupi/Huellas de la Sal: Procedimiento de Consulta y Consentimiento Previo, Libre e Informado para las Comunidades Indígenas de la Cuenca de Salinas Grandes y Laguna de Guayatayoc.*

Ferradás Abalo, E., Lobo, A. L., and Lucero, M. J. (2016). *Conflicto socioambiental en Salinas Grandes: neoextractivismo, resistencias y nociones de desarrollo en el nuevo escenario político regional.* Villa María.

Flores, C. (2017). Entre el Litio y la Vida: Comunidades Originarias y la Lucha por la Conservación del gua y de su Cultura. In FARN (Ed.), *Informe Ambiental Anual 2017.* Buenos Aires: FARN.

Giorgi, A. (2018). *Religioni di Minoranza tra Europa e Laicità Locale*. Milano: Mimesis Edizioni.
Jonas, H., Bavikatte, K., and Shrumm, H. (2010). Community Protocols and Access and Benefit Sharing. *Asian Biotechnology and Development Review*, 12(3), 49–77.
Marchegiani, P., Hoglund Hellgren, J., and Gomez, L. (2018). *Lithium Extraction in Argentina: a Case Study on its Social and Environmental Impacts*. Buenos Aires: Fundacion Ambiente y Recursos Naturales (FARN).
Marchegiani, P., Morgera, E., and Parks, L. (2020). Indigenous Peoples' Rights to Natural Resources in Argentina: The Challenges of Impact Assessment, Consent and Fair and Equitable Benefit-Sharing in Cases of Lithium Mining. *The International Journal of Human Rights*, 24(2–3), 224–240.
McAdam, D., Tarrow, S., and Tilly, C. (2001). *Dynamics of Contention*. Leiden: Cambridge University Press.
Meyer, D. S. (2004). Protest and Political Opportunities. *Annual Review of Sociology*, 30, 124–145.
Morgera, E., Tsioumani, E., and Buck, M. (2014). *Unraveling the Nagoya Protocol*. Leiden: Brill | Nijhoff.
Nelson, F. (2010). Democratizing Natural Resource Governance: Searching for Institutional Change. In F. Nelson (Ed.), *Community Rights, Conservation and Contested Land: The Politics of Natural Resource Governance in Africa*. London: Earthscan, 310–333.
Parks, L. (2020). *Benefit-Sharing in Environmental Governance Local Experiences of a Global Concept*. Abingdon: Routledge.
della Porta, D., and Diani, M. (2005). *Social Movements: An Introduction*. Second ed. Oxford: Wiley-Blackwell.
de Sousa Santos, B. (2014). *Epistemologies of the South: Justice Against Epistemicide*. London: Paradigm Publishers.
Tarrow, S. (1998). *Power in Movement*. Cambridge: Cambridge University Press.
Tobin, B. (2013). Bridging the Nagoya Compliance Gap: The Fundamental Role of Customary Law in Protection of Indigenous Peoples' Resource and Knowledge Rights. *Law, Environment and Development Journal*, 9(13), 142–162.
Uggla, Y. (2010). What Is this Thing called 'natural'? The Nature-Culture Divide in Climate Change and Biodiversity Policy. *Journal of Political Ecology*, 17(1), 79–91.
Vanhala, L. (2018). Is Legal Mobilization for the Birds? Legal Opportunity Structures and Environmental Nongovernmental Organizations in the United Kingdom, France, Finland, and Italy. *Comparative Political Studies*, 51(3), 380–412.
Vermeylen, S. (2017). Materiality and the Ontological Turn in the Anthropocene: Establishing a Dialogue Between Law, Anthropology and Eco-philosophy. In L. Kotzé (Ed.), *Law and Governance for the Anthropocene*. Oxford: Hart Publishing.
Viveiros de Castro, E. (2004). Perspectival Anthropology and the Method of Controlled Equivocation. *Tipití: Journal of the Society for the Anthropology of Lowland South America*, 2(1), 3–22.

8

BIOCULTURAL RIGHTS AND PROTOCOLS IN THE PACIFIC

Miri (Margaret) Raven and Daniel Robinson[1]

Introduction

The Nagoya Protocol to the United Nations Convention on Biological Diversity encourages academics, government bureaucrats and Indigenous peoples to consider the utilisation of customary laws within state law frameworks to improve and resolve access and benefit-sharing (ABS) processes. The Nagoya Protocol (Article 12.1) specifically calls on parties to "take into consideration indigenous and local communities' customary laws, community protocols and procedures, as applicable, with respect to traditional knowledge associated with genetic resources". Article 12 encourages states to support the development of community protocols by Indigenous and local communities. The community of scholars and practitioners dedicated specifically to ensuring the implementation of Article 12.1 has grown significantly over the past decade (see, for example: Arjjumend, 2018; Goagoses et al., 2020; Halewood et al., 2021; Malsale et al., 2018; Parks, 2018; Su, 2018; Suvanto, 2020). This chapter contributes to that growing body of knowledge.

In this chapter, we report on some of our work under the Australian Research Council (ARC) Discovery Project (DP180100507): *Indigenous knowledge futures: protecting and promoting indigenous knowledge* (2018–2022). This is a participatory action research project involving research and work with Aboriginal peoples and enterprises in Northern Australia, as well as Indigenous communities in Vanuatu and the Cook Islands. The research project focuses on patent-landscaping activities to profile patents relating to traditional uses of plants used for food and medicines that may originate from Indigenous communities in these countries; and the development of biocultural community protocols (BCPs) which reflect and respect local customary laws and norms. The project specifically seeks to support the implementation of the Nagoya Protocol in these countries.

DOI: 10.4324/9781003172642-10

An important aspect of implementation is understanding the nature of Indigenous knowledge and its associated customary laws and beliefs. To this end, one of our recent papers (Robinson & Raven, 2020) examines and reviews legal, anthropological and historical texts relating to biodiversity and associated knowledge to explore Aboriginal and Torres Strait Island peoples' customary laws and governance. Understanding the broader place of Indigenous customs, laws and beliefs, as sitting on the oral-written continuum and expressing through the Dreaming,[2] provides the foundation for understanding Indigenous Australian customary laws as they relate to plants and animals. Plants and animals may be regarded as totemic species that have specific relational significance, and which have specific customary laws, and rights for specific individuals or families.

Biopiracy and Technological Innovation: Why Protocols Are Needed

Documenting Indigenous knowledge of plants in Australia began from at least the late eighteenth century (Clarke, 2003). It was part of the wider processes of colonisation of Australia based on the notion of discovery (Miller et al., 2010). This ethnobotanical research involved the documentation and collection of Australian native plants that Indigenous peoples were known to use. This information has found its way into monographs, books, journal articles and herbariums and museums around the world, with a vast majority of the collections ending up in England. The first phases of collecting were "primarily driven by practical needs, with the plants observed in use by Indigenous Australians in the 'new' land being investigated for their potential economic benefit to eventual settlers" (Clarke, 2003, p. 21). These benefits included the use of native plants for life stock feed and to help settlers supplement their food between supply ships. The types of collectors (such as settlers, explorers, biologists, pharmacologists, herbalists, anthropologists, geographers, medical specialists and ethnologists) (Clarke, 2003) indicate the wide and diverse value of Indigenous knowledge across various specialist fields.

Similarly, in the Pacific, a colonial history exists whereby European explorers and early settlers exploited natural resources, and collected specimens for scientific interest for botanical gardens, herbariums and then later in genebanks. In more recent times, there have been collections by a wide range of ethnobotanists and anthropologists such as Paul Cox (1991) and Art Whistler (1992), which have seen many traditional medicines and useful plants from parts of Polynesia and Melanesia documented and published widely. This, in turn, has resulted in concern about research continuing based on those earlier collections, and patents relating to the research, as discussed by several Indigenous authors in the collection by Mead and Ratuva (2007).

The rising recognition of the value of biodiversity, since the early 1990s, and the influence of technology has meant that plants and animals are considered an important biological resource for the development of products in the pharmaceutical, beauty products and food industries. Past and current practices of collecting biological resources and associated traditional knowledge share some similarities, but also differ in significant ways. Collecting Indigenous knowledge of plants and placing them in the public domain, via journal articles and other mediums, remain. The difference, for some public domain placements, is that they are put in the public domain by Indigenous peoples themselves. These are defensive approaches to ward off the "theft" of their knowledge, with the idea being that public domain information cannot be claimed by others under intellectual property laws. But some collecting is undertaken without necessary biodiversity approvals (such as permits and free prior informed consent), or where approvals do not protect Indigenous peoples' rights to their own knowledge. These instances of "biopiracy" are difficult to track and monitor as they are often known about through word of mouth, approvals are subject to confidentiality agreements and getting access to these approval documents is difficult because they are subject to commercial-in-confidence or government regulations.

Additionally, some collections, particularly those for universities and other research institutions, now have an ethical research framework which stipulates the requirements for prior informed consent for the collecting of this knowledge. Some of these, although updated in more recent years, pre-date the Nagoya Protocol. In Australia, this includes the Australian Institutes for Aboriginal and Torres Strait Islander Studies' (AIATSIS) *Code of Ethics for Aboriginal and Torres Strait Islander Research* (and what was previously the AIATSIS Guidelines 2012; AIATSIS, 2020) and the National Health and Medical Research Council's (NHMRC) *National Statement on Ethical Conduct in Human Research* (NHMRC, 2018b) and their *Ethical conduct in research with Aboriginal and Torres Strait Islander Peoples and communities: Guidelines for researchers and stakeholders* (NHMRC, 2018a). They all stipulate the requirements for researchers, that works with Aboriginal and Torres Strait Islander people and their data, to seek ethics review through a human research ethics committee. While these mechanisms can protect Indigenous knowledge of biological resources, it only does so for Australian research institutions, or those who access funding through the NHMRC Australia. Even with the existence of such mechanisms, there are research organisations which do not require ethics to operate in Australia (as their funding is sourced elsewhere).

Confounding this is that ethical frameworks (just like international law making) can't always keep up with the pace of technology. Biodiversity and biotechnology research have seen an uptake in technologies such as phytochemical screening (Al Rashid et al., 2019) and next-generation genomic sequencing (NGS) (Escalante et al., 2014), and encompass various methods such as high-throughput DNA sequencing (HTS) (Gao et al., 2012) and environmental DNA sequencing (eDNA) (Huerlimann et al., 2020), which are being clustered together in policy CBD policy circles as "digital sequencing information" (DSI) (Laird & Wynberg, 2018).

Additionally, NGS and DSI technology is no longer confined to the laboratory and can be undertaken through portable screening (Watsa et al., 2020). While

Indigenous peoples hold biodiversity knowledge, NGS' new portable capabilities mean that existing knowledge, and vast amounts of associated biological resources, can be quickly screened. This may pick up other associated biological materials to which the initial plant or animal of interest is dependent. It may help to map plant and animal communities that would otherwise have not been known by outside researchers.

Protocols, or BCPs, provide a process for Indigenous peoples to determine whether to allow access to their knowledge and lands for research and biodiversity conservation purposes, among potentially other uses (such as ecotourism). Establishing clear processes for free prior and informed consent to establish ABS procedures to Indigenous knowledge associated with biological resources ensures that the rapidly changing technological landscape can be included in discussions about what access entails now, and in the future. The research we are undertaking seeks to recognise BCPs in various locations across the Pacific as well as working with Aboriginal peoples and enterprises in Australia.

The Geographical Contexts

The research occurs in Vanuatu, Cook Islands and in various locations across Northern Australia. These countries are geographically located in the "Pacific". Vanuatu is located in the Melanesian sub-region and Cook Islands is located in the Polynesian sub-region. The geographical delineation of Melanesia, Micronesia and Polynesia was coined by French navigator Jules Dumont d'Urville to designate ethnic and geographic regions in the Pacific. While we use the Melanesian and Polynesian terminology in this chapter, we understand that for some Indigenous peoples, these groupings may be problematic and that there are overlaps between culture and history. Additionally, the location of Australia, as part of the Pacific, is also a difficult geographic distinction for Indigenous peoples located on the western and southern parts of the continent which borders the Indian Ocean. Notwithstanding this, we, as geographers, need some kind of word to geographically located research. We have chosen to use these terms, but acknowledge the difficulties with them. Vanuatu and Cook Islands were included in the project so as to advance further research and development of BCPs.

To date, in the communities in Vanuatu and Cook Islands, the development of BCPs has been a highly practical process, with the people that we have worked with preferring simple BCP tools and documents – including posters and signs – that can be posted in airports or at the entrance to protected areas to indicate their customary laws, rules and processes.

We have begun the development of community protocols in some specific communities that utilise traditional medicines for customary purposes (in Cook Islands), as well as some communities that are heavily involved in "biotrade"[3] of traditional skin-care oil products derived from tree nuts (in Vanuatu) and use and sale of kava.

Vanuatu

Vanuatu is an archipelago of over 80 islands in the Melanesian Pacific. The Melanesian grouping also includes the Nation States of Solomon Islands, Fiji and Papua New Guinea, and the French Territory of New Caledonia. Vanuatu is located 1,750 kilometres east of Australia, and 1,200 kilometres west of Fiji. The Ni-Vanuatu (Indigenous peoples of Vanuatu) make up over 98% of the population. Vanuatu is also extremely culturally and linguistically diverse with an estimated 138 languages spoken in a population of 300,000 people (Van Trease, 1987). A series of European colonial powers visited, what was then called by Captain James Cook the New Hebrides in 1774, with both the United Kingdom and France claiming ownership in the 1880s which eventually led to the "condominium" division of control of the country between the two (Van Trease, 1987). This caused disruptions to *kastom* (customary/traditional) systems of governance and law, as well as land tenure, throughout the colonial period. Vanuatu has become independent in 1980 and its Constitution has recognised *kastom* rules, rights and responsibilities in regard to land and natural resources (Tonkinson, 1982).

We have worked in the islands of Malekula and Santo in the North, Tanna and Aneityum in the South, to discuss whether those communities have existing research or commercial biotrade activity that could benefit from clearly established community protocols. This has also involved discussions with the Vanuatu Government to establish the possibility of government recognition of those protocols within formal permit approvals processes under the Nagoya Protocol.

Cook Islands

Cook Islands are a group of 15 islands located in Polynesia. It is an independent state in free association with the "Realm of New Zealand". The Polynesian grouping includes 15 nation states such as New Zealand, Hawaii, Tahiti, Tonga, Samoa and Tuvalu. Cook Islands are located about 3,500 km northeast of New Zealand. The Maori (Cook Island Maori people) make up the majority of the population; however, a large number of Cook Islands people reside in New Zealand either permanently or as dual residents. Named after Captain James Cook who landed in the islands in 1773, the islands received British missionaries in the 1820s, became a British Protectorate in 1888 and was then annexed to New Zealand in 1901. The islands have been independent but in "free association with New Zealand" since 1965 (Stone, 1966).

Australia

The Australian continent is composed of a plateau (with deserts and rangelands in the northern, southern, western and central parts), and mountain ranges in the east and southeast. The west tropics (with forests in eastern Australia) and the southwest (in Western Australia) are recognised as two of the 35 international global biodiversity hotspots in the world.

While Australia had colonial contact with the French and Dutch, it was the English who established the landmass, and its peoples, as a colony of England. Australia is an island nation state, which federated as a nation in 1901, with six states, three internal territories and seven external territories. At the time of Australian colonisation in 1788, there were between 250 and 750 languages spoken by Indigenous peoples, and more than 500 first nations (Arthur & Morphy, 2019). Despite the existence of Indigenous peoples prior to the arrival of the British, Australia was founded on the idea of *terra nullius* ("land belonging to no one"). Unlike in Vanuatu and Cook Islands, colonisation specifically sought to remove people from their land. Australian government policies and procedures to remove children from their families and country (known as the Stolen Generations) have substantially impacted Indigenous peoples' relationships with their own lands and seas and families. Indigenous peoples, in the earlier days of the federation, were excluded from the national census and voting in elections until 1967.

The Legal Contexts

While Australia considers itself a leader in the Pacific, it lags behind other nation states in the region in implementing the Nagoya Protocol. Australia, a signatory to the protocol, is yet to ratify it. Vanuatu and Cook Islands have both introduced legislation to implement the Nagoya Protocol.

Vanuatu

Vanuatu ratified the Nagoya Protocol in 2014 and is now implementing it through operation of the Biodiversity Advisory Council (BAC), established under sections 29–34 of the amended Environmental Protection and Conservation Act (Vanuatu Government, 2014)[4]. The ABS processes under the Nagoya Protocol operate under this Act, and the BAC issues permits to researchers who intend to collect and study biological resources (and associated knowledge). Benefit-sharing conditions are also required to be negotiated with provider communities for both academic and commercial research, with academic research typically sharing non-monetary benefits, while commercial research would usually include monetary benefits.

The government has also recently developed and passed the *Protection of Traditional Knowledge and Expressions of Culture Act No. 21 of 2019*[5]. This Act does not specifically mention the use of cultural protocols. However, the scope of the Act (Article 4(1)) is such that traditional *kastom* (the Bislama term for custom) owners have exclusive rights to traditional knowledge to *inter alia* control, exploit and use; grant prior informed consent to use traditional knowledge and to access it on mutually agreed terms; and prevent misappropriation and granting of unauthorised intellectual property rights. Additionally, the establishment of "The Traditional Knowledge and Expressions of Culture Authority" (TKECA), under the Act, provides an avenue for use of community customary protocols (or

BCPs). BCPs may play a role in assisting the Authority in its deliberations related to Article 14(1) on the granting of bio-prospecting licences; to "consider and determine applications that have elements of traditional knowledge and expressions of culture, for registration of" intellectual property rights and research based on this knowledge. Within the legislation, there is also a scope for the use of BCPs to provide the basis through which communities can determine their own process for granting consent and access. BCPs can potentially provide the framework for this conversation and soft law setting.

Cook Islands

The *Cook Islands Act* of 1915 (Article 422) gave recognition to customary laws and rules and made the following provision in relation to land: "Every title to and interest in customary land shall be determined according to ancient custom and usage of the Natives of Cook Islands" (Boer, 1996, p. 30; Robinson & Forsyth, 2016). The state in Cook Islands engages with custom partly through the establishment of state-recognised customary institutions. For example, in Cook Islands, the House of Ariki was established to represent the paramount or high chiefs in 1966, followed by the Koutu Nui, a house for the sub-district chiefs comprising mataiapo and rangatira, in 1972 (House of Ariki Act 1966, as amended in 1972).

The government of Cook Islands has enacted the Traditional Knowledge Act 2013 to "give legal recognition to the rights in the traditional knowledge of the traditional communities of Cook Islands" (preamble). The Act encourages the registration of traditional knowledge by knowledge-holders, and its written documentation although this practice is not something that would traditionally have occurred. It provides a range of rights, among them one that only rights-holders of registered traditional knowledge have the right to use, transmit, document or develop the knowledge in any way, whether commercial or not (Article 7(1)). Registered traditional knowledge is protected in perpetuity, with the rights being inalienable, and purporting not to limit or affect other intellectual property rights (Forsyth & Farran, 2015). The Traditional Knowledge Act 2013 anticipates multiple and overlapping registrations. If two or more registrations for ostensibly the same knowledge are made, then each applicant is able to view the other application and must come to an agreement before registration is accepted for protection under the law. This is an interesting legal adaptation of mediation replacing or paralleling what would have occurred through customary laws and norms (Robinson, 2014; Robinson & Forsyth, 2016).

Australia

As a federated state, aspects of biodiversity conservation are distributed across national and State and Territory legislation. In 2002, the "Nationally consistent approach for access to and the utilization of Australia's native genetic and biochemical resources" was endorsed to promote consistency of regulations (Department of Environment, 2014). ABS requirements of the Convention on Biological Diversity are included in the Environmental Protection and Biodiversity Conservation Act (EPBC) 1999 Regulations, Part 8A. These regulations apply to Commonwealth land and sea areas only, limiting the scope of the regulations significantly (Department of Environment, 2014). Importantly, these regulations require PIC where access is sought to Indigenous people's land (Article 8A 10(1), EPBC Regs) and also stipulate (Article 8A.08, EPBC Regs) that PIC and benefit-sharing are required for access where Indigenous knowledge is used for research and development (R&D) (Robinson & Raven, 2017). States and Territories in Australia, following the development of the Commonwealth EPBC regulations, gradually implemented ABS provisions. Further legal and policy developments are likely if Australia ratifies the Nagoya Protocol (Robinson & Raven, 2017).

Indigenous peoples' rights to land are recognised through native title (through the Native Title Act 1993) and land rights legislation at the State and territory jurisdiction, which includes, for example, the Northern Territory Land Rights Legislation 1976. Native title is the federal Australian mechanism for recognising Indigenous or customary laws in Australia related to land, waters, and by association, plants and animals. Although highly flawed, native title is one of the mechanisms through which Aboriginal and Torres Strait Islands peoples are given recognised rights in land and resources, and which is therefore relevant for the governance of biological resources. Land rights acts (administered by the states) are the other mechanisms. As we explain in another article (Robinson & Raven, 2020), native title determinations and Aboriginal land rights have relevance for establishing whether native title holders (or claimants), or land holders, might be genetic resource "access providers" as recognised by the Northern Territory Biological Resources Act (2006) and the Commonwealth Environmental Protection and Biodiversity Conservation Regulations (2000), Part 8A.

Protocols: What They Are and Can Do

While the term "biocultural community protocols" has become the norm for referring to the types of arrangements, standards and rules in biodiversity conservation, protocols are used in various other contexts. Through Indigenous research, protocols are positioned as a mechanism to remedy the misappropriation of Indigenous knowledge. They have been conceptualised as etiquettes or guides (Argumedo et al., 2011; Bowrey, 2006; Garwood-Houng, 2005; Nakata et al., 2005); rules or rules of engagement; standards; prescriptive tools; rights-based approaches to affirm self-determination; agents of change; and form and source of private laws (Australia Council for the Arts, 2007a, 2007b; Carter, 2010; Dunbar & Scrimgeour, 2005; Janke, 1998; Janke and Dawson, 2012; Jonas et al., 2010).

Through our preliminary research where they are used for biodiversity conservation goals, community protocols appear as tools that expand the regulatory toolbox for implementing the Nagoya Protocol, and assist with connecting Indigenous law and state law relating to the governance and management of Indigenous knowledge and totemic species. They can be derived from existing cultural norms and customs or be newly constructed to serve a particular purpose. In Australia, our work on "protocols" has been heavily focused on Aboriginal enterprises that are actively "biotrading" plant species that have been of interest to many larger companies for R&D, and in which there have been issues of potential "biopiracy" (Robinson, 2010). At the request of these Aboriginal enterprises and communities, we are advising on protocols and agreement-making as part of the PIC and ABS process.

We are working with the Vanuatu and Cook Islands governments to leverage better protections for Indigenous knowledge, and to include legal recognitions of customary laws and community protocols. However, we note that in Vanuatu and Cook Islands, some progress is being made towards implementing the Nagoya Protocol. Vanuatu has ratified the protocol and introduced the *Traditional Knowledge Act* (2020). While Cook Islands is yet to ratify the Nagoya Protocol, they are developing ABS legislation that will support implementation once they have ratified. Both the ABS and TK legislation could be utilised to recognise and respect BCPs as part of the ABS PIC process. Importantly, Cook Islands has signed and been active in ABS processes with its own UNDP-GEF ABS programme[6] relating to the CIMTECH agreement (bone healing based on traditional medicines).

BCPs have the appearance of hybrid instruments in biodiversity conservation. Their hybridity derives from incorporating a mix of traditional customary law, research ethical principles and ABS framing related to notions of consent and customary rights. Their origin in customary law means that protocols are able to transmit customary rules and regulations into spaces and places that would ordinarily exclude or dismiss customary law. The very nature of protocols, as hybrid flexible instruments, is what actually enables this to occur. In Vanuatu, for example, which is attempting to codify aspects of customary law into protocol formats, it means that customary law over biological resources will be recognised by state law. While protocols are non-binding tools, the way that Vanuatu legislation recognises them arguably turns them into quasi-binding legal instruments. The extent to which they are also able to operate as ethical instruments, in formal ethical research processes, remains to be seen.

Indigenous knowledge systems are often defined as including the idea that Indigenous peoples have relational obligations to each other and nature (Berkes & Berkes, 2009; Pierotti & Wildcat, 2000; Whyte, 2013). Relational obligations derive from customary rules, laws, norms and protocols which act as governance arrangements over Indigenous knowledge (Christie, 2006; Janke, 2008; Mackay, 2009). In Australia, for example, customary law has figured in the high-profile copyright dispute through the *Bulun Bulun & Anor v R & T Textiles Pty Ltd* case (the Bulun Bulun case[7]) over the reproduction of a waterhole design by John Bulun Bulun, a Ganalbingu man of Arnhem Land, which was printed in Indonesia and imported into Australia (Janke, 2008; Mackay, 2009). In this case, Judge Von Doussa found that a fiduciary relationship (one based on trust) existed between John Bulun Bulun and the Ganalbingu clan, and that customary laws influence what the artist can do with the work embodying the Indigenous knowledge "in a way that he had to discuss and negotiate use of traditional knowledge with relevant persons in authority within his clan" (Janke, 2008, p. 19). What the *Bulun Bulun* case highlighted and is often discussed in relation to defining the use of Indigenous knowledge, and thus to customary law and protocols, is the tension between individual and communal ownership of this knowledge and the obligations related to this ownership.

Relational obligation is often captured through the terminology of "guardianship" or "stewardship" (Argumedo et al., 2011; Bavikatte & Robinson, 2011; Carpenter et al., 2009; Kolig, 2002; Lai, 2014; Lotz, 2002; Stephens et al., 2007; Wade, 1999; Whyte, 2013). Guardianship and stewardship lay the foundations on which protocols sit. This was clearly articulated by Argumedo, Asociación ANDES, Potato Park Communities and IIED in their "Community Biocultural Protocol", where they stated:

> The fact that biocultural approaches [...] are now emerging as useful concepts is testament to the inadequacy of reductionist, disciplinary methods that continue to be the *modus operandi* of conservation and development research, policy formulation, and action. Mainstream perspectives do not value the role Indigenous Peoples have played historically – and continue to play today – as stewards and guardians, innovators and developers, of their eco- and knowledge systems.
>
> *(Argumedo et al., 2011, p. 19)*

The very nature of protocols, as being based on customary law, encompasses an "ethics of stewardship". Stewardship is a concept used to encompass a duty of care ethics that extends beyond people towards nature (Carpenter et al., 2009). The notion of "stewardship", as Carpenter et al. (2009) point out, is used elsewhere in management literature (see, for example: Albanese et al., 1997; Arthurs & Busenitz, 2003; Davis et al., 1997; Hernandez, 2008, 2012; Kuppelwieser, 2011; Le Breton-Miller & Miller, 2009; McArthur, 2012; Preston, 1998; Schillemans, 2013; Van Slyke, 2006). Across this literature, stewardship theory focuses on intrinsic motivation and rewards as a contrast to self-interested individuals (Kuppelwieser, 2011).

It is useful to consider label BCPs and Indigenous peoples through "stewardship" because both of them fulfil communal obligations and responsibilities to support the conservation of biological diversity and Indigenous rights to self-determination. However, overly prescriptive uses of the term risk simplifying the everyday lived realities of Indigenous peoples. Indigenous knowledge "is based on the premise that humans should not view themselves as responsible for nature, i.e., we are not stewards of the natural world, but instead we are a part of that world, not greater than any other part" (Pierotti & Wildcat, 2000, p. 1336). Similarly, guardianship has its roots in the idea of the "noble savage" which essentialises Indigenous peoples and risks Indigenous peoples being wed to a static identity which is unable to legitimately respond to changes in the broader global political economy (Wade, 1999). The association of Indigenous peoples with "guardian" also obfuscates Indigenous peoples' reality who use, whether through choice or necessity, technological innovations from the industrial and post-industrial eras in their farming and livelihood practices (for example, pesticides and firearms[8] (Wade, 1999)).

Challenges in Developing and Implementing Community Protocols

In our own research and engagement with communities, we have noted a range of challenges in developing and implementing community protocols. First, there are challenges surrounding the scope of the protocol and the scope (or coherence) of the "community". Communities are heterogeneous and so there are often competing interests and agendas about the use of natural and biological resources. There are often different interpretations of customary law, and different ideas about what should go into a protocol, or how it might best be used. In some of the communities we worked with, at the village level in Vanuatu, for example, there were individuals who were much more active in collecting and trading the biological resources (bush foods and medicines) than others. So, these specific individuals may have had more vested interests in the activities and in ensuring their voices would be heard in community meetings. There are also questions that can be asked about whether the scope of the community protocols should just focus on ABS and biotrade, or if it should be broader and include cultural heritage concerns, protection of intangible cultural heritage and folklore. In other circumstances, it seemed relevant that protocols might also be wider in scope so as to also deal with tourism and its potential impacts on biological resources and ecosystems. Because we have been focused on Indigenous knowledge relating to biological resources, the Nagoya Protocol clause on "community protocols" has guided our project down a particular path. But that path could certainly be wider if our project had been broader in scope.

Second, there were different ideas about what the protocol would look like. We had originally envisaged BCPs along the lines of those produced and facilitated by Natural Justice. These often had been influenced by and included statements about the relevant international laws, such as the Nagoya Protocol and the United Nations Declaration on the Rights of Indigenous Peoples (UNDRIP, 2007). However, many of the communities we met in the Pacific countries (Vanuatu and Cook Islands) did not know about these conventions and international laws. Nor did it seem important to them to be included in our drafting of the protocols, which were highly grounded in local customary law (or *kastom*) and local contexts of biotrade, beliefs, stories and practicalities. This meant that most communities asked us to work on highly practical protocols that in some cases would be like guidelines in both English and local language, as well as posters, flow charts and similar representations about what rules people accessing the community's knowledge (or a specific individual's knowledge) must follow. While these are still a work in progress (interrupted by COVID-19), there were many requests that the protocols be as simple and practical as possible – and so our drafts have been developed with this in mind. The community rules were perceived by government as practically linked to the Nagoya Protocol ideas of PIC. Additionally, for the national governments, there seemed a willingness to consider the protocols as supporting their national legislation through local PIC processes.

Third, the scale of protocol development was something that we have often thought about. For our five- to six-year project (interrupted by COVID-19), we have reached a protocol drafting stage for several communities in the three countries (in Australia, most of our support has been for bush foods and medicines enterprises and has focused more on biotrade agreements and ABS clauses in contracts). The process of working with remote communities involved in biotrade is highly labour-intensive and time-consuming. This means that we may potentially miss out on developing protocols with other communities that may be interested. It is not clear if communities in these three countries will self-generate their own protocols unless encouraged or supported to do so in some way. In Australia, bush foods enterprises are quite actively establishing contracts and agreements for their trading with clients, but this is somewhat different to what we had envisaged as a community protocol. In Vanuatu and Cook Islands, it is unclear if there is legal, governance and administrative capacity for local NGOs or government entities like the Vanuatu Cultural Centre (VKS) to develop community protocols on a wide scale. It is also not clear if there is enough funding, or if having a few protocols "here and there" in these countries is sufficient. For example, researchers who are not acting ethically might circumvent the protocols by seeking out other communities that are less well organised and willing to readily provide access to the knowledge and resources. As our project continues in the next three years and as borders re-open for travel and for our fieldwork, we will continue to consider these challenges and gaps in the development of community protocols.

Conclusion

Countries such as Vanuatu and Cook Islands, and indeed many of the Pacific Island countries, have strong legal and customary systems for the recognition of custom, making them ideal places to work on strengthening custom systems and/or developing BCPs. The existence of these pluralist legal orders aligns well with the broader agenda globally for better recognition of biocultural jurisprudence, respectful and responsive to self-policy or self-governance, and bottom-up decision-making.

Protocols are non-binding tools that are a hybrid of legal, ethical, political and cultural norms. They work and operate because they are, paradoxically, a tool that is both flexible and fixed (Raven, 2010). Because they have strong relational obligations, protocols also mirror stewardship. However, caution must be used in extending the use of this term in ways that curb self-determination. In Vanuatu and Cook Islands, protocols create interlinkages between national legislation and community practices; as such, they form the basis of a counter-narrative that embeds Indigenous ownership of knowledge in biodiversity conservation. Overall, we have found them useful tools for allowing communities to self-express their rules, beliefs and to manage access processes to community resources and knowledge. However, community protocols are also somewhat experimental and are not without a range of challenges that we need to continue to consider, monitor, evaluate, report on and discuss with the relevant communities, governments, researchers and enterprises working with biological resources and associated knowledge.

Notes

1 This research is supported by the Australian Research Council (ARC) Discovery Project (DP180100507): *Indigenous knowledge futures: protecting and promoting indigenous knowledge* (2018–2022). *All URLs retrieved on 1 September 2021.

2 The Dreaming term derives from the Australian anthropologist Stanner (1953), who used the term to describe creation beliefs, lores and stories for Aboriginal peoples in Australia.

3 In the general sense, biotrade refers to when a product or service sourced from biodiversity is commercialised and traded. Under certain initiatives, BioTrade refers to when a product or service sourced from biodiversity is commercialised and traded *in a way that respects people and nature, as per The BioTrade Principles and Criteria developed by UNCTAD* (see UNCTAD BioTrade Initiative at: https://unctad.org/topic/trade-and-environment/biotrade, accessed 9 June 2021). There is also a similar membership-based initiative that can certify the ethical and sustainable aspects of supply chains, called the "Union for Ethical Biotrade" (UEBT) (see https://www.ethicalbiotrade.org/).

4 DEPC https://environment.gov.vu/index.php/environment-conventions-and-agreements/laws/other-environmental-laws/94-policy-legislation, accessed 9 July 2021.

5 Protection of Traditional Knowledge and Expressions of Culture Act No. 21 of 2019 (2019). https://wipolex.wipo.int/en/text/546121

6 UNDP-GEF Project Document Strengthening the Implementation of the Nagoya Protocol on Access to Genetic Resources and Benefit Sharing in the Cook Islands, see: https://procurement-notices.undp.org/view_file.cfm?doc_id=32749, accessed 9 June 2021.

7 Bulun Bulun Case: *John Bulun Bulun & Anor v R & T Textiles Pty Ltd* [1998] IndigLawB 87; (1998) 4(16) Indigenous Law Bulletin 24

8 In Australia, for example, the *Yanner v Eaton* case in 1999 (201 CLR 351) recognised that under the *Native Title Act 1993*, the use of firearms was allowed for traditional hunting practices (Weir, 2012).

References

AIATSIS. (2012). *Guidelines for Ethical Research in Australian Indigenous Studies*. http://aiatsis.gov.au/sites/default/files/docs/research-and-guides/ethics/gerais.pdf

AIATSIS. (2020). *AIATSIS Code of Ethics for Aboriginal and Torres Strait Islander Research*. https://aiatsis.gov.au/sites/default/files/2020-10/aiatsis-code-ethics.pdf

Al Rashid, M. H., Majumder, S., Mandal, V., Mandal, S. C., & Thandavarayan, R. A. (2019). In Search of Suitable Extraction Technique for Large Scale Commercial Production of Bioactive Fraction for the Treatment of Diabetes: The Case Diospyros Melanoxylon Roxb. *Journal of Traditional and Complementary Medicine, 9*(2), 106–118.

Albanese, R., Dacin, M. T., & Harris, I. C. (1997). Agents as Stewards. *The Academy of Management Review, 22*(3), 609–611.

Argumedo, A., Asociación ANDES, Potatoe Park Communities, & IIED. (2011). *Community Biocultural Protocols. Building Mechanisms for Access and Benefit Sharing among the Communities of Potato Park basd on Customary Quechua Norms*. International Institute for Environment and Development (IIED). http://pubs.iied.org/G03168.html

Arjjumend, H. (2018). Recognition of Customary Law and Institutions and Community Protocols of Indigenous People in Domestic ABS Legislation or Policies in Accordance with the Provisions of Nagoya Protocol. *Journal of Vasyl Stefanyk Precarpathian National University, 5*(2), 67–82.

Arthur, B., & Morphy, F. (Eds.). (2019). *Macquarie Atlas of Indigenous Australia*. Macquarie University.

Arthurs, J. D., & Busenitz, L. W. (2003). The Boundaries and Limitations of Agency Theory and Stewardship Theory in the Venture Capitalist/Entrepreneur Relationship. *Entrepreneurship, Theory and Practice, 28*(2), 145–162.

Australia Council for the Arts. (2007a). *Music: Protocols for Producing Indigenous Australian Music*. http://www.australiacouncil.gov.au/about/protocols-for-working-with-indigenous-artists/

Australia Council for the Arts. (2007b). *Visual Arts: Protocols for Producing Indigenous Australian Visual Arts*. http://www.australiacouncil.gov.au/about/protocols-for-working-with-indigenous-artists/

Bavikatte, K., & Robinson, D. (2011). Towards a People's History of the Law: Biocultural Jurisprudence and the Nagoya Protocol on Access and Benefit Sharing. *Law, Environment and Development Journal*, 7(1), 35–51. http://www.lead-journal.org/content/11035.pdf

Berkes, F., & Berkes, M. K. (2009). Ecological Complexity, Fuxxy Logic, and Holism in Indigenous Knowledge. *Futures, 41*(1), 6–12.

Boer, B. (Ed.). (1996). *Environmental Law in the South Pacific*. South Pacific Regional Environment Programme and IUCN Environmental Law Centre. IUCN – The World Conservation Union.

Bowrey, K. (2006). Alternative Intellectual Property?: Indigenous Protocols, Copyleft and New Juridifications of Customary Practices. *Macquarie Law Journal, 6*, 65–95. http://www.austlii.edu.au/au/journals/MqLJ/2006/6.html

Carpenter, K. A., Katyal, S. K., & Riley, A. R. (2009). In Defense of Property. *The Yale Law Journal, 118*(6), 1022–1125.

Carter, J. L. (2010). Protocols, Particularities, and Problematising Indigenous 'engagement' in Community-Based Environmental Management in settled Australia. *The Geographical Journal, 176*(3), 199–213.

Christie, M. (2006). Transdiscplinary Research and Aboriginal Knowledge. *The Australian Journal of Indigenous Education, 35*, 78–89.

Clarke, P. A. (2003). Australian Ethnobotany: An Overview. *Australian Aboriginal Studies, 2*, 21–38.

Cox, P. (1991). Polynesian Herbal Medicine. In P. Cox & S. A. Banack (Eds.), *Islands, Plants, and Polynesians: An Introduction to Polynesian Ethnobotany* (pp. 147–168). Dioscorides Press.

Davis, J. H., Schoorman, D. F., & Donaldson, L. (1997). Toward a Stewardship Theory of Management. *Academy of Management Review, 22*(1), 20–47.

Department of Environment. (2014). *Access to Biological Resources in States and Territories.* Commonwealth of Australia. Retrieved Accessed 1 July 2021, from http://www.environment.gov.au/topics/science-and-research/australias-biological-resources/access-biological-resources-states-and-0.

Dunbar, T., & Scrimgeour, M. (2005). Ethical Assessment of Indigenous Health Research: A Review of the Literature. *Discussion Paper (Onemda VicHealth Koori Health Unit), No. 14.* Onemda VicHealth Koori Health Unit.

Escalante, A. E., Jardón Barbolla, L., Ramírez-Barahona, S., & Eguiarte, L. E. (2014). The Study of Biodiversity in the Era of Massive Sequencing. *Revista Mexicana de Biodiversidad, 85*(4), 1249–1264.

Forsyth, M., & Farran, S. (2015). *Weaving Intellectual Property Policy in Small Island Developing States.* Intersentia.

Gao, Q., Yue, G., Li, W., Wang, J., Xu, J., & Yin, Y. (2012) Recent Progress using High-Throughput Sequencing Technologies in Plant Molecular Breeding. *Journal of Integrated Plant Biology, 54*(4), 215–227. doi: 10.1111/j.1744–7909.2012.01115.x. PMID: 22409591

Garwood-Houng. (2005). Protocols: Meeting the Challenges of Indigenous Information Needs. *Australian Academic and Research Libraries, 36*(2), 143–151.

Goagoses, N., Winschiers-Theophilus, H., & Zaman, T. (2020). Community Protocols for Researchers: Using Sketches to Communicate Interaction Guidelines. *AI and Society, 35*(3), 675–687.

Halewood, M., Bedmar Villanueva, A., Rasolojaona, J., Andriamahazo, M., Rakotoniaina, N., Bossou, B., Mikpon, T., Vodouhe, R., Fey, L., Drews, A., Kumar, P. L., Rasoanirina, B., Rasoazafindrabe, T., Aigbe, M., Agbahounzo, B., Otieno, G., Garforth, K., Kiene, T., & Nnadozie, K. (2021). Enhancing Farmers' Agency in the Global Crop Commons Through Use of Biocultural Community Protocols. *Agriculture and Human Values.* https://doi.org/10.1007/s10460-020-10164-z

Hernandez, M. (2008). Promoting Stewardship Behavior in Organizations: A Leadership Model. *Journal of Business Ethics, 80*(1), 121–128.

Hernandez, M. (2012). Toward an Understanding of the Psychology of Stewardship. *Academy of Management Review, 37*(2), 172–193.

Huerlimann, R., Cooper, M. K., Edmunds, R. C., Villacorta-Rath, C., Le Port, A., Robson, H. L. A., M., S. J., Burrows, D., & Jerry, D. R. (2020). Enhancing Tropical Conservation and Ecology Research with Aquatic Environmental DNA Methods: An Introduction for Non-environmental DNA Specialists. *Animal Conservation, 23*(6), 632–645.

Janke, T. (1998). Don't Give Away Your Valuable Cultural Assets: Advice for Indigenous Peoples. *Indigenous Law Bulletin, 4*(11), 8–11. http://www.austlii.edu.au/au/journals/ILB/1998/35.html

Janke, T. (2008). Indigenous Knowledge & Intellectual Property: Negotiating the Spaces. *The Australian Journal of Indigenous Education, 37*(Supplement 1), 14–24.

Janke, T., & Dawson, P. (2012). New Tracks: Indigenous Knowledge and Cultural Expression and the Australian Intelletual Property System. Terri Janke and Company Pty Ltd. Retrieved 1 July 2021, from https://apo.org.au/sites/default/files/resource-files/2012-05/apo-nid38273.pdf

Jonas, H., Shrumm, H., & Bavikatte, K. (2010). Biocultural Community Protocols and Conservation Pluralism. In *Policy Matters 17* (pp. 102–112). IUCN-CEESP. http://community-protocols.org/wp-content/uploads/documents/BCPs_and_conservation_pluralism-policy_matters.pdf

Kolig, E. (2002). Guardians of Nature or Ecologists of the Stomach? The Indigenous Cultural Revival in New Zealand, Resource Use and Nature Conservation. In E. Kolig & H. Mückler (Eds.), *Politics of Indigeneity in the South Pacific: Recent Problems of Identity in Oceania* (pp. 95–118). LIT Verlag.

Kuppelwieser, V. G. (2011). Stewardship Behavior and Creativity. *Management Revue, 22*(3), 274–295.

Lai, J. L. (2014). *Indigenous Cultural Heritage and Intellectual Property Rights. Learning from the New Zealand Experience?* Springer.

Laird, S., & Wynberg, R. P. (2018). *A Fact-Finding and Scoping Study on Digital Sequence Information on Genetic Resources in the Context of the Convention on Biological Diversity and the Nagoya Protocol.* CBD/DSI/AHTEG/2018/1/3. https://www.cbd.int/doc/c/e95a/4ddd/4baea2ec772be28edcd10358/dsi-ahteg-2018-01-03-en.pdf

Le Breton-Miller, I., & Miller, D. (2009). Agency vs. Stewardship in Public Family Firms: A Social Embeddedness Reconciliation. *Entrepreneurship, Theory and Practice, 33*(6), 1169–1191.

Lotz, M. (2002). Colliding Worlds: Indigenous Rights, Traditional Knowledge, and Plant Intellectual Property. *Business & Professional Ethics Journal, 21*(3/4), 71–94.

Mackay, E. (2009). Indigenous Traditional Knowledge, Copyright and Art-Shortcomings in Protection and an Alternative Approach. *UNSW Law Journal, 32*(1), 1–26. http://www.austlii.edu.au/au/journals/UNSWLJ/2009/1.html

Malsale, P., Sanau, N., Tofaeono, T. I., Kavisi, Z., Willy, A., Mitiepo, R., Lui, S., Chambers, L. E., & Plotz, R. D. (2018). Protocols and Partnerships for Engaging Pacific Island Communities in the Collection and Use of Traditional Climate Knowledge. *Bulletin of the American Meteorological Society, 99*(12), 2471–2489.

McArthur, M. (2012). *The Meaning and Practice of Stewardship.* University of Calgary.

Mead, A. T. P., & Ratuva, S. (2007). *Pacific Genes and Life Patents: Pacific Indigenous Experiences and Analysis of the Commodification and Ownership of Life.* Call of the Earth Llamado de la Tierra and the United Nations University Institute of Advanced Studies.

Miller, R. J., J. Ruru, L. Behrendt & T. Lindberg. (2010). *Discovering Indigenous Lands: The Doctrine of Discovery in the English Colonies.* Oxford University Press.

Nakata, M., Byrne, A., Nakata, V., & Gardiner, G. (2005). Libraries, Indigenous Australians and a Developing Protocols Strategy for the Library and Information Sector. *Australian Academic and Research Libraries, 36*(2), 185–199.

NHMRC. (2018a). *Ethical Conduct in Research with Aboriginal and Torres Strait Islander Peoples and Communities: Guidelines for Researchers and Stakeholders.* https://www.nhmrc.gov.au/about-us/resources/ethical-conduct-research-aboriginal-and-torres-strait-islander-peoples-and-communities#block-views-block-file-attachments-content-block-1

NHMRC. (2018b). *National Statement on Ethical Conduct in Human Research.* https://www.nhmrc.gov.au/research-policy/ethics/national-statement-ethical-conduct-human-research

Parks, L. (2018). Challenging Power from the Bottom Up? Community Protocols, Benefit-sharing, and the Challenge of Dominant Discourses. *Geoforum, 88*, 87–95.

Pierotti, R., & Wildcat, D. (2000). Traditional Ecological Knowledge: The Third Alternative. *Ecological Applications, 10*(5), 1333–1340.

Preston, L. E. (1998). Agents, Stewards, and Stakeholders. *The Academy of Management Review, 23*(1), 9–13.

Raven, M. (2010). Protocols: Devices for Translating Moralities, Controlling Knowledge, Defining Actors in Indigenous Research, and Critical Ethical Reflection. *Australian Aboriginal Studies, 2*, 33–47.

Robinson, D. (2010). *Confronting Biopiracy: Challenges, Cases and International Debates.* Earthscan/Routledge.

Robinson, D. F. (2014). *Biodiversity, Access and Benefit-sharing: Global Case Studies.* Routledge.

Robinson, D., & Raven, M. (2017). Identifying and Preventing Biopiracy in Australia: Patent Landscapes and Legal Geographies for Plants with Indigenous Australian Uses. *Australian Geographer, 48*(3), 311–331.

Robinson, D. F., & Forsyth, M. (2016). People, Plants, Place, and Rules: The Nagoya Protocol in Pacific Island Countries. *Geographical Research, 54*(3), 324–335.

Robinson, D. F., & Raven, M. (2020). Recognising Indigenous Customary Law of Totemic Plant Species: Challenges and Pathways. *The Geographical Journal, 186*(1), 31–44.

Schillemans, T. (2013). Moving Beyond the Clash of Interests: On Stewardship Theory and the Relationships between Central Government Departments and Public Agencies. *Public Management Review, 15*(4), 541–562.

Secretariat of the Convention on Biological Diversity. (2011). Nagoya Protocol on Access to Genetic Resources and the Fair and Equitable Sharing of Benefits Arising from their Utilization. To the Convention on Biological Diversity. Retrieved 9 September 2014, from http://www.cbd.int/abs/doc/protocol/nagoya-protocol-en.pdf

Stanner, W. E. (1953). *The Dreaming.* ABC Press.

Stephens, C., Parkes, M. W., & Chang, H. (2007). Indigenous Perspectives on Ecosystem Sustainability and Health. *EcoHealth, 4*(4), 369–370. https://doi.org/10.1007/s10393-007-0140-2

Stone, D. (1966). Self-Government in the Cook Islands 1965. *The Journal of Pacific History, 1*(1), 168–178.

Su, C. (2018). Towards New Ethics Protocols for Community-Based Research. *URBAN Research Network Working Paper 2018*, 1–18. http://urbanresearchnetwork.org/wp-content/uploads/2018/06/community-protocols-urban.pdf

Suvanto, S. (2020). *The Protection of Traditional Knowledge Associated with Genetic Resources by using Community Protocols.* University of Helsinki. http://urn.fi/URN:NBN:fi:hulib-202008263871

Tonkinson, R. (1982). National Identity and the Problem of Kastom in Vanuatu. *Mankind, 13*(4), 306–315.

Van Slyke, D. M. (2006). Agents or Stewards: Using Theory to Understand the Government-Nonprofit Social Service Contracting Relationship. *Journal of Public Administration Research and Theory, 17*(2), 157–187.

Van Trease, H. (1987). *The Politics of Land in Vanuatu: From Colony to Independence*. University of South Pacific.

Wade, P. (1999). The Guardians of Power. Biodiversity and Multiculturality in Colombia. In A. Cheater (Ed.), *The Anthropology of Power. Empowerment and Disempowerment in Changing Structures* (pp. 73–87). Routledge.

Watsa, M., Erkenswick, G. A., Pomerantz, A., & Prost, S. (2020). Portable Sequencing as a Teaching Tool in Conservation and Biodiversity Research. *Plos Biology, 18*(4). https://doi.org/https://doi.org/10.1371/journal.pbio.3000667

Weir, J. (2012). Country, Native Title and Ecology. In J. Weir (Ed.), *Country, native title and ecology* (pp. 1–19). Canberra: ANU E Press.

Whistler, W. A. (1992). *Tongan Herbal Medicine*. University of Hawaii Press.

Whyte, K. P. (2013). On the Role of Traditional Ecological Knowledge as Collaborative Concept: A Philosophical Study. *Ecological Processes,* 2(7), 1–12. https://doi.org/10.1186/2192-1709-2-7

9

THE KHOIKHOI COMMUNITY'S BIOCULTURAL RIGHTS JOURNEY WITH ROOIBOS

Leslé Jansen and Rayna Sutherland

Introduction: The Khoikhoi, the San and Rooibos

The Khoikhoi have been documented as African Indigenous peoples, along with the San, who traditionally roamed and stewarded Southern Africa: the Khoikhoi as nomadic pastoralists and the San as hunter-gatherers. The Khoikhoi people are made up of historical groupings which include the (i) Griqua, (ii) Nama, (iii) Koranna and (iv) Cape Khoi, each of which has further subgroupings. There are also Indigenous farming communities, as descendants of the original Khoikhoi, living mainly in the Cederberg region in the Western Cape province of South Africa. In this area, where the cultural heritage of the Khoikhoi is rich, the Indigenous farming communities continue to practise their traditional knowledge (TK) of Rooibos to steward the plant. In their land-based cultures and economies, the Khoikhoi are "known for their spiritual connectedness to land" (National Khoi and San Council & Cederberg Belt Indigenous Farmers Representatives, 2019, p. 24). Tied to their beliefs that land is a sacred gift from nature, rather than an economic commodity, was their shared commitment of a sacred duty to care for nature, as it cared for them (Natural Justice, 2019). Expressions of these ways of knowing, embedded in values of sharing, neighboroughly love, respect for the environment and nature, and compassion, can be found in community-held myths (ibid.). Not only do these stories carry these values, but storytelling of them was a custom of education to ground these values across generations.

Within these cosmologies, land was believed to "belong to all living creatures that live on it" (National Khoi and San Council & Cederberg Belt Indigenous Farmers Representatives, 2019, p. 24). Therefore, while the Khoikhoi claimed no individual right to land, as individualised ownership was outside of their ways of knowing, they maintained deeply held connections to land (Boezak, 2017). The

DOI: 10.4324/9781003172642-11

Khoikhoi's heritage and identity, intimately tied to land, is still beautifully etched into the environment today, in the form of rock art. Though the Khoikhoi's rock art is comparatively rare to that of the San, it may be described as follows:

> The art of painting on rocky surfaces in caves and in open lands is a unique and defining characteristic of our distinct identity and heritage. Our paintings date back thousands of years and serve as a testimony to our right to land and its resources in South Africa. Khoikhoi rock art is made up of different designs, finger dots and handprints common amongst the Khoikhoi people. The designs were applied with fingers, making a striking contrast to the work of the San.
>
> *(National Khoi and San Council & Cederberg Belt Indigenous Farmers Representatives, 2019, p. 24)*

As documented in colonial British, Dutch, German and Portuguese archival records, the Khoikhoi held and shared vast wealth inclusive of cattle, sheep, TK, customary resources and diverse languages, constituting a rich heritage (National Khoi and San Council & Cederberg Belt Indigenous Farmers Representatives, 2019). Just as the Khoikhoi expressed their stories with land through their respective forms of rock art, they were also connected to their environment through TK of its flora and fauna, including of Rooibos (Natural Justice & the ABS Capacity Development Initiative, 2018). Rooibos is endemic to South Africa and is plentiful specifically in the Cederberg Mountains of the Western Cape province, as well as in some areas of the Northern and Eastern Cape (National Khoi and San Council & Cederberg Belt Indigenous Farmers Representatives, 2019). Through "oral tradition, storytelling and teaching by doing" (ibid., p. 60), the Khoikhoi's ancestors passed down their TK to share the knowledge to use and care for the plant across generations. The Khoikhoi expressed their belief that they "have a transgenerational link to the transmission of the traditional knowledge and that this is evident in the knowledge that [they] hold and share" (ibid.). While there is no internationally accepted definition of TK, the World Intellectual Property Organization (WIPO) speaks to it as follows:

> Traditional knowledge (TK) is knowledge, know-how, skills and practices that are developed, sustained and passed on from generation to generation within a community, often forming part of its cultural or spiritual identity.[1]

Rooibos, known to the Khoikhoi in Afrikaans as "*die arm man se tee*" (the poor man's tea) or "*bossie tee*" (bush tea) and known scientifically as *Aspalathus linearis*, is plentiful in uses, intimately known by the Khoikhoi (National Khoi and San Council & Cederberg Belt Indigenous Farmers Representatives, 2019, p. 6). For generations, it has been prepared for its benefits of high antioxidants as well as its uses to relieve allergy systems, boost energy, heal damaged skin, stimulate appetites, boost immune systems, facilitate relaxation and treat maladies such as hypertension and cardiovascular health, hypotension, stomach-related ailments, blood circulation, kidney ailments, stress relief, health skin and boost polyphenol/micro-nutrient levels (ibid.). When paired with other herbs, it may also be used to treat diabetes and improve oral health. As it has been incorporated as an ingredient in skincare products, Rooibos is used to treat eczema and minor skin injuries. Beyond its health benefits, it also has agricultural uses such as to be used in soil mixtures to assist with mulching and soil fertilisation (ibid.).

In their culture of sharing, the communities shared their knowledge of Rooibos. This began with the story of Tryntjie Swarts, a Khoikhoi woman living in the Cederberg who, in the 1920s, shared her "ecological-cultural knowledge of the Khoikhoi about how to locate the 'golden nests' of Rooibos seeds" (ibid., p. 60). This knowledge was a critical catalyst of the expansion of the present Rooibos industry as Rooibos tea, in itself, became popular the world over as well as its processing into health and cosmetic products. As will be explored, despite the foundational role of the Khoikhoi's TK upon which the development and commercialisation of the industry were built, the communities were never recognised as the knowledge holders nor received intellectual property rights and never gave free, prior or informed consent (FPIC) to the use of their knowledge (ibid.).

Historical Context

BOX 9.1 TIMELINE OF SOUTH AFRICAN HISTORY WITH SPECIFIC POINT RAISED IN RELATION TO THE EXPERIENCES OF THE KHOIKHOI

Historical Timeline

Below is a brief timeline of South African history with specific points raised in relation to the experiences of the Khoikhoi:

Pre-Colonial Era:

Pre-1400s: The Khoikhoi and San roamed Southern Africa as hunter-gatherers and nomadic pastoralists, respectively. Deoxyribonucleic Acid (DNA) confirms that the Khoi and San are first nations in South Africa.

1488: First contact with colonial forces as Vasco de Gama rounded the Cape as the Portuguese searched for the "New World".

Colonial Era:

1652–1800: Dutch colonial occupation of South Africa.

1800–1910: British colonial occupation of South Africa.

- *Hottentots Code of 1809* and the *Masters and Servants Act of 1856* were among a series of laws which legalised the restriction of the Khoikhoi's movement, land rights and further bound their livelihoods to that of commercial farm labourers.
- The *Apprentice Act of 1812* legalised the forced removal of Khoikhoi children from their parents to serve as farm labourers.

Apartheid Era:

1948–1990: White minority rule under the apartheid era.

- *Native Land Act of 1913* further legitimised the dispossession of lands as it affirmed the white minority's ownership to approximately 90% of the land and confined the Black majority rights to the remaining 10% of the land.
- *Race Classification and Population Registration Act of 1950* forcibly classified the Khoikhoi and San as "Coloured" along with several other communities.
- *Group Areas Act of 1950* once again forcibly removed the Khoikhoi from their homes as per their racial classification.

1990–1994: Negotiations to end apartheid and form new democratic government.

Post-Apartheid Era:

1994: First democratic election of South Africa and new democratic constitution is drawn.

1994: *Restitution of Land Rights Act 22 of 1994* is formed to restore land and provide other remedies to people dispossessed by racially discriminatory legislation and practice. The 1913 cut-off date for restitution claims discriminated against Khoikhoi as their historical claims dating from the start of European colonisation in 1652 were not seen admissible.

1996: President Mandela promulgates the 1996 Constitution of South Africa; yet, the Khoikhoi and San are left out of legislative recognition and governance inclusion within the institution of traditional leaders.

1999: The National Khoi and San Council is established as the official forum to negotiate with government regarding constitutional accommodation of the Khoikhoi and San.

1999: San, Cape Khoi, Griqua, Korana and Nama, five major Khoi and San groupings, were written into a Status Quo report.

Along this history of roaming and stewarding the lands of Southern Africa, the Khoikhoi faced enduring forces of colonialism beginning in 1652. Genocidal wars, widespread land dispossession and civilising mission's intent on cultural erasure were layers of violence which not only physically harmed the Khoikhoi, but sought to strip them of their freedom, dignity and means of living their land-based cultures, governance systems and identities as Indigenous peoples. This historical period was seen as the onset of what would become 300 years of systematic oppression seeking to relegate the Khoikhoi's Indigenous beliefs, ways of life, systems and overarching cosmologies "to that of mere savagery" (National Khoi and San Council & Cederberg Belt Indigenous Farmers Representatives, 2019, p. 28). The colonial era further systematically oppressed the Khoikhoi through several colonial government Codes. The *Hottentots Code of 1809* (also known as the Caledon Code) as well as the *Masters and Servants Act of 1856* were among a series of laws which legalised the restriction of the Khoikhoi's movement and further bound their livelihoods to that of commercial farm labourers (ibid.). In this way, they were fixed to commercial farms or mission stations as a "place of abode" to ensure commercial labour following the abolishment of slavery in that era. In addition, any and all movements beyond this abode had to be legitimised with a pass from their master or local official (ibid.). In this way, the Khoikhoi's traditional livelihoods of roaming nomadic pastoralists and land-based peoples were not only restricted in their landless state, but criminalised. The Khoikhoi state this as "legal and moral justification for the colonial dispossession of [their] sovereign indigenous people from their lands and resources" (ibid., p. 30). Community cohesion and cultural continuity were further hindered with the *Apprentice Act of 1812* which legalised the forced removal of Khoikhoi children from their parents to serve as farm labourers with little regard for their freedom from servitude as they aged into adults (ibid.). With this Act, Khoikhoi children were removed from their communities, subject to the trauma of servitude and stripped of their right to grow up learning their culture, language and TK. On these farms, trauma was further inflicted, and dignity further denied through the "dop system" (drink system), prevalent in the Western Cape, in which menial wages were offset by giving farm labourers cheap wine as a "fringe benefit" (ibid.). In the context of the extent of physical violence as well as violence in the form of stolen freedom and resources and denigration of identity, culture, livelihoods and cosmologies, this system further exacerbated trauma to incite and encourage substance abuse.

In the apartheid era (1948–1990), the Khoikhoi's resilient ties to land and identity were further eroded. With the *Race Classification and Population Registration Act of 1950*, racial categories of (i) White, (ii) African, (iii) Indian and (iv) Coloured were created and systematised. The Khoikhoi were forcibly classified as "Coloured" along with several other communities, obscured under this homogenising racial category. For the Khoikhoi, this imposed classification under this label "further dispossessed [them] from [their] African beingness by severing the last connections to [their] customary resources and knowledge" (National Khoi and San Council & Cederberg Belt Indigenous Farmers Representatives, 2019, p. 30). This violent apartheid legacy was confirmed in this effect by the UN Special Rapporteur, Prof. Rodolfo Stavenhagen (2005) in his UN Mission Report on the situation of Indigenous peoples in South Africa. Finally, the Khoikhoi were once again forcibly removed with the onset of the *Group Areas Act of 1950* which facilitated systemic spatial disruption "clearly along racial lines" (National Khoi and San Council & Cederberg Belt Indigenous Farmers Representatives, 2019, p. 30). Many of these communities were forcibly removed from the foothills of sacred mountains, such as Hoerikwagoo, now known as Table Mountain, to the Cape Flats. Since 2004, Table Mountain remains a renowned UNESCO Natural World Heritage Site and demarcated protected area. In contrast, the Cape Flats is rich in culture; yet, as it is ongoingly neglected by government development priorities, it is also a "crime ridden, impoverished wetland area with extreme social and economic challenges" (ibid., p. 30).

Contemporary Socio-political Challenges

Colonial and apartheid legacies of these forms of violence live on today as many Khoikhoi people continue to serve as commercial farm labourers on lands in which they "hold deep ancestral claim to" (National Khoi and San Council & Cederberg Belt Indigenous Farmers Representatives, 2019, p. 29). Yet, they "hold no real land and resource rights on these farms, other than some residential rights and informal employment" (ibid., p. 29). Following the early colonial criminalisation of the Khoikhoi and the later facilitated prison slave labour for Afrikaner commercial farmers under apartheid, the Khoikhoi continue to be disproportionately incarcerated 200 years later (ibid.). The legacies of the dop system live on to still affect the health of Khoikhoi as foetal alcohol syndrome is seven times higher in their communities in comparison to the world average.

Since the colonial era, the African Indigenous identity of the Khoikhoi has been disregarded or marginalised to where the Khoikhoi continue to experience a lack of social and political visibility in the new democratic state. As an ongoing apartheid legacy, the Khoikhoi continue to be homogenised under the category of "Coloured" in official state statistics which still mirror apartheid typologies of race which never reflected the existence of the Khoikhoi as Indigenous peoples (Le Fleur & Jansen, 2013). For example, 2011 census data was disaggregated by racial categories of (i) Black South African, (ii) White, (iii) Coloured and (iv) Indian/Asian (Alexander, 2018). Similarly, while there are 11 official languages represented in South Africa, none of the Khoikhoi people's Indigenous languages are included among them. As a result, to advocate for the human rights and fundamental freedoms of the Khoikhoi as Indigenous peoples, the UN Special Rapporteur, Radolfo Stavenhagen, recommended in his 2005 report "that indigenous communities be recognized as such constitutionally and that legal institutions maintaining the stigma of their classicisation as 'Coloured' by the apartheid regime be removed" (Stavenhagen, 2005, p. 3). The South

African Human Rights Commissioner (2016) affirmed the reality of a "cultural genocide" as follows:

> According to representatives for indigenous communities, the perpetuation of what they deem to be 'cultural genocide' continues as a result of the lack of recognition of indigenous peoples, together with inadequate measures to promote protect, and preserve indigenous culture and tradition. The prevailing lack of recognition and invisibility of Khoi-San groups exacerbates their vulnerability and marginalisation, fuelling their sense of an identity crisis and contributing to the assertions in submissions that indigenous persons are unable to access their full entitlement of rights contained in the Bill of Rights.
>
> *(ibid., p. 36)*

Up until recently and throughout the BCP process, the Khoikhoi communities were not constitutionally recognised by the state, in terms of national legislation, as a customary/Indigenous community (Jansen, 2019). However, with the signing of the *Traditional and Khoi-San Leadership Act 3 of 2019* into law and its commencement date of 01 April 2021, a new moment has since been heralded for state affirmation of the Khoikhoi as self-identifying African Indigenous communities. With this legislation, the Khoikhoi will be included in South Africa's traditional leadership and governance system on par with other customary communities (Jansen, 2019). Not only is this symbolically key for recognition of self-identification and dignity, but it is also significant in relation to conferring formal representation at different levels of government. This will pragmatically facilitate greater access to justice "as communities who have so far been formally left outside of the South African rule of law as it pertains to their cultural recognition, customary communities, Indigenous languages and ancestral lands" (Jansen, 2021, p. 135).

While the Khoikhoi were not previously recognised in South Africa's constitution system, the communities have been represented by the National Khoi and San Council (NKSC) since 1999. In 1999, the state commissioned the Status Quo Reports of 1999, to research the leadership claims of the Khoikhoi and San. Former President Nelson Mandela then established the NKSC based on the reports' recommendation to do so, given the absence of a national body to govern the communities' interest (National Khoi and San Council & Cederberg Belt Indigenous Farmers Representatives, 2019). The NKSC continues to act as a

> national, non-statutory and voluntary organization with its own legal personality, set up as a negotiating body for the Khoikhoi and San peoples of South Africa to represent their constitutional interests and to accommodate the Khoikhoi and San mandates within government departments.
>
> *(ibid., p. 46)*

Following independent research into Khoikhoi and San communities and membership in 1999 and an election of further leadership in 2012, 30 members of the NKSC now represent the historical Khoikhoi and San groups across South Africa: the San, Griqua, Nama, Cape Khoi and Koranna (ibid.).

Finally, just as the Khoikhoi and San were so long unrecognised as Indigenous peoples and excluded from post-apartheid development agendas, they were also excluded from land reform processes, mainly, land restitution. Given the vast inequity of contemporary land holdings along racial lines in South Africa, as a legacy to apartheid policies, land reform is a critical three-pronged process to redress systematic land dispossession and restore equity, justice and dignity. In 2018, the South African Parliament endorsed a report proposing an amendment to Section 25 of the Constitution to allow expropriation without compensation for the sake of accelerating a critical, yet acutely delayed land reform process. Thereafter, the President of South Africa, Cyril Ramaphosa, appointed an Expert Advisory Panel on Land Reform and Agriculture to analyse and give input on this process. However, questions remain on the inclusion of the Khoikhoi Indigenous representation on this panel. Despite the importance of this report, its output recommendations offered little and unclear considerations to meaningful inclusion and accommodation of the Khoikhoi's needs in the advised implementation planning. This oversight in recommendation is likely tied to the associated critique that the report's analysis did not sufficiently represent the specific land concerns of the Khoikhoi.

One central land concern for the Khoikhoi is their unique positioning in land restitution policies. As the first peoples of Southern Africa, along with the San, the Khoikhoi have stewarded and remained connected to these lands since time immemorial but lost the vast majority of their lands during the colonial era. Therefore, they hold land claims dating back to 1652. However, *Land Restitution Act 22 of 1994* only permits claims for land lost due to racially discriminatory apartheid legislation post-1913. Therefore, this restricts the Khoikhoi's land claims, despite the evidenced importance of land to their identity, culture and livelihood as traditionally land-based Indigenous peoples. The Advisory Panel on Land Reform and Agriculture (2019) affirmed this injustice in that the Act "will not and has not delivered substantive justice for those persons that lost land long before 19 June 1913" (ibid., p. 27). In response to the Advisory Panel's report, Prof. Stanley Peterson, a member of the NKSC and Rooibos ABS Negotiations Team, stated that communities are left with questions and without assurance as to whether and how justice will be facilitated for the Khoikhoi following the report, which was intended to offer clarity in this regard (Jansen, 2021). In this way, once again, the Khoikhoi continue to be disregarded and left behind in the name of post-modern development initiatives. These processes intended to redress land dispossession and historical violence for collective dignity and rebuilding, and continue to insufficiently include equity considerations of the Khoikhoi and their due needs and rights of restorative justice.

Overall, the post-apartheid era, hoped to mark a shift towards equity and justice, brought an unexpected challenge for the Khoikhoi. As the nation looked to rebuild in the conceptualisation of the new democratic South Africa, the Khoikhoi were all but excluded from envisioning their right to self-determination on par with the rest of South African society. Within this broad challenge of unrecognition, the Khoikhoi were equally left out of processes of development and rebuilding from land restitution to recognition and protection of their languages. With layers of justice and healing due to their communities, in 2012, the NKSC decided to develop a community protocol with the objective of articulating the Khoikhoi communities' struggles as a non-recognised African Indigenous community in South Africa (National Khoi and San Council & Cederberg Belt Indigenous Farmers Representatives, 2019). As aligned with the historical and living contemporary struggles outlined, the identified priority issues for the protocol were as follows: constitutional recognition, recognition of their Indigenous language, land rights, intellectual property and Access and Benefit-Sharing (ABS), women and youth, and healing and economic development.

The BCP Process

The Rooibos Restitution Journey

In 2010, shortly before the Khoikhoi began their community protocol process, a biopiracy case linked to Rooibos came to light in South Africa (National Khoi and San Council & Cederberg Belt Indigenous Farmers Representatives, 2019). At that time, the multinational company, Nestlé, applied for patents to the uses of Rooibos and honeybush without the consent of the resource provider, South Africa (Natural Justice & the ABS Capacity Development Initiative, 2018). In the same vein, consent was also neither sought nor freely given from the knowledge holders, the Khoikhoi and San. As a response, the pan-African non-profit organisation (NPO), Natural Justice and Swiss NPO, The Berne Declaration (now known as the Public Eye), successfully launched a high-profile campaign against Nestlé, for contravening South African Law and the UN Convention on Biological Diversity and its Nagoya Protocol on ABS (National Khoi and San Council & Cederberg Belt Indigenous Farmers Representatives, 2019). This advocacy campaign rendered the patent application unsuccessful. Thereafter, in 2013, Nestlé approached both South Africa and its knowledge holders of Rooibos to seek consent for the development of a Rooibos tea product (Natural Justice & the ABS Capacity Development Initiative, 2018). This opened the development of a benefit-sharing agreement between Nestlé, the South African San Council and the NKSC.

Parallel to the unfolding of the Nestlé biopiracy case and then benefit-sharing process, the roots of a broader industry-wide benefit-sharing negotiations process began to take hold. First, the South African San Council wrote to the Minister of the then Department of Environmental Affairs (now Department of

Environment, Forestry and Fisheries) requesting negotiations with the Rooibos industry in accordance with the South Africa's legislative requirements: the National Environmental Management: Biodiversity Act 10 of 2004 (Natural Justice & the ABS Capacity Development Initiative, 2018). Two years later, the NKSC joined them in what would become a shared journey to seek recognition as the knowledge holders to the uses of Rooibos. The industry-wide negotiations were then collectively initiated by both councils in their first meeting with the South African Rooibos Council (SARC) that same year, in 2012.

As this journey further progressed and took shape as a priority, intellectual property and ABS rose as the key relative focus for the community protocol. As the ABS negotiations, and more deeply, the struggle for recognition as knowledge holders to the uses of Rooibos intensified, it was decided to shift attention onto this process (National Khoi and San Council & Cederberg Belt Indigenous Farmers Representatives, 2019). The significance of the issue at hand was recognition of the Khoikhoi as the TK holders of Rooibos, along with the San, and uniting the diversity of Khoikhoi communities to organise collectively, including the unrepresented Cederberg belt Indigenous farming communities.

In terms of the struggle for recognition, part and parcel to this process was the South African government commissioned TK study to enquire whether the Khoikhoi and San were the undisputable knowledge holders to the uses of Rooibos and honeybush (DEA, 2014). Similarly, the Rooibos industry submitted their official notice to conducting an independent TK study that they planned to commission for their own research on the same issue. Although they commissioned their report as intended, their results were never made public. However, the South African government's study found that there was "no evidence that disputes" the communities' claim to being the TK holders (DEA, 2014, p. iii).

With the publication of the study, affirming their TK status, and the successful signing of the first ABS agreement for Rooibos, between Nestlé, the NKSC and the San Council, the momentum of this journey continued as the NKSC formed an ABS subcommittee for ABS/Intellectual Property, which would become the central ABS Negotiations Team for Rooibos as well as other plant ABS agreements (Natural Justice & the ABS Capacity Development Initiative, 2018). The agreement with Nestlé was also strategically used to insert a clause stating that all Rooibos sourced should be ethically done in compliance with the law (ibid.). A year later, the NKSC began to outreach with the Cederberg belt communities to include them in the Rooibos industry negotiations and eventually, the Cederberg belt community joined the NKSC ABS negotiations committee in 2015. The process continued to develop as, in 2016, the South African Rooibos industry, through the SARC, came to the negotiations table overseen by, what was then known as, the Department of Environmental Affairs. At this stage, a specific Rooibos Biocultural Community protocol was opened by the NKSC. The goal of the negotiations was an industry-wide agreement, including a TK levy on the use of TK associated with Rooibos. The goal of the Rooibos BCP was to document the Khoikhoi as the TK holders to Rooibos and as will later be explored, organise the communities collectively, with due representation, for resource governance and benefit-sharing.

In 2019, negotiations concluded, and the finalised ABS agreement brought the Rooibos Industry into compliance with their benefit-sharing obligations as per international and South African bioprospecting law and regulations (Natural Justice & the ABS Capacity Development Initiative, 2018). In terms of international law, the ABS agreement is grounded by the UN Convention on Biological Diversity (CBD) and Nagoya Protocol on Access and Benefit-Sharing (Article 7 and Article 12), a supplementary agreement to the CBD. Specifically, one of the fundamental objectives of Article 1 of the CBD is the fair and equitable sharing of the benefits arising out of the utilisation of Indigenous biological resources. In this way, the agreement centrally incorporates FPIC for access to resources and TK (Natural Justice, n.d.). In terms of South African Law, the National Environmental Management: Biodiversity Act 10 of 2004 and the Bioprospecting, Access and Benefit Sharing Regulation of 2008 (No. R. 138 of 2008) were enacted in order to provide a regulatory framework for ABS (Natural Justice, n.d.). The South African legislative framework goes beyond the Nagoya Protocol's standard of regulating "genetic components" of resources to regulating the utilisation of the whole of the Indigenous biological resource (Natural Justice, n.d.). This policy approach has been crucial for the Khoikhoi with regard to Rooibos and social justice.

These benefit-sharing obligations are upheld in the form of a TK levy, a percentage of the profits from the marketing of Rooibos by the South African Rooibos industry, as a benefit to be shared with Khoikhoi and San communities. The first levy payment by the Rooibos industry was expected to be due and payable to Khoikhoi and San communities in June 2020. However, the industry reported that the payment would be delayed due to governmental administrative lags and the impact of COVID-19 on their own farming. While the first Rooibos levy was not paid by the industry to the South African Biodiversity Fund, the Fund should, in turn, pay the levy over to the two groups. Both the Khoi and San have formed their representative community trusts for the purposes of distributing the projected benefits to their communities, once the levy is received.

Significance of Restitution for Landless Knowledge Holders

Given the unredressed colonial and apartheid legacies the Khoikhoi continue to experience, inclusive of their enduring landlessness and, at the time of signing of the agreement, their constitutional unrecognition, this landmark ABS agreement was not only historically precedent setting for Indigenous peoples, but incredibly significant and symbolic for the Khoikhoi. The Khoikhoi have lived as landless people, made to work as farm labourers on others' commercial farms or lease the land upon which they stewarded their precious Rooibos. While they labour on these lands, without constitutional recognition of their customary community as the first peoples of these same lands, they are instead still obscured under the imposed classification of "Coloured". In their BCP, the communities express this as living as a "landless proletariat and de-Africanised, with no access to [their] resources, no communal land, stripped of [their] cultural and collective identity" (National Khoi and San Council & Cederberg Belt Indigenous Farmers Representatives, 2019, p. 39). While the Rooibos industry grew from the misappropriation of their knowledge, the "legacy and demeaning stigma" of the racial classification of "Coloured" continued to "effect the image of Rooibos Indigenous farming communities as being poor 'Coloured' labourers, as opposed to traditional knowledge holders central to the growth and development of the Rooibos industry" (ibid., p. 54). One Indigenous farming community leader, Alida Afrika (as cited in National Khoi and San Council & Cederberg Belt Indigenous Farmers Representatives, 2019), expressed that the compounding challenges of the "Coloured" stigma and lack of land ownership lead to a lack of freedom for self-governance and determination to shape their own community development. In this way, they are hindered in their capacity to offer their youth prosperous futures in their home communities and within their knowledge systems and livelihoods. It is expressed that this challenge leads many youths to desire to leave their communities and farming practices to migrate to urban settings. In this vein, cultural continuity in terms of opportunities to intergenerationally pass down TK as well as confer feelings of pride in one's valuable Indigenous knowledge, identity and community is significantly hindered.

Challenges of community farming are compounded in the ongoing absence of support from the state. Specifically, the absence of support is felt in terms of means to mitigate and overcome the challenges of the dry, drought-prone area, exacerbated by climate change-driven higher temperatures, droughts and extreme rainfalls (National Khoi and San Council & Cederberg Belt Indigenous Farmers Representatives, 2019). Moreover, communities which so long stewarded these lands sustainably are now disproportionately left to face the brunt of climate change. It was expressed by one Indigenous farming community member, Alida Afrika, that, "[t]hrough [their] cooperative, [they] try to develop systems to help [their] members to expand their existing traditional knowledge to improve their organic practices, but it remains a challenge without sufficient funds available" (National Khoi and San Council & Cederberg Belt Indigenous Farmers Representatives, 2019, p. 55).

As landless peoples with little avenue for redress, given the limits to land restitution, access to land as well as plant, marine and wildlife resources is "near impossible" (ibid., p.32). The Khoikhoi continue to face challenges in that their customary law dictates that they are well within their rights to hunt, fish and gather for their subsistence needs and yet, they face barriers such as having to pay "expensive rentals to commercial farmers and municipalities to harvest these species that [they] consider [their] ancestral resources" (ibid., p. 32). In other cases, communities face criminalisation for accessing resources as per their subsistence needs. Amidst the challenges of COVID-19 in the Northern Cape, a lack of infrastructure such as economic centres in close proximity to the Khomani

San, a San community, and price inflation of basic food stocks in relation to the market pressure of the pandemic has led a surge of food insecurity (Jansen, 2021). In tandem, communities' incomes, driven mainly by tourism, became even more precarious, exacerbated by their already limited access to sufficient food and nutrients (ibid.). As a result, members of the community hunted wildlife to address their hunter challenges. In times of needs, they were met with criminalisation for their acts of self-reliance and efforts to meet their most basic needs.

It is within this context of the struggle for recognition as Indigenous peoples, access to their land and resources and the issue of dignity embedded within both that this journey for recognition as TK holders carries such significance. The Khoikhoi have always stewarded Rooibos and intimately carried a culture tied to it. One Cederberg Belt Indigenous farming community leader, Barend Salomo, expressed this as follows: "My culture connects me with rooibos, it is unthinkable to have a culture without rooibos involved in it" (Natural Justice & the ABS Capacity Development Initiative, 2018, p. 48). Their TK of Rooibos is so deeply linked to their culture as Indigenous people; yet, it was misappropriated the world over without affirmation of their status as TK holders. In this way, the absence of recognition in tandem with the exploitation of their TK parallels ongoing patterns of unrecognition of the Khoikhoi and San. Therefore, their newfound recognition as the TK holders and agreement to grant a traditional levy for the use of this knowledge represents "a form of restitution, as redress for past injustices. It is the symbolic restoration of [their] beloved Rooibos" (National Khoi and San Council & Cederberg Belt Indigenous Farmers Representatives, 2019, p. 14). Through this two-pronged process of ABS negotiations and the parallel documentation of the BCP, the community was able to "document how [they] brought justice to [their] cultural resources, what [they] call 'onse moedersmelk' (our mother's breast milk)" (ibid., p. 18). At large, the process brings a shared sense of healing and restoration of dignity for Khoikhoi peoples. For communities who endured centuries of dismissal and denigration of their indigeneity, to then be internationally recognised as Indigenous peoples with valuable TK, known to be the backbone of an international industry, marks a powerful shift. This is a long-desired step towards recognition and affirmation of the value of their cultural identity and knowledge, so long forgotten. Prof. Stanley Peterson expressed it as such: "Our dignity is being restored in the land of our forefathers by recognising our Khoikhoi ancient traditional knowledge on rooibos. Our people are happy and our land healed" (Natural Justice, n.d.).

Organising for Resource Governance

As the BCP process shifted to focus on Rooibos, it became a critical concurrent component to the ABS negotiations in its role of bringing together once disparate Khoikhoi communities to organise them for the Rooibos negotiation process and later fair sharing of benefits from the successful ABS agreement. Given the Khoikhoi's de-Africanising apartheid experience, in being forcibly classified as "Coloured", coupled with their centuries of land dispossession and forced removals through both colonialism and apartheid, a sense of unity and community cohesion in their Indigenous identity was also stolen. As a result, the negotiations and associated BCP process had to ensure community outreach and inclusion to overcome this historic legacy of community cultural disruption. In the process, it worked to bring healing, pride and restoration to unity and community identity, along this journey of seeking recognition and benefits collectively.

The NKSC had represented, and continued to represent, 30 Khoi communities in their governance writ large, inclusive of the Nestlé ABS negotiations and early Rooibos industry negotiations process (Natural Justice & the ABS Capacity Development Initiative, 2018). However, it became clear that rural farming communities in the Cederberg Mountains, where Rooibos is endemic and grows most fruitfully, needed to also be included in this process (ibid.). Within this mountain range are the communities of Wupperthal, Nieuwoudtville and Suid Bokkeveld, as well as further surrounding farming areas, all of which are regions in which original Khoikhoi descendants live as Indigenous farming communities, many of which steward and farm Rooibos (National Khoi and San Council & Cederberg Belt Indigenous Farmers Representatives, 2019). In this collective process to overcome apartheid cultural erasure, and for many, to shed the imposed classification of "Coloured" and resurge with Indigenous self-identification, it was important to recognise communities who had been "deliberately unrecognized" and note their indigeneity, including the indigeneity of the farming community in the Cederberg (ibid., p. 50).

Joining this process for due recognition of Indigenous farming communities, in this collective journey of struggle for recognition, was an important process of healing for these farming communities in their own particular histories and living legacies of apartheid. Wupperthal, specifically, is a region which hosts the most densely populated concentrations of wild Rooibos. However, it is also a region that lives with a stark apartheid legacy in relation to land ownership. Wupperthal, once known as *Reitmond* and home to the *Gouri-Grika* (Griqua in English), was occupied as a mission station during the colonial era (ibid.). However, this land dispossession was further compounded and systemised as the land was later registered in the name of the Moravian church rather than ever being returned to Khoikhoi communities. Presently, the Moravian church still owns 36,000 ha of Khoikhoi ancestral land around Wupperthal (ibid.). As a result, the Indigenous farming communities in Wupperthal lease their ancestral land directly from the church to farm Rooibos (ibid.). Therefore, resurging from persistent landlessness and systemic discrimination, to recognise their rights as TK holders and affirm their heritage through the Rooibos ABS process, was a fundamental journey of restoration for the Wupperthal community. In their own right, the communities of Nieuwoudtville and Suid Bokkeveld, living amidst rich Fynbos[2] and heritage as Khoikhoi descendants, joined in this journey of resurgence to shed the stigma of their imposed "Coloured" identity and reconnect with fellow

Khoikhoi communities in this journey of self-determination, affirmation and recognition (ibid.).

As historical descendants of the Khoikhoi community and in their distinct role as the traditional custodians of Rooibos, an elected representative from each of the three farming communities joined the NKSC in this shared journey to represent the Khoikhoi TK holders to Rooibos in these Rooibos negotiations and associated BCP. In this way, they represented due beneficiaries within the Rooibos benefit-sharing model on the Rooibos Access and Benefit-Sharing Negotiations Team along with members of the NKSC, as well as members of Natural Justice who had signed a Memorandum of Understanding with the NKSC to support their advocacy, negotiations and BCP process. For the purpose of organising benefit distribution, the Indigenous farming community representatives also served on the Khoikhoi Peoples Rooibos and Biodiversity Trust which was registered in November of 2019.

Prior to joining in collaboration with the Cederberg Belt Indigenous farming communities, the NKSC had signed a 50/50 Benefit Sharing Agreement with the South African San Council around four key plant species (Natural Justice & the ABS Capacity Development Initiative, 2018). In their joint negotiations around benefit-sharing agreements for these various Indigenous biological resources, the agreement was for the benefit to be shared equally. In this way, as the Cederberg Belt Indigenous Farming Communities joined with the NKSC, the latter would represent the 30 Khoikhoi communities, as outlined in the 1999 Status Quo reports, together with the farming communities. The Khoikhoi Peoples Rooibos and Biodiversity Trust was then responsible for distributing the funds, accruing to the trust, as per the Rooibos benefit-sharing model (National Khoi and San Council & Cederberg Belt Indigenous Farmers Representatives, 2019). Respectively, the San communities were represented by the San Council, given that they first initiated this negotiations process, in the case of Rooibos, and had already been concluding benefit-sharing agreement on behalf of San communities over the course of the last decade, since the Hoodia case (Chennells, 2012.). Beyond negotiations, the San decided to document, assert and protect their TK through the South African National Recordal System, a national database which "records, documents, preserves and protects Indigenous knowledge for the benefit of the communities of South Africa" (National Khoi and San Council & Cederberg Belt Indigenous Farmers Representatives, 2019, p. 18; Wynberg, 2017, p. 205). Their uses of Rooibos were officially recorded in 2017. In their own right, the Khoikhoi communities decided to produce their BCP as facilitated by the NKSC inclusive of representatives from the Cederberg Belt Indigenous farming communities.

The BCP serves the purpose of providing a platform for the Khoikhoi to set out the relationship of their communities with the Rooibos plant, in their role as TK holders and "safeguard [their] resources and associated TK from being misappropriated" (ibid.). For one, given the colonial and apartheid legacies of challenges to access and exercise cultural traditions, this collective documentation of knowledge on traditional uses of plants, animals and other resources is, as the community expresses, "vital to maintaining [their] heritage for future generations" (ibid., p. 18). In addition to these internal community benefits of the BCP process, it is also an outward articulation of how the Khoikhoi communities are organised for the purposes of FPIC and ABS in relation to Rooibos. The Khoikhoi communities self-determined and expressed who they are and their current organisation, aims and priorities in the post-apartheid era. In self-determining their identity, their customary Indigenous resources (as set out in their Cultural Biodiversity Register), their intellectual property rights related to Rooibos and its commercialisation, and their terms of consensual engagement, the communities are able to assert their rights in all such regards. Therefore, the BPC "provides an interface between [their] communities that want to engage in ABS (including current and future intellectual property rights), and the ethical users of [their] traditional knowledge" (ibid., p. 18).

Importantly, the BCP is also a resource for the communities as TK rights holders and beneficiaries of the Rooibos industry ABS agreement. The BCP outlines how the Khoikhoi Peoples Rooibos and Biodiversity Trust operates as a governance structure in the form of an NPO in public interest. The distinct purpose of the trust is to "utilize the capital and income paid into the Trust Fund to support and benefit the Beneficiaries (Khoikhoi peoples inclusive of the Cederberg Indigenous farming communities)" (ibid., p. 90). The BCP outlines the representative membership of the Board of Trustees, its duty to host Annual General Meetings of which any beneficiary community is welcome, the aims, objectives and protocol by which to distribute funds, and annual reporting processes. Most importantly, the BCP clearly explains through text and diagrams, in both English and Afrikaans, how benefit-sharing works in the case of Rooibos and how community beneficiaries may access their benefits. In this way, in seeking to meet its objective of uniting and serving Khoikhoi and San communities, to both affirm and protect their rights inclusive of due benefits to uses of their TK, the BCP is able to do so accessibly. This is a critical success, given the challenges of uniting communities who have experienced a deep disruption to their culture and community cohesion in light of apartheid legacies of the "Coloured" classification and layers of dispossession.

Analysis

The damage left by apartheid and colonialism remains something that cannot be calculated or sufficiently articulated on paper. The documented wealth the Khoikhoi communities owned with the advent of colonialism, reflected in modern-day economics, would have equated to extreme wealth. The present reality of these landless TK holders, however, is a stark contrast. Many of these communities live on what is, today, called the Cape Flats, a wetland alongside Table Mountain, a World Heritage Site (called Hoerikwagga in their Khoekhoegoewab language). One sees nearly no material or physical semblances of their once wealth. The Western Cape province holds 18 prisons where a vast majority of Khoikhoi communities are living under a forcibly labelled identity of "Coloured". The evidence of genocide, slavery, cultural assimilation and land dispossession across a 400-year period is layered by apartheid spatial planning where the wealthy owns 70% of the land in South Africa. It remains a miracle that this community was able to secure this victory of the Rooibos agreement.

On top of these layers of oppression, academics, advocating in the name of these communities, were silently seeking to ensure this victory never saw the light of day. Yet, this divine miracle did indeed occur through the sweat of a long journey after nearly a decade of hard work.

Natural Justice lead the first crack at this miracle victory by running an international campaign, with the Berne Declaration, against Nestlé, for their illegal patent applications which were in violation of the Nagoya Protocol. Their sourcing of the plant material was done in violation of South Africa's bioprospecting regulatory framework. This international campaign set the global stage for Rooibos to emerge as an ABS case. This was critical work at the time, but it fell short in so far as recognising the communities as the TK holders. While the campaign opened the door to start addressing the ethical sourcing of Rooibos as a biotrade resource, their campaign did not address the fundamental issue, namely the recognition of the Khoikhoi communities, as the TK *rights holders*.

Parallel and unconnected, in 2011, the South African San Council wrote a letter to the government informing them of their claim as TK holders of Rooibos. A certain synchronicity of events aligned, a Khoikhoi Indigenous lawyer, joining Natural Justice at that time, a second Indigenous lawyer, heading Open Society's Indigenous Peoples Program, and the Chairperson of the Khoikhoi national community body, all met by divine coincidence. Once together, they then could form an alliance, with Natural Justice, to support the National Khoisan Council (NKSC) to ensure the second claim to Rooibos, as TK holding communities, included them, as the NKSC, alongside that South African San Council. This now meant that two Indigenous communities, namely, the Khoikhoi and the San, had staked their claim as TK holders of the uses of Rooibos. And so, the eight-year journey could unfold to ensure the SA Rooibos industry would be held accountable for benefit-sharing in the context of the Nagoya Protocol, on the basis that the Khoikhoi and San were the first inventors of the uses of Rooibos.

The Rooibos BCP process was a delicate process that could have fallen apart at any point. The Rooibos case was fought in the most unlikely circumstances, namely, being that none of these communities are documented as Indigenous or cultural communities in South Africa. They have no form of constitutional recognition on par with other cultural communities, their languages are not official languages, they have no access to their natural resources and as a result of extreme land dispossession, they are no longer able to show a connection to the land. They comprise different alliances, namely, the San's four historical communities, the 40 Khoikhoi communities and the Cederberg belt Indigenous

Rooibos farming communities (who fall under the larger Khoikhoi component). A division therefore existed between the cultural communities and the resource-based communities living with Rooibos. Managing this division, between these different groups, needed constant attention. What connects them is that they are all bound by the apartheid typology of being forcibly labelled "Coloured". This was in addition to apartheid's cruelty and systemic violence, a massive de-Africanisation process. The fundamental identity and existence as an organised Indigenous nation are so far altered; it made their claim as TK holders to the uses of Rooibos a most unbelievable claim to both government and industry alike. Their traditional institutions faced constant battles within fights that were not theirs, but they had to pay for through the proposed Traditional and Khoi-San Leadership Act 3 of 2019 (TKLA). This piece of legislation finally recognised their communities and their leadership as part of the state system, on par with other cultural communities. Despite the significance of this legislative change for the Khoikhoi and San, it was met with widespread controversy. It is argued that the TKLA premises governance as aligned with colonial and apartheid laws, in how it presents "jurisdiction over land for traditional authorities in the former Bantustans" and thus retains "apartheid created structures and boundaries" which structure traditional leadership's accountability to the state as opposed to its people (Mzwakali, 2019). However, this struggle to rectify imperfect legislation was beyond the fight of the Khoikhoi and instead, the controversy threatened the necessary inclusion the Act offered the Khoikhoi communities who had long been excluded from South Africa's constitutional system.

What complicated their claim further was the fact that they were not on the land. Confusion surrounded the idea of what it meant to want to claim one's rights as TK holders. It was largely assumed to mean a farming issue; however, it was not. Therefore, communities were being critiqued for saying that they do not live in the Cederberg, on the land where Rooibos grows. If it were the case that it was indeed a farming issue and a claim tied to being on the land in which Rooibos is endemic, it would only be the Cederberg communities that could make this claim. The question became: why would communities, living in different parts of South Africa, all have a claim as TK holders? They are not on the land; the Cederberg farming communities are looking after Rooibos.

Brewing beneath it all, the fundamental difficulty with the claim was the communities' assimilationist identity of being labelled "Coloured". The communities, labelled as "Coloured", have been the victims of a deep de-Africanising process. Consequently, they became viewed as not "legitimately" African. Their place at the South African table was not secured and they were left at the margins of the mainstay of South Africa. Their history, like that of many others, is fraught with complexities. One of those complexities is that the Khoikhoi people were forcibly labelled "Coloured". The communities coming forward as part of the Khoikhoi Indigenous resurgence movement are viewed as opportunistic and charlatans. However, there's more to the story, so much more. One piece of this story is that many people labelled "Coloured" were actively part of the anti-apartheid movement and regarded themselves as Black African. That was the basis on which the anti-apartheid struggle was fought; they were united in their cause. Yet, when the new South Africa was designed, these communities were told, "Sorry, you're not African". It forced this collective to go deep within themselves to find out, then, who exactly they are. The result was a resurgence with an indigeneity that held a great amount of wealth, stewarded the environment and its natural resources, and held genetic markers linking them to people, over 200,000 years prior, at the tip of Africa. Thirty years into South Africa's dispensation, these communities' resurgence movement is forcing cracks into the modern-day development space to find their place at the South Africa table. Rooibos lit the torch to help make those linkages. One of the knowledge holders, Oom Barend, said, "You don't have to teach them, just dust the peoples' memory a little and they will remember". And remember they did. The Rooibos BCP process facilitated a shift which forced them to relook at their traditional institutions and what they needed to do to respond to benefit-sharing amidst this complicated context.

It was as if Rooibos gave a clarion call, to these communities, about all their other resources. Rooibos shed light on the lie of biotrade. It forced South Africa to have a different conversation on their high value plant species. It now has to speak about TK.

Notes

1 https://www.wipo.int/tk/en/tk/. *All URLs retrieved on 1 September 2021.
2 Fynbos is a small belt of distinctive, heather-like tree and shrub vegetation endemic to the Western and Eastern Cape Provinces in South Africa.

References

Advisory Panel on Land Reform and Agriculture. (2019). *Final Report of the Presidential Advisory Panel on Land Reform and Agriculture*. Republic of South Africa. https://www.gov.za/sites/default/files/gcis_document/201907/panelreportlandreform_0.pdf

Alexander, M. C. (2018, July 9). South Africa's Population. *South Africa Gateway*. https://southafrica-info.com/people/south-africa-population/

Boezak, W. (2017). The Cultural Heritage of South Africa's Khoisan. In A. Xanthaki, S. Valkonen, L. Heinämäki, & P. K. Nuorgam (Eds.), *Indigenous Peoples' Cultural Heritage* (pp. 253–272). Brill | Nijhoff. https://doi.org/10.1163/9789004342194_013

DEA. (2014). *Traditional Knowledge Associated with Rooibos and Honeybush Species in South Africa*. Siyanda Samahlubi Consulting for Department of Environmental Affairs. Pretoria, https://www.dffe.gov.za/sites/default/files/reports/traditionalknowledge_rooibosandhoneybushspecies_report.pdf

Jansen, L. (2019). South Africa. In D. N. Berger (Ed.), *The Indigenous World 2019* (pp. 543–550). IWGIA.

Jansen, L. (2021). South Africa. In D. N. Berger (Ed.), *The Indigenous World 2021* (pp. 160–167). IWGIA.

Le Fleur, A., & Jansen, L. (2013). *The Khoisan in Contemporary South Africa*. Konrad Adenauer Stiftung Country Report. www.kas.de/suedafrika/en

Mzwakali, S. (2019). Draft traditional laws reimpose colonial logic. *New Frame*, Tuesday, 10 June 2022. https://www.newframe.com/draft-traditional-laws-reimpose-colonial-logic/

National Khoi and San Council, & Cederberg Belt Indigenous Farmers Representatives. (2019). *The Khoikhoi Peoples' Rooibos Biocultural Community Protocol.* https://naturaljustice.org/wp-content/uploads/2020/04/NJ-Rooibos-BCP-Web.pdf

Natural Justice. (n.d.). The Rooibos Access and Benefit-sharing Agreement. *Natural Justice.* Retrieved 21 April 2021, from https://naturaljustice.org/the-rooibos-access-and-benefit-sharing-agreement/

Natural Justice. (2019). *Indigenous Land Rights Report.* [unpublished draft].

Natural Justice, & the ABS Capacity Development Initiative. (2018). *Community Protocols In Africa: Lessons Learned for ABS Implementation.* https://naturaljustice.org/wp-content/uploads/2018/11/2018_Community-Protocols-in-Africa_Lessons-Learned_Natural-Justice.pdf

South African Human Rights Commission. (2016). *Report of the South African Human Rights Commission. National Hearing Relating to the Human Rights Situation of the Khoi-San in South Africa.* South African Human Rights Commission.

Stavenhagen, R. (2005). *Human rights and indigenous issues: Report of the Special Rapporteur on the Situation of Human Rights and Fundamental Freedoms of Indigenous People, Rodolfo Stavenhagen. Addedum. Mission to South Africa* (E/CN.4/2006/78/Add.2). Economic and Social Council, Commission on Human Rights.

Wynberg, R. (2017). One Step Forward, Two Steps Back?: Implementing Access and Benefit-sharing Legislation in South Africa. In C. R. McManis & B. Ong (Eds.), *Routledge Handbook of Biodiversity and the Law* (pp. 198–218). Routledge.

10

BIOCULTURAL COMMUNITY PROTOCOLS AND BOUNDARY WORK IN MADAGASCAR

Enrolling Actors in the Messy World(s) of Global Biodiversity Conservation

Fabien Girard and Manohisoa Rakotondrabe[1]

Introduction

The following reflection is the result of an ethnographic survey and a theoretically informed study of the Biocultural Community Protocol (BCP) of the farming communities of Analavory (Madagascar).[2] Focused on the development of the protocol, the analysis draws on the concept of "boundary work" in a bid to describe how the various actors involved strove to negotiate the boundaries between different social worlds; and why this fell short of producing any anticipated or expected result. We believe that attention should be paid to how the various stakeholders, including funding entities, non-governmental organisations (NGOs) and State agents, introduce "scripts" and how, through their practical implementation and the resulting unintended competition, these scripts may compromise the successful completion of the boundary work and further the conciliation of worlds.

The BCP in question was named after the rural commune of Analavory, which is located in the district of Miarinarivo, almost 100 km from Antananarivo, the capital of Madagascar. The commune is home to around 68,000 inhabitants and has always been an area of high immigration, due in part to its geographical location and agricultural potential. The area is located at the crossroads of the RN1, a major, well-maintained road leading to the capital, and the RN23, which provides access to the tourist area of Ampefy and the irrigated areas suitable for rice cultivation in the district of Soavinandriana. Analavory is a productive farming area supplying Antananarivo with rice and market garden produce (see Map 10.1).

The development of the Analavory BCP was part of a project funded by the Darwin Initiative, a UK government funding scheme aimed at countries which are rich in biodiversity but poor in financial resources. The project was entitled

DOI: 10.4324/9781003172642-12

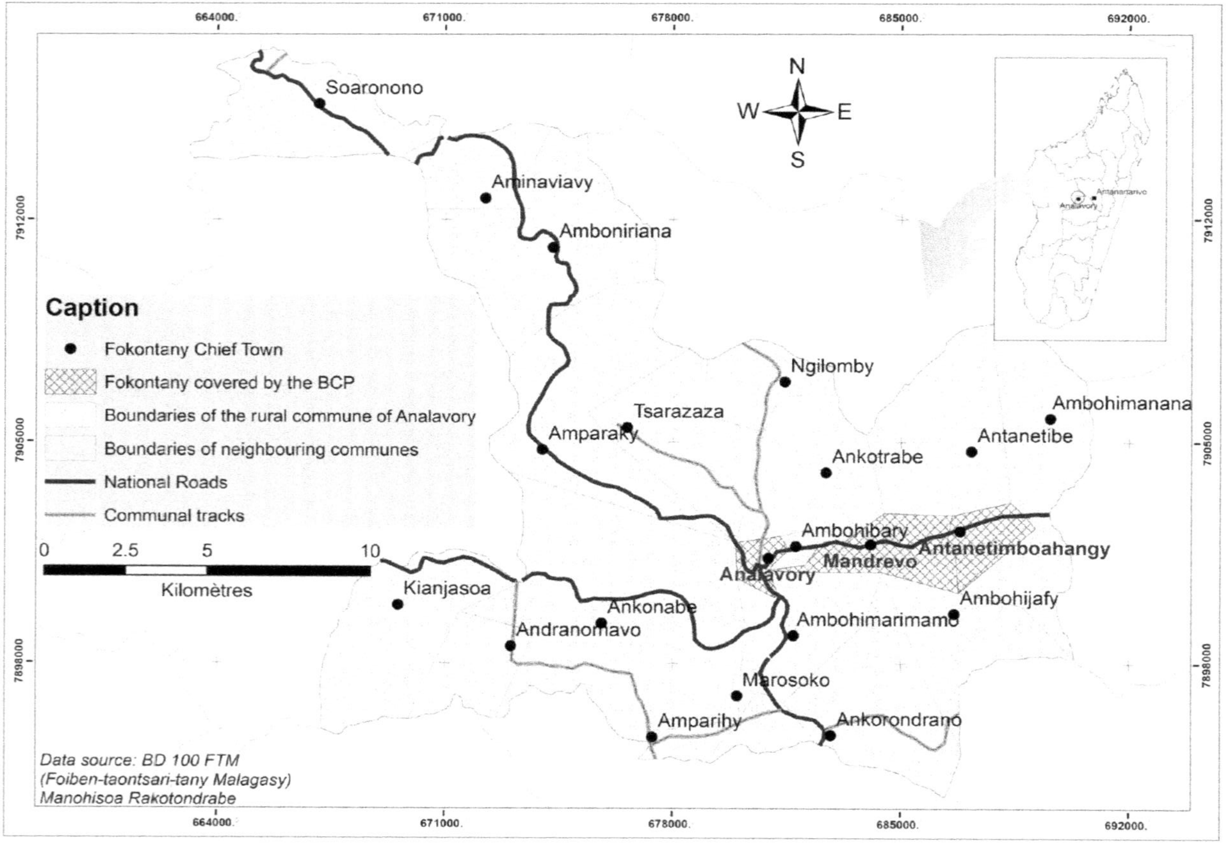

MAP 10.1 Rural Commune of Analavory and location of the three *Fokontany* covered by the Analavory BCP (Manohisoa Rakotondrabe –

"Mutually supportive implementation of the Plant Treaty and the Nagoya Protocol in Benin and Madagascar", hereinafter referred to as the "Darwin project". Bioversity International was the contract holder, with funding covering a period of three years (from 1 April 2015 to 31 March 2018). It took two years to negotiate and draft the Analavory BCP (2016–2017), which was signed on 27 December 2017 by the deputy Mayor of the commune and the two regional heads of the Ministries directly concerned: the Ministry of Agriculture, Livestock and Fisheries (MAEP) and the Ministry of the Environment and Sustainable Development (MEDD). The development of the BCP was outsourced to Natural Justice (the facilitator).[3]

As the title indicates, the project initially aimed at mutually supportive implementation of the International Treaty on Plant Genetic Resources for Food and Agriculture (ITPGRFA or Plant Treaty) and the Nagoya Protocol in Benin and Madagascar. The Nagoya Protocol,[4] which complements the legal framework established by the Convention on Biological Diversity (CBD), was ratified by Madagascar on 3 July 2014.[5] It significantly strengthens the obligations of States Parties regarding access to genetic resources and traditional knowledge (TK) held by indigenous peoples and local communities (IPLCs). Importantly, all parties are bound to take measures with a view to ensuring that TK and genetic resources are accessed with the prior and informed consent (PIC) or approval and involvement of IPLCs, and that mutually agreed terms (MAT) have been established. States are also obliged to take measures to ensure that benefits arising from the use of this TK and these genetic resources are shared in a fair and equitable way with the communities concerned (Morgera et al., 2014). In Madagascar, one outcome of this component of the Darwin project (together with the GIZ-PAGE programme)[6] was Decree No. 2017–066 of 31 January 2017 on the regulation of Access and Benefit Sharing (ABS) arising from the use of genetic resources, which mandates community PIC from "local custodians (*gestionnaires*) of natural resources" and "holders of TK" associated with genetic resources and also encourages the development of BCPs, while making compliance with their provisions mandatory.[7] The Plant Treaty[8] was also ratified by Madagascar on 11 June 2006.[9] It should be recalled that this treaty only covers a sub-category of genetic resources, referred to as "plant genetic resources for food and agriculture" (PGRFA). Central to the Plant Treaty is the Multilateral System (MLS) of ABS, whereby contracting parties agreed to "virtually" pool 35 food crops and 29 forages that are listed in Annex 1 of the Treaty, in order to facilitate access to these resources and to fairly and equitably share the benefits arising from their use (Frison, 2018).

It should be stressed that the general theme of this chapter sits at the interface between the global regime of biodiversity governance and the regulation of access to genetic resources and associated TK "held" by local communities. Resituating the debate in this way and specifying the spatio-temporal scales are critical: the interface between global and local has gone through a sea change in recent decades, which has led to international recognition of IPLCs as the "guardians" or "stewards" of biodiversity and to the foundations being laid for the implementation of community or local PIC (Morgera et al., 2014, p. 145) and procedures for benefit sharing with IPLCs (Morgera et al., 2014, p. 117) being incorporated into domestic legislation. The ground-breaking nature of the Nagoya Protocol in this respect is indisputable. However, the true extent of its provisions only emerges against the background of the recent "rediscovery" of the pivotal role of IPCLs and farmers/peasants in the breeding, maintenance and improvement of plant varieties and agricultural ecosystems, after many decades of their innovative agency being downplayed and their local practices around biodiversity conservation being made invisible (Bonneuil, 2019, p. 12). There is now much debate about the "socio-natural" (Watts & Scales, 2015) or "bio-cultural" dimension of landraces and folk varieties, which certainly frames the current questions around the role of IPLCs in biodiversity maintenance in the context of political ontology (Demeulenaere, 2014).

Within this new framework, therefore, community PIC is not only about guaranteeing genuine involvement and "true" consent from IPLCs every time a bio-prospector wishes to access their genetic resources and/or TK. It is also about making sure that consent is obtained and potential benefit sharing arrangements are negotiated following culturally appropriate procedures which are capable of offsetting power asymmetries and, most importantly, of securing "the market-inalienability" (Bavikatte et al., 2010, p. 316) of certain aspects of the community's heritage (i.e. all that which should be kept *extra commercium*, outside the realm of private rights, such as language, traditional rules regulating the exchange of goods, and sacred sites) (Banerjee, 2003, p. 144; Hughes & Lamont, 2018, p. 162). BCPs, enshrined in the Nagoya Protocol as "community protocols" (Arts. 12.1, 12.3(a) and 21(i)), have been heralded as ground-breaking tools to achieve this goal.

As the introductory chapter of this book recalls,[10] BCPs were probably born out of the "indigenous protocols" and "cultural protocols" (Bannister, 2009; Gray, 2004) developed in Australia and Canada as ethical instruments to protect indigenous peoples' cultural heritage. While there are still some doubts as to how these cultural protocols morphed into ABS-related instruments, both share common characteristics. Significantly, cultural protocols and BCPs both build on "procedures" or "protocols" – commonly referred to as ancient, traditional or immemorial – through which a community manages its resources and defines and regulates access to them within and outside the community (see Bannister, 2009, p. 288). As written documents, BCPs strive, if only partially, to codify these internal norms and procedures (Tobin, 2013, p. 142), with a view to providing clear guidance to would-be bioprospectors on the terms of the negotiations for accessing their resources and TK (Shrumm & Jonas, 2012, p. 13).

As a result, BCPs are generally featured as *technical* tools that enable "a community to prepare in advance for negotiations of an ABS arrangement" (Morgera & Tsioumani, 2010, p. 157) and to set up internal benefit sharing mechanisms, thereby preventing any intracommunity conflicts (ibid., p. 158) that may follow monetary and non-monetary arrangements with bioprospectors.

While none of the above is inaccurate, a cursory examination of the few protocols established throughout the world (Delgado, 2016; Parks, 2019) indicates that BCPs cannot be reduced to mere technical tools. They commonly encapsulate recurring discursive patterns which extend the instrument's scope beyond its procedural dimension. For instance, one may commonly find strong calls to tradition, pervasive references to a holistic way of life and, above all, an emphasis on the "stewardship" role of IPLCs in biodiversity conservation, itself systematically linked to "traditional" ways of life.[11] These recurring patterns and tropes can certainly be read as a form of "strategic essentialism" (Banerjee, 2003, p. 144; Hughes & Lamont, 2018, p. 162) which is already apparent in the way representations of "stewards" and "ecological natives" have been used, with some success, by indigenous peoples "to transform nonindigenous peoples' ideas of their identities not only within the nation-state, but also in transnational arenas" (Ulloa, 2005, p. 215).

As was the case for the discourse around biodiversity, which was wrought globally by dominant players but "transformed and re-inscribed into other knowledge-power constellations" as it circulated through local networks (Escobar, 1998, p. 56), so too are BCPs re-inscribing into new local constellations the discourses that are being framed in international arenas as "nature-in-the-making" (Harvey & Haraway, 1995, p. 517). Furthermore, as alluded to while referring to strategic essentialism, BCPs are part of a political project which is ontological in nature. This means that BCPs are not about promoting and securing the acceptance of different *representations* of seed, plant and TK, but more radically about recognising and conciliating different "worlds" (Cadena & Blaser, 2018; Demeulenaere, 2014). In other words, BCPs touch upon fundamental "assumptions" about "what kinds of things do or can exist, and what might be their conditions of existence, relations of dependency" (Scott & Marshall, 2009, quoted by Blaser, 2009, p. 887).

For this reason, we argue that BCPs should be approached as processes (rather than textual sources) that call upon and engage all the tangible and intangible elements of community life and should be analysed through the conceptual lens of "boundary work". We draw here upon theoretical insights from anglophone human geography, which itself largely borrows from science and technology studies and actor-network theory. In what follows, the concept of "boundary work" is used to describe the questions of interactions and interfaces between communities of researchers or developers and IPLCs in the governance of biodiversity ("place-based governance"). Boundary work covers those "arrangements" that ensure "the creation and transformation of boundaries between different social worlds that are inhabited by specific communities of actors" (Koehrsen, 2017, p. 2), so as to navigate knowledge systems and even conciliate "diverging ontological claims" (Löfmarck & Lidskog, 2017, p. 28). In addition to the concepts of mediation and translation (sometimes referred to as "boundary concepts", Mollinga, 2010), boundary work generally involves a whole range of methods intended to support the sharing and co-creation of knowledge between partners (Robinson & Wallington, 2012).

The Analavory BCP is used as a case study to examine how and to what extent the current global negotiations on seeds, genetic resources and TK are being translated locally into practice. How did the boundary work manage to articulate, if at all, different worldviews by mobilising boundary concepts such as sustainable development (and its new incarnation, deployed against a modernist background, of "sustainable intensification" – Ministère de l'agriculture et de l'élevage, 2018), biodiversity, nature stewardship and "biocultural diversity" (Bridgewater & Rotherham, 2019; Graddy, 2013)? More specifically, can it be said that the Analavory BCP has opened, as promised by advocates of BCPs, a new discursive space aiming at a better considering the lifestyles and worldviews of the communities involved (Bavikatte et al., 2015)? Conversely, should it be concluded that this BCP has achieved nothing but the deployment of a "site of promises, hopes, fears and hypotheses which itself sets new relations in motion" (Hayden, 2003, p. 75), at the expense of local institutions, peasant solidarity and traditional rules of seed exchange, lending credence to the idea that protocols are nothing more than new forms of instrumentalisation of local practices and the notion of "community" to serve powerful institutional interests?

What the Analavory case shows is that attention must be paid to how the various actors, including funding entities, NGOs and State agents, introduce "scripts" or "scenarios" into negotiations (on the concept of scripts, see this volume, Chapter 11); and, above all, how some scripts, due to the power of the players driving them and through their mode of circulation, end up constituting a "canvas of interpretation of the world" (Bonneuil, 2019, p. 10). Built on "hypotheses about the entities that make up the world", scripts "define actors with specific tastes, competences, motives, aspirations, political prejudices, and the rest" (Akrich, 1992, pp. 207–208). A script is akin to a "film script" which sets "a framework of action together with the actors and the space in which they are supposed to act" (Akrich, 1992, p. 208) – in other words, the script works to "enrol" or "enscript" the players (i.e. "conquer and discipline", Fujimura, 1992, p. 171) and in this way "manufactures" identities.

In Analavory, the boundary work fell short of opening up a new discursive space for the conciliation of worlds. One piece of evidence of this is that the Natural Justice's community facilitator and the contract holder (from Bioversity International) continue to disagree as to the public release of the text of the BCP. Above all, it lacks clarity when it comes to the categories of resources that it purports to cover, while doubts remain as to whether these categories are, in their own right, intelligible to the peasants. Focused on the development process, this chapter strives to explain these misalignments or equivocations.

For this purpose, it first identifies and traces the dominant scripts as they unfolded locally through "development brokers" (Bierschenk et al., 2000), brokers or "intermediaries between development institutions and peasants" (Lewis & Mosse, 2006, p. 13), such as Bioversity International in this case, which are still able to "mediate multiple worlds" (Demeulenaere, 2016, pp. 14–15), but whose "local expertise" is now detached from a strong local foothold (ibid.).

Consequently, they need to go through various "chain aids" (Sharma, 2016), and it is exactly this which opens the way to alternative scripts, destabilises the "enrolment" process and puts a strain on the boundary work.

There is little doubt that all this competition between scripts creates a new political space on the ground which benefits some community members. It remains to be seen, however, whether it stands as a "mobile space" of empowerment and emancipation (Tsing, 1999, p. 157) or, on the contrary, as an "elite capture" phenomenon which is potentially disruptive of group's cohesion in the long run.

In the next section, we offer an overview of the two competing public scripts that the Darwin project brought in its wake. It then goes on to show the tension that this competition created, allowing notably the Ministry of Agriculture to push for a hidden script and some community members to exploit the competition between scripts. The final section investigates the effect that the competing scripts and the hindered boundary work which resulted had on the content of the BCP itself.

Two Competing Public Scripts

Quite apart from the establishment of an ABS legal framework in Benin and Madagascar, the Darwin project pursued the development of BCPs within several communities in these same countries. To this end, two communities with different profiles were to be selected within each country: first, a community living in high diversity area that would act as providers of resources under ABS agreements; and second, a community that would be selected on the basis of its possible need for adapted germplasm (Bioversity International, 2014). In Madagascar, farmers in the rural commune of Analavory fell within the second category, while the *Fokontany* (a basic administrative subdivision covering several villages) of Ampangalantsary fell within the first. In total, four BCPs were developed and consultants from Natural Justice acted as facilitators (Bioversity International, 2018).

Turning now to the development of the Analavory BCP, the case study shows a seeming tension, if not competition, between two scripts that is discernible throughout the working documents and internal reports produced over the lifespan of the project, a tension that we were able to confirm through semi-directed interviews with the project leader, the facilitator from Natural Justice, staff of the Ministry of Agriculture and public officials from the Ministry of Agriculture's decentralised departments.

To understand this tension, we need to refer back to the Darwin project as it was designed and submitted by Bioversity International. Immediately striking to any reader versed in international biodiversity governance is the project's alignment with the "Grand Bargain" narrative (Wynberg & Laird, 2009) and adherence to the "ILPCs as small entrepreneurs" script which is analysed at length in Chapter 11 of this volume. On closer examination, however, this sense of familiarity and simplicity evaporates. The script is in fact Janus-faced and offers two different, albeit interrelated, versions of the function of ABS in biodiversity conservation. The tension stems from this duality.

Something familiar remains, though, as foundational in both cases is the notion that ABS is a "win-win" scenario whereby advanced economies (the "gene-poor North") can maintain their access to plant genetic resources (PGRs) located in the tropics, while the Global South, through the benefits flowing from intellectual property rights (IPRs) on "biodiscovery", is supposedly better equipped to tackle the erosion of biodiversity. Key to this is the understanding, based on economic attributes (low excludability and high rivalry), of PGRs as "impure public goods" (Halewood, 2013; Halewood et al., 2021).[12] In the reports submitted to the funding agency, this idea features prominently in the theoretical underpinnings, along with scattered, but strong references to New Institutional Economics, collective action and the "Commons". The main threat to the conservation of PGRs is underuse, not overexploitation (Schmietow, 2012, p. 82). In addition, while the biodiversity maintained and enriched by farmers and IPLCs generates positive externalities (in the form of "use value" or "option value", Jarvis et al., 2016), none of these values are appropriated on the local level by those responsible for its maintenance due to the lack of "property rights associated with the public good genetic resources" (Sedjo, 1992, p. 200). It follows from this that, in order to ensure that farmers and local communities continue to use, trade in and access their seeds and plants, economic incentives need to be developed.

Here, the first version of the script begins to materialise: ABS agreements and BCPs are tools to incentivise[13] and drive changes in IPLCs' attitudes towards the conservation of PGRs, building on the rather narrow and contested premise that farmers and IPLCs have always "operationalized the management of [PGRs]" (Eyzaguirre & Dennis, 2007, p. 1492) and make decisions regarding their lands and the maintenance of diversity according to wealth maximisation criteria (or economic efficiency) alone.[14] We refer to this as the **"economic incentives" script**.

In its second version, the script continues to be about designing appropriate incentives, i.e. on ways to change how "humans interact with their environment and how they use natural resources", as well as "patterns of behaviour and traditions that have emerged over long periods of time, and have, as a result, become enshrined in law or social custom".[15] However, in the wake of North's work on institutional changes (North, 1990), incentives are more broadly conceived of as a blend of "formal constraints" (economic and legal instruments, regulations and public investment), "social constraints" (such as cultural norms and social conventions) and "levels of compliance", all of which constitute "institutional incentives". We refer to this script as the **"institutional incentives" script**.

In its institutional version, the script takes on very distinctive features: incentives are no longer about fine-tuned property rights over resources or knowledge,[16] or Coasean contracts (Sedjo, 1992, p. 204),[17] as is the case of the economic incentive script, but rather about acting on local institutions and social norms in order to reinforce enabling social constraints and drive necessary changes.[18] In other words, risks of underuse are not to be blamed on the lack of economic incentives (benefit sharing) but, as the project holder for Bioversity International recently put it, on institutional and capacity weaknesses that prevent farmers from taking advantage "of the technological and organizational developments that have changed the shape and functioning of the global crop commons over recent decades" (Halewood et al., 2021, p. 12).[19] Here, the function of the BCP is radically different, and appears to make plenty of sense in the context of Analavory. Indeed, although Madagascar is a biodiversity hotspot, other than a very few crops, it is not a centre of crop diversity (see Table 18.1 in Kamau, 2013, pp. 380–381).[20] As a consequence, for farming communities, such as that in Analavory, endowed only with PGRFA, by nature of little interest to potential bioprospectors, the BCP can hardly be conceived of as anything other than a tool for accessing PGRFA through the MLS or other sources, as well as including landraces in the MLS.[21]

For Bioversity International, it was clearly the second version of the script ("institutional incentives") which supported the development of the BCP. The goal was, unquestionably, to roll out a new institutional framework for communities to engage in transnational germplasm transfers, test new cultivars through participatory plant breeding and adopt new varieties (see Halewood et al., 2021). Yet, nothing was more at odds with the agenda of Natural Justice and with the facilitator's professional habitus. The facilitator therefore clung on to the first script ("economic incentives") and pushed to shift the protocol away from PGRFA and to rearticulate it around genetic resources and the Nagoya Protocol.

There seems little doubt that the facilitator believed that, with greater time and effort spent on bioprospection, the BCP would deliver. This would, furthermore, only be fair from the peasants' perspective – in that he clearly claimed to be able to decipher their will and cater to their needs. And yet, he was all too aware of having been "recruited" (just as much as the farmers) (Lewis & Mosse, 2006, p. 13) into a separate script – linking the BCP with ITPGRFA and achieving exchanges of germplasm with centres of research and partner communities. As one of our semi-directed interviews with the facilitator shows, the BCP, in relation to ABS, only exists because he overstepped the limits set for his mission, something which, as he is keenly aware, injects a first level of precariousness into the BCP:

> That's something that was a little bit difficult for us, as Natural Justice that actually facilitated the whole process, and that I think is something that we're going to really incorporate into the overall community protocols guide. It's this relationship, well, this mediation between the local communities on the one hand and those who fund the whole project on the other. Because the institutions that are involved, like for example the Darwin project, they get into their head "here we are going to develop a community protocol for such and such an end!" [...] and on the other hand, there are the communities that have their own priorities, which may

> be other than those of the donors, technical and financial partners. If you look at the reports as well in fact, if we had to add budget to the Darwin project, it was because of the community protocols [laughs], because I was stubborn actually [...].[22]

From the viewpoint of Bioversity International and the project leader, there is no doubt that deepening the boundary work with the community and refocusing the BCP on ABS and bioprospection were misguided. As the project leader told us in a semi-directed interview, although he was eventually able to secure additional funds, at the request of the facilitator, to be invested in further developing the protocol and capacity-building activities such as a community seed bank and a community registry, he had no doubt that the process was headed in the wrong direction:

> I mean the other problem with all these protocols from the Nagoya point of view is you can build a fence around all kinds of things and that doesn't create a demand for it and that doesn't make it useful. You just build a fence around it and it remains ignored, and it is allowed to continue to drop off the face of the Earth and a lack of investment in it. [...]
>
> You know where in communities where it's purely agricultural, they're working on crops, from other centres of origin and put a lot of effort in developing a protocol but if there's no interest in getting stuff from them [...] it's another thing.[23]

The Emergence of a Hidden Script

The boundary work undertaken by Natural Justice on the ground was further hindered by the emergence of another script, this time a hidden one, that the Ministry of Agriculture was able to deploy on account of the indeterminacy that affected the aims of the projects. This script is that of **"IPLCs as premoderns"** and is further described in Chapter 11 of this volume.

In terms of concrete effect on the boundary work, the competing public scripts yielded a certain indeterminacy or blurriness, which itself opened up a hermeneutic space that was first used to influence the choice of sites for the implementation of the Darwin project. While the choice of the Analavory commune was officially presented as being a collective decision of the Expert Guidance Group[24] which oversaw the whole project, our work shows that it was decisively influenced by the former heads of the National Agency for the Official Control of Seed and Plant Material (now known as the Official Service for Certification of Seeds and Plant Material, SOC). Importantly, the choice of Analavory, apart from it being close to the capital and served by well-maintained roads, offers benefits that are threefold, and incidentally ensured a triple convergence of interests:

- the Darwin Initiative has supported several projects in the area and is familiar with Analavory;

- as the commune is a rich farming area, Bioversity International can work on the linkages between the BCP, the Plant Treaty and MLS;
- and last but not least, the municipality is made up of several dynamic Seed Producer Groups (SPGs),[25] which are considered exemplary for the rest of the country.

This third aspect is critical for the Ministry of Agriculture and fostered its opportunistic behaviour. While seemingly fitting into the second script (i.e. "institutional incentives" and the MLS), the major impetus behind the choice of site was in fact the Malagasy government's willingness to advance its own public policies. It is no coincidence that the predecessor of the SOC pushed for Analavory: the model SPGs play a critical role in the multiplication of basic seeds developed by the National Centre for Applied Research on Rural Development (FOFIFA). From the Ministry's point of view, the Mitsinjo and Santatra SPGs, in particular, play two essential roles. First, for what is still a minority of farmers who are "market-oriented", certified seed must be more easily accessible (MAEP, 2008). For example, one of the priorities of the national seed strategy, as well as the national rice development strategy (Ministère auprès de la Présidence en charge de l'agriculture et de l'élevage, 2017), is to increase the use of certified seeds, to boost the growth of agricultural production and, thus, to ease land pressure. Second, for the great majority of farmers who practise subsistence farming, and who therefore depend on landraces and farmers' seed networks for seed supplies, SPGs are confidently seen as able to bring about a change in practice, notably by providing local services and implementing incentives, such as swapping seeds for grain or selling seeds on credit (MAEP, 2008).

This interpretative twist, in other words the move from the "institutional incentive" script to that of farmers in need of modernisation, has largely gone unnoticed. The first reason for this is that, as the result of a fundamental shift away from colonial and Cold War patterns, major development agencies now incorporate managerial tools (such as "log-frames" and "theories of change") and long-term impact indicators (such as the fight against poverty and sustainable development goals). As a consequence, partners are given greater leeway in the kind of actions and means they can deploy (which can also be modified part way through the process), and the criteria for assessing the success of the project are significantly broader.[26] In light of this, it is easy for the Ministry of Agriculture to surreptitiously integrate its objectives into the project. For instance, in a confidential pre-report prepared by SAGE for Bioversity International's second annual report (Bioversity International, 2017), significant parts of the text insist on the role of new varieties that have "a multiplier effect on the economy by increasing the efficiency and total volume of agricultural production contributing to food security" (SAGE, confidential). In any event, the Ministry can showcase the role of the MLS and ABS mechanisms (and thus the BCP) without losing sight of its own hidden agenda. Following extensive fieldwork, financed through additional funds, ecologists, agronomists, technicians and farmers carried out a rich diagnosis of climate change adaptation issues at both sites (MAEP et al., 2017; Ministère de l'environnement et al., 2016, p. 18). The study helped identify challenges of climate adaptation and opened the way for discussions on the development of climate-resilient agricultural and seed systems. At this juncture, it is not hard for the Ministry to (re)align its agenda to the aims of the Darwin Initiative project; it is enough to conveniently cast the BCP as a tool for accessing, through the MLS, "climate-resilient varieties".[27]

Infrapolitics: From the Art of Resistance to the Risk of Subversion

The reference to "infrapolitics" (Scott, 2009) is meant to convey three main ideas: subaltern actors, who are the target of environmental policies, do not have scripts[28] – they are "actors", i.e. they are subject to scripts and, as such, exposed to "enrolment"; but scripts never succeed in fully "conquering". At best, they "enrol" through encountering resistance, and this resistance – which is the politics of the subaltern – is unobtrusive, unorganised, unstructured and is therefore difficult to detect.

At this juncture, two further caveats are needed. The first is that there is a fine line between genuine resistance and elite capture (Platteau, 2004, p. 228), and there may be doubts as to the proper meaning of subaltern manoeuvres, as will shortly be illustrated. The second is that the fuzzier the aims, the greater the odds of communities escaping the grip of the scripts.

With this in mind, we may now turn our attention back to Analavory and address the main argument of this section: as the development process was never geared towards a clear objective (e.g. the creation of a value chain around a shared resource and the strengthening of community preparedness to bioprospecting), the "community" in the community protocol is somewhat "caught in the making". In other words, the community is unable to self-identify as a community and is constantly compromised by disengagements and defiance. Ultimately, a collective ends up forming around a local elite who takes ownership of the development process and seizes the opportunity to roll out a cooperative that gradually takes control of the community seed bank and the community registry.

In practice, the Analavory development process was encumbered by issues pertaining to the scope of the BCP, in particular those touching on the precise contours of the communities involved in the protocol. The process was first deployed at the rural commune level, but all boundary work was directed towards the *Fokontany*, as the smallest administrative divisions in Madagascar, that make up the commune of Analavory. That the whole project revolved around the *Fokontany* should not come as a surprise: these are the administrative embodiments, as it were, of what has always been considered in Madagascar as the flesh and bones of "communities": the *Fokon'olona* (Condominas, 1961).[29] The first working meeting at the community level thus began with a general assembly of the *Fokontany*, held on 7 September 2016. This meeting also involved leaders of the Analavory commune and members of the SPG. Subsequently, the number of participants attending the work meetings constantly decreased (see Figure 10.1). When our fieldwork was carried out, only three *Fokontany* out of the 22 in the rural commune of Analavory were still involved in the development process. These *Fokontany* were Analavory, Antanetimboahangy and Mandrevo (see Map 10.1).

One possible reading would be to recognise this as a salutary "creaming-off process" which swiftly excludes any free riders looking for immediate opportunities, and which singles out those members eager to commit themselves to long-term objectives. Alternatively, the obvious disinterest or disengagement could be explained in terms of opportunity cost, as attending meetings represents lost workdays and may have an impact upon subsistence activities.

We feel, however, that these explanations give too much weight to "outward trajectories" – disengagement – at the expense of "inward pulls" – engagement – that offer much greater insights into the community-making process and infrapolitics. In fact, one major determinant in the disorderly operation of boundary work and the constant disruption in the community-making process was the seeming appropriation of the BCP development process by a small local elite gathered in a so-called "*kômity mpanorona*" or drafting committee (*ad hoc* as far as can be ascertained) which comprised about 20 peasant leaders. It is most unsettling that there is no mention at all of this committee, either in the summary documents or in the workshop reports issued during the lifespan of the project. Quite the contrary, documents and reports refer to a far-reaching "community committee made up of representatives from each *Fokontany* concerned" tasked with making proposals to be validated by all farmers (Ministère auprès de

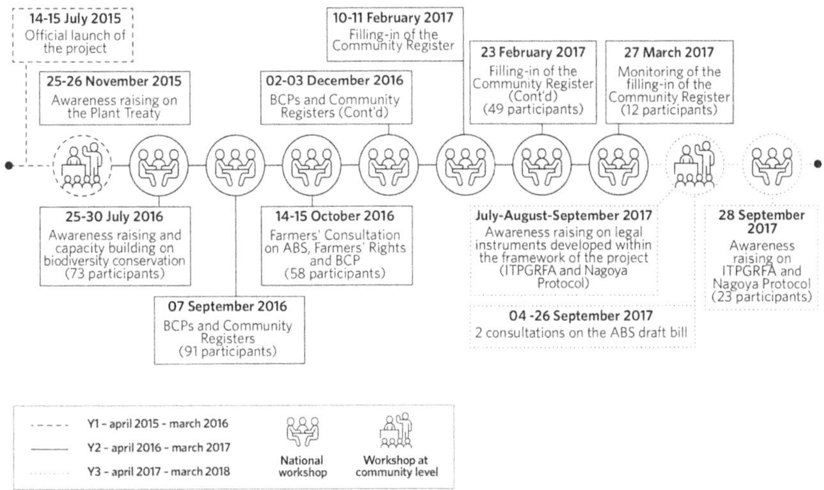

FIGURE 10.1 Timeline of the Darwin Initiative project and the boundary work in Analavory (Manohisoa Rakotondrabe, original material for the book)

la Présidence en charge de l'agriculture et de l'élevage, 2016). It is not clear how much work this committee carried out, but it is known for a fact that the "*kômity mpanorona*", which came to life with the blessing of the facilitator (who probably saw in it an expedient for agreeing on a draft in a timely manner), contributed decisively to the content of the BCP.

A similar confusion permeates the other entity that was also born out of the implementation of the Darwin project at Analavory: the Fa.M.A. seed-producing cooperative that brings together the two SPGs and unaffiliated community members interested in the production and sale of seeds (rice, maize, groundnuts and sorghum). Quite apart from the fact that the cooperative increases uncertainty around community delineation (its members are more or less all those who makes up the "*kômity mpanorona*"), the *Koperativa Fa.M.A.* appears to have taken hold of the outcomes of the development process in Analavory, namely the BCP, the seed registry and the community seed bank. In terms of the community seed bank, one cannot help but note a striking confusion as to who owns it. It is notable that the cooperative and the seed bank share the same premises, and the inscription carved on the main facade gives the impression, through the direction of the script (left-to-right) and the juxtaposition of Malagasy and English, that the latter is now subjected to the authority of the former (see Figure 10.2).

Evoking the leading roles taken by the *kômity mpanorona* and the cooperative, the facilitator admits his "mistake" in this regard, which has created tension

FIGURE 10.2 Community seed bank and head office of the cooperative Fa.M.A., Fokontany Mandrevo (Manohisoa Rakotondrabe)

within the community and has had a dramatic impact on non-members of the cooperative. As one of these farmers told us: "*Andrizareo kôperativa io!*" (all this belongs to the cooperative!).

Certainly, this string of equivocations, misunderstandings and disruptions can be accounted for by the proliferation of competing scripts that hinder the implementation of boundary work, to such an extent that the facilitator himself evokes a protocol developed "in a vacuum".[30] Yet, we would also like to add a layer of political reading, if only to underline the power games at play: it is the emergence of the hidden script of the Ministry of Agriculture, made possible by the competition between the public scripts, that enabled SPG members to weave their way into the Darwin project and to bring to life a cooperative project that they had seemingly been maturing for a long time.

On the surface, everything occurred as though the members of the cooperative were aligning themselves with the Ministry's script. After all, the model of the cooperative was looked upon favourably by the administration (Ministère de l'Agriculture et de l'Élevage, 2002), and when the Fa.M.A. cooperative has taken over the commune's SPGs, it will be able to provide for the multiplication of R1-certified seeds produced from the basic seeds developed by FOFIFA (Bioversity International, 2017). And yet, behind this apparent convergence of views, there is also an ongoing process of subversion whereby the Fa.M.A. cooperative is dramatically altering the function of the seed bank: henceforth, the cooperative will be able to use the premises as a retail outlet for the distribution of fertilisers, a valuable source of additional income that will help to fill the hunger gap (particularly tough at the start of the school year). No less important is the transfer of part of the decision-making power regarding resources to the members of the cooperative, a point that cannot be discussed here, but which is quite apparent upon reading the Analavory BCP to which we now turn.

A Glance at the BCP's Content: Lost in In-scription

That the boundary work fell short of its goal in Analavory is clearly reflected in the content of the BCP itself. Out of respect for the facilitator and the community members who continue to object to the dissemination of this 31-page document (in its second version), written in Malagasy and translated into French, we reproduce here only the outlines of the protocol (see Figure 10.3, Annex). However, this is probably enough to discern an intriguing paradox: while Bioversity International aimed to develop a BCP revolving around PGRFA and geared towards farming communities as beneficiaries of germplasm (the "institutional incentives" script), only three pages out of the 31 that make up the document deal with PGRFA and the MLS.[31] The bulk of the BCP revolves around bioprospection, bilateral negotiations and benefit sharing (the "economic incentives" script).

At this point in the chapter, the explanation for this is straightforward. While necessarily aware of Bioversity International's goal, the facilitator invested almost all the resources and energy supporting the boundary work in laying down rules for community PIC and ABS and connecting the BCP to the Nagoya Protocol. The multiplication of scripts circulating in the preliminary documents, in addition to the configuration of modern brokering, allowed the boundary work to be refocused on the Nagoya Protocol, which was more in tune, for that matter, with the facilitator's ethos and professional culture. An additional point stressed by the facilitator was the presence in the area of endemic species used in traditional pharmacopoeia (SAGE & MAEP, 2016), resources that might be exploited to attract bioprospectors and foster benefit sharing arrangements.

In summary, in Analavory, the boundary work remained torn between two objectives and hung on two competing public scripts which never succeeded in connecting with one another or embodying themselves in the daily life of the community, despite the fact that the BCP clearly hinges on the two fundamental categories of the international regime of biodiversity: the PGRFA on the one hand and "genetic resources **other than** the Plant Genetic Resources for Food and Agriculture (PGRFA) used for food and agriculture" (our emphasis) on the other hand.

Over one of our "group analyses" sessions,[32] a lengthy discussion touched on the term "*fototarazo*", the Malagasy translation of "genetic resources". In the limited space of this text, we cannot report the exchanges verbatim, but a few extracts give an overview of the way the peasants remain outside the categories of the international regime on biodiversity, while seemingly being able to navigate seamlessly through their inherent complexity.

MR R. (PEASANT): For an old man like me, I don't know how to explain the topic. When you talk to me about "*fototarazo*", it's all Greek [French] to me. [Laughter from the audience]. I call it "traditional crops", I didn't know it had another name, "*fototarazo*"!

[...]

MR T1. (DRAEP): I have a question. I'd like to know how far you have got [he is addressing the members of the cooperative] with the other genetic resources, because as of this morning, we're still on the first type of resource [PGRFA], while you said that there are two categories of resources?

MRS T. (PEASANT): We are currently in the process of adding in the BCP the genetic resources of medicinal plants and genetic resources needed for environmental protection. This is one of our projects: reforestation.

MRS B. (PEASANT): Protecting these medicinal plants is essential; because of the bush fires... if this continues, these plants will disappear in a few years. And it is very difficult to find seeds of these plants, such as "*talapetraka*" for example [...]. But nowadays, as there are more and more bushfires, then, in my opinion, it is essential to collect these different resources [...] if only to put them in our gardens. And I say this, not only for the Fa.M.A. cooperative, but for everyone... because we have to think about what will happen tomorrow, if all this continues... it will be sad for us Malagasy... especially since these are real medicines, we don't need to buy them from the pharmacy... we grow them a little bit everywhere, here and there, and they become medicines for colds, coughs, etc.

In my opinion, it is of utmost importance to protect them. We can plant them in our garden to start with [...]. They are real medicines, and we don't need to spend money to get them.

FACILITATOR (NATURAL JUSTICE): What resources do you think are covered by this BCP? Is that only PGRFA or all Genetic Resources?

MRS B.: For the time being, it is only genetic resources for food and agriculture.

FACILITATOR (NATURAL JUSTICE): So, if there are people who want to access medicinal plants in our area, will you or won't you use this BCP?

MRS B. (PEASANT): In my opinion, these medicinal plants are not yet covered by this BCP.

MR T2 (PEASANT): In my opinion, we should not have put in the title of the BCP that it only concerns PGRFA, because these are plants that are not grown but which are very useful to us [...]. So, in my opinion, we should change the title of the BCP.

FACILITATOR (NATURAL JUSTICE): But what is the current title of the BCP?

MR H. (DRAEP) [HE READS THE TITLE]: "*Arofenitra Iombonan'ny Tantsaha Analavory*" [...]. Biocultural Community Protocol of the peasants of Analavory.

MRS M. (MAEP): It is clear from the title that the BCP concerns all Genetic Resources including PGRFA.

FACILITATOR (NATURAL JUSTICE): You've probably forgotten what we put in it [laughter].[33]

The very first part of the exchanges stumbles over a translation issue which extends the space of communicability to breaking point. Even when translated into Malagasy, the foreignness or "otherness" of the word *fototarazo* persists ("it's all Greek [French] to me").[34] Certainly, Mr R, the first speaker, is able to grasp the meaning of the word through a referent that appeals to vernacular representations. But this comes at a cost: the protocol is stripped of half of its substance, for it is then limited to "traditional crops", i.e. PGRFA (or "*fototarazo famboly sy fanao sakafo*" in Malagasy). Through a series of deceptively naive questions, the facilitator scrambles to bring "genetic resources other than" PGRFA back into the BCP.

And this raises a new challenge. As we shall see, PGRFA are almost consistently referred to as "traditional knowledge" which, by extension, causes a mismatch between the vernacular definition and the international and legal sense of the expression. A subsequent sequence of exchanges illustrates this point. It begins with a question raised by a ministry representative: "On the other hand, what I don't find in the BCP, but maybe I'm reading it too quickly, is the traditional knowledge dimension". Following a lengthy silence, one member of our team brings the issue up again. Here are the replies:

MRS B. (PEASANT): There was a meeting, it was even in this room where we are meeting right now, and we discussed traditional knowledge [pause]. So, I am quite surprised and I wonder why we talked about this subject during the meeting if it was not with a view to adding it to this protocol. In addition, we had spent a total of two days on this topic and group assignments had been done on it.

MR N. (PEASANT): Perhaps it is an oversight [laughs].

MR R. (PEASANT): Because these are not things that we grow [Laughter from the audience]. These plants grow like that, in the fields. We just know that a plant is used to treat a certain disease if a person asks us. None of us has made the decision to grow these plants yet.

MR T1 (DRAEP): My apologies, it is not my turn to speak yet, but it is a matter of understanding. When you talk about "traditional knowledge", is it knowledge as it should be or is it traditional plants of our ancestors? [Assembly laughter].[35]

The last question is particularly important as it shows most clearly the incommunicability of worlds and the fragility of the category used in the BCP with regard to vernacular classifications. Even more striking is the two-fold limit that the peasants set on the legal category of "genetic resources": apart from Ms B (who alludes to "seeds" nonchalantly and haphazardly sown in the garden), genetic resources are nothing but "medicines", "medicinal plants" and "traditional medicines". Strictly speaking, these plants are neither planted, nor grown, at least not in the sense of traditional crops. Equally important is the fact that these resources appear to belong to a separate universe: as one peasant stated, "these concern those with healing gifts", i.e. the *mpitaiza olona* (healers) or *mpanao tambavy* (herbalists) (Lefèvre, 2008, paras 12–14).

Some may argue that, subject to minor efforts at translation and readjustment, vernacular categories could be mapped onto legal (and Western) ones. We disagree: the way each category is embodied in farming *praxis* should not be overlooked. That what we refer to as "genetic resources" other than PGRFA constantly appears to be on the verge of morphing into knowledge and therapeutic practices are proof that these "resources" are in fact foreign to them and belong to separate realm dominated by healers and herbalists. Importantly, this explains why they are often forgotten. This also provides clues as to the appropriate interpretation to be made of peasants' attitudes towards the facilitator. Throughout the group analyses and semi-directed interviews, the peasants did not hesitate to join the facilitator in deploring the lack of bioprospectors, apparently sharing with him the hope that in the near future bioprospecting contracts on aloe vera and *talapetraka* (*Centella asiatica (L)*) could reverse course. The sincerity of these statements is, however, doubtful; it was as if peasants were playing a scene in which their lines do not quite fit with the rest of the screenplay.

***

Lost in the scripts, the peasants and the facilitator strove to resume the boundary work with a new starting point. The expression "traditional plants of our ancestors", as we heard one state agent said, should give some guidance as to what the further work of negotiation and translation should look like. The centrality of ancestry in peasants' cosmovisions cannot be overstated and is confirmed by one thorny issue our ethnographical workshop revealed, but which remains unaddressed in the protocol: "'*Sakafo ve dia asiana contrat*?' – Do we really need to have a contract for food?". As the Malagasy people say: "*sakafo masaka, tsy mba manantompo*" – "prepared food has no owner"; this is about sharing.

The "*sakafo*" is embedded in a system of complex idea which interweaves numerous aspects: tomb, filiation and ancestry, with the tomb being, as Bloch demonstrated, "the ultimate criteria of membership" (Bloch, 1994). Identity stems from this hub and continues to be linked to the *tanindrazana* (the land and soil of the ancestors) where the family tomb is located. The land of the ancestor is the legacy (*lova*) of the deceased, which comes with responsibilities in perpetuity and the *famadihana* ("double funerals" or the ceremony of the turning of the bones). A range of obligations include an obligation to plant the "*angady*" (a traditional spade) in the ancestral rice fields to ensure the family's food security and, above all, to rapidly collect wealth to worship the ancestor (Deliége, 2012, p. 132). However, the ritual never wipes out the debt, because the descendants are not satisfied simply with giving it back. They have to "re-give", so to speak, by calling for a new gift and then tracing a never-ending circle, which ensures the continuity of the vital flow or *aina*, focused on the *razana* (the ancestor) (Deliége, 2012, p. 202).

This echoes the importance of the "growth process" which, as Eva Keller recalls, is the first criterion of success for the Malagasy farmer (Keller, 2008, p. 659). Farm, cultivate and grow, in the hope of being able to fastidiously honour the *razana* and thus capture additional vitality. Yet, this vitality can only be achieved by following the "normal course of things" which is set by God (*Zanahary*), i.e. by scrupulously and manifestly respecting the intermediate divinities (the "*lolon-kazo*" or the "tree spirit", the Chthonian powers, etc.) and the *razana* themselves (Beaujard, 1995, pp. 251, 255; Rakoto Ramiarantsoa & Lemoigne, 2014, p. 99). This is the subject of offerings, prayers and even *ody andro* or *ody avandra*, i.e. spells against (poor) weather ("*andro*") or hail ("*avandra*") which aim to ward off climatic risks (Grandidier, 1932, p. 161).

In such a system, which constantly weaves together the origin of death and that of crops, fecundity, the world of the dead (Beaujard, 1995, p. 260), agricultural production and ancestry, some exchanges necessarily bring into question the continuity of the group and constitute supports for identity (Gudeman, 2012, p. 27). They respond to their own logic, as is the case of seeds-food. When foodstuffs are shared (raw or prepared), the Analavory farmers say this means that there is trust among the guests. If a "contract" has to be signed, this means that the trust is in doubt, and trust has not been acquired, which changes the status of the food (the "*safako*" is also a vector for poison or witchcraft "*vorika*") (Rakotomalala, 2006). Any negotiation or commitment (*fifanekena, fifampiraharahana, fifandaminana, fifanarahana*) takes place orally. No written contract is prepared. Belief in the "*tsiny*" ("blame" or the "readiness to answer" – Genard, 1999) and "*tody*" (the retaliation brought by fate) is enough to guarantee that each party respects the commitments. In contrast, written agreements represent the State/the *Fanjakana*, requiring signatures and leading to distrust.

The conclusion is self-evident: if the boundary work is to ensure the mediation between different social worlds and to achieve a BCP as a bridge of dialogue between diverging ontologies, the importance of traditional crops in the peasants' cosmological system cannot be glossed over, just as it cannot be ignored that medicinal plants and traditional pharmacopeia belong to a universe of which Analavory peasants are not part.

With this in mind, recentring the protocol on the MLS of the Plant Treaty and strengthening farmers' collective action, which was Bioversity International's original aim, could be an option. But this would have to be done keeping in mind the specific mode of rationality which governs the on-farm conservation of crop genetic resources and farmers' seed exchanges (see, in general, Altieri et al., 1987; Badstue et al., 2006; Garine et al., 2018; Toledo, 1990), in particular the cosmological system which presides over the circulation of traditional seeds in Analavory. In other words, reflections on "biocultural heritage" (Girard & Frison, 2021; Toledo, 2001) seem to be a fundamental prerequisite for resuming the boundary work in Analavory, a way to recall, in passing, that the epithet "biocultural" in BCPs is not pure affectation or symbolism. A not-to-be-forgotten lesson for the "boundary workers" of Natural Justice.

Annex

Preface

- Presentation of the community, the wealth of its natural resources;
- Definition of the protocol: "guide" to strengthen and inform about the local organisation for the conservation, sustainable use, exchange and sharing of benefits arising from the use of genetic resources and associated traditional knowledge;
- Origin of the protocol;
- Addressees of the protocol;
- Aims of the protocol;
- Reminder of the national and international instruments supporting the implementation of the BCP;
- Protocol outline

I. Methods of genetic resource management followed by the farmers of Analavory

- References to the "Community Register of Biodiversity" and the seed bank;
- References to the holders of TK. Role of the "*Ray Aman-dReny*" (ancestors) and the "*mpanao tambavy*" (traditional phytotherapists).

II. Challenges encountered in the management of genetic resources and associated TK in Analavory

- Disruption of the cropping calendar due to the degradation of the environment in general; disappearance and rarefaction of animal and plant species in Analavory;
- Forgetting and loss of TK held by the *Fokonolona* due to the modernisation of medicine. Loss of TK associated with resources due to the rarefaction of these resources;
- Decreasing interest of the youth in TK;
- Discrepancy between Analavory's local resource potential and the low standard of living of peasants;
- Problems linked to the involvement of farmers in decisions that may affect their lives (practices of exploitation of local resources at odds with local customs);

III. Modalities of exchanges of Plant Genetic Resources for Food and Agriculture (PGRFA) between farmers in Analavory

- Peer-to-peer seed swapping as the main modality of seed exchange;
- Mention of "*firaisan-kina*" (solidarity) as a fundamental basis for exchange and the role of plant genetic resources in everyday social life.

IV. Modalities of exchange of, and access to, PGRFA from farmers outside the commune or abroad

- Role of the SML;
- Role of the general assembly of Fa.M.A. members and the rest of the community in decisions on the inclusion of a resource in the Multilateral System;
- Steps to apply for PGRFA from outside (Multilateral System or other research centres): mention of participatory plant breeding.

V. How to apply for genetic resources located in Analavory

- Prior and informed consent of the entire community for access to genetic resources and associated TK;
- Steps to be taken when using resources for research purposes or when using resources and associated TK for commercial purposes;
- Description of the 6 steps to be followed for access and benefit sharing.

VI. Preservation and development of TK

- Recognition of the link between the preservation of TK and the preservation of genetic resources;
- Reminder of the conditions of access to TK (mutually agreed terms and benefit sharing)

VII. Benefit sharing

- Monetary and non-monetary benefits;
- Preference is given to benefits that promote the agricultural sector, strengthen the production capacity of farmers or increase the competitiveness of products on the market.

VIII. Resolution of potential conflicts and grievances

Preference given to friendly settlement;
Failing which, resort to "higher instances".

IX. Call for the respect and promotion of the rights of local communities and farmers

Reminder of the international and national legal framework:

1. International Treaty on Plant Genetic Resources for Food and Agriculture (ITPGRFA) and Order No. 11.567 /2017 of 27 May 2017 on interim measures related to requests for access to PGRFA and benefit sharing within the framework of the Multilateral System of the ITPGFA;

2. Nagoya Protocol on Access to Genetic Resources and the Fair and Equitable Sharing of Benefits Arising from their Utilisation, and Decree n° 2017-066 of 31 January 2017 on the regulation of access and benefit-sharing arising out of the use of genetic resources;

3. Texts governing TK: Art. 8(j) of the Convention on Biological Diversity, Decree n° 2017-066 of 31 January 2017 on the regulation of access and benefit-sharing arising out of the use of genetic resources, and Law 2013-017 of 20 February 2014 for the safeguarding of national intangible heritage;

4. United Nations Declaration on the Rights of Peasants and Other People Working in Rural Areas (Arts. 3.5; 5 and 10)

Signature page

- Top of the page: Logos of the two ministries: Ministry of Agriculture and Environment.
- Signatures and stamps of the Director of the regional directorate of Agriculture and Animal Husbandry for the Itasy Region, the Director of the Regional directorate of Environment, Ecology and Forestry for the Itasy Region, and the Deputy Mayor of the Rural Commune of Analavory, dated 27/12/2017;
- At the bottom of the page: logos of Darwin Initiative, Bioversity International and ABS Initiative.

Last page

- Contact details of individuals and their institutions in case of questions related to the use of the PBC.

FIGURE 10.3 Outline of the Biocultural Community Protocol of the peasants of Analavory (version as of 21 October 2019) (Manohisoa Rakotondrabe, Fabien Girard, original material for the book)

Notes

1 *All URLs retrieved on 1 September 2021.
2 In the local dialect: "*Arofenitra Iombonan'ny Tantsahan'Analavory*". The fieldwork was conducted between October and November 2019.
3 On this NGO, which was founded in 2007 in Cape Town, see Chapter 1.
4 Nagoya Protocol, Nagoya, 29 October 2010.
5 Ratification authorised by Act No. 2013-010 of 31 October 2013.
6 The *Programme d'Appui à la Gestion de l'Environnement* (PAGE) is a support programme for the Ministry of the Environment.
7 "Local custodians of natural resources" are defined as "groups of inhabitants who legally and/or traditionally manage the resources for which access is requested and whose way of life is of relevance to the conservation and sustainable use of biodiversity". The text adds that for the local custodians of natural resources and holders of associated TK, consent shall be given in a written contract. This contract is established according to customary rules, traditional values and practices as prescribed in the locality and must not be contrary to the statues and regulations in force (Art. 14). Finally, a paragraph specifies that in cases where traditional values and practices are already documented in an instrument implemented by the communities, the instrument shall be consulted and embedded in the contract (Art. 14).
8 ITPGRFA, Rome, 3 November 2001.
9 Ratification authorised by Act No. 2005-042 of 15 December 2005.
10 See, this volume, Chapter 1.
11 See, this volume, Chapter 11.
12 In the field of the conservation of crop genetic resources, the blueprint was created by the economist Swanson and his colleagues (1994) in the first Background Study Paper commissioned by the FAO's Commission on Genetic Resources from 1994 onwards (see also, Cooper et al., 1994).
13 On incentives, see Chapter 1 (this volume).
14 In truth, there is mounting evidence that "[f]armers carry out both unconscious and deliberate selection through their various agronomic practices (for example, thinning of emerged seedlings, transplanting, rogueing of off-types" (Jarvis et al., 2016, p. 264). This is something that is progressively accounted for in policies and strategies on the *in situ* management of biodiversity. In fact, as recently argued, along a continuum ranging from "intentionality by default" to "conscious intentionality" (Almekinders et al., 2019, p. 122), small-scale farmers "do not typically choose agrobiodiversity for its own sake but rather because it fits with underlying farming rationales or trait preferences" (ibid.). Intentionality is more visible at the natural landscape level where subsistence farmers, driven by their needs, operate as "multiuse strategists" and seem to "play the game of subsistence through the manipulation of ecological components and processes (including forest succession, life cycles, and movement of materials)" (Toledo, 2001, p. 460).
15 CBD, COP, Characteristics Specific to Biological Diversity and Suggestions to Funding Institutions on How to Make their Activities more Supportive of the Convention. A Preliminary Consideration. Note by the Executive Secretary, UNEP/CBD/COP/3/7 (22 September 1996), para. 7.
16 CBD, COP, Sharing of Experiences on Incentive Measures for Conservation and Sustainable Use. Note by the Executive Secretary, UNEP/CBD/COP/3/24 (20 September 1996), para. 68.
17 SBSTTA, Sustainable Use of the Components of Biological Diversity. Note by the Executive Secretary UNEP/CBD/SBSTTA/5/13 (12 November 1999), para. 5.
18 UNEP/CBD/COP/3/24, paras 13, 68-69.
19 As far as we grasp the meaning, "underuse" is understood here in a radically different sense. The assumption is that, from the perspective of the "global crop commons" (total crop genetic diversity available), small-scale farmers, as "diversity managers",

cannot play an active role "in conserving, using, and improving crop diversity" as long as they cannot access the greatest crop genetic diversity available. This is a fairly expansive understanding of underuse that draws attention away from the much more problematic underuse resulting from the disruption of farmer seed networks (Girard & Frison, 2018).

20 For the main food crops, Madagascar's level of dependence on genetic resources from other primary regions of diversity is extremely high (94%–100%) (Palacios, 1998, p. 13).

21 In a recent study conducted on this BCP by the contract holder, the focus is almost exclusively on access to and use of crop genetic diversity, and there is only a brief mention of "standards for access to genetic resources managed by the communities, including crop genetic resources" (Halewood et al., 2021, p. 9).

22 Interview, Natural Justice, Facilitator, 5 March 2019 (in French).

23 Interview, Bioversity International, project holder, 23 May 2019 (in English).

24 Comprising the National Focal Points (NFPs) of ITPGRFA and the Nagoya Protocol both in Madagascar and in Benin, the representatives of Secretariats of the Plant Treaty and of the CDB, ABS Initiative, and Bioversity International in its capacity as a project holder. In Madagascar, the project also involved two main Ministries: the MAEP for the implementation of the ITPGRFA and the MEDD for the Nagoya Protocol.

25 SPGs are groups of farmers who specialise in the production of seeds for marketing. To be able to multiply and sell seeds, farmers have to follow training courses provided by the FOFIFA. There are about ten SPGs at the rural commune level.

26 A further reason is that brokers operate in complex and unstable settings. There are no pre-existing arrangements for development. As "assortment of interests, expertise, technologies and disciplines" (Sharma, 2016, p. 10), development projects are "always unforeseeable" and "become real through the work of generating and translating interests, creating context by tying in supporters and so sustaining interpretations". As a result, "heterogeneous actors in development are constantly engaged in creating order through political acts of *composition*" (Lewis & Mosse, 2006, p. 13; Latour, 2004).

27 Building climate-resilient seed systems was clearly at the core of the project in Madagascar and Benin (Halewood et al., 2021).

28 As James Scott himself stressed, the forms of domination he was studying contained a "strong element of personal rule". He deemed therefore its analysis less relevant to forms of "*impersonal* domination" (J. C. Scott, 2009, p. 21), i.e. "technologies of self" and "intimate government", what Arun Agrawal called "environmentality" (Agrawal, 2005) (or eco-governmentality).

29 Originally patrilineal and patriarchal, the *Fokon'olona* used to bind together, in the same territory (*Fokontany*), "the descendants of the same ancestor whose tomb constitutes the mystical pole where the group comes to find its cohesion" (Condominas, 1961, p. 25). Subsequently, the family dimension of the *Fokon'olona* faded away. It is now the territorial attachment which is decisive and makes it possible to consider as being related to all those who obey common rules (Delteil, 1931, p. V-VI). Furthermore, the *Fokon'olona*, once an autonomous administrative unit, lost most of its prerogatives in 1994 with the advent of the Third Republic. The new decree n° 2004-299 of 3 March 2004 on the *Fokontany* no longer recognises the *Fokon'olona* as anything other than the inhabitants of the *Fokontany*. It is the *Fokontany* alone that has legal personality and is considered as "a basic administrative subdivision at the commune level".

30 Interview, Natural Justice, Facilitator, 20 February 2020 (in French).

31 As the contract holder told us most clearly, "one of the objectives of the project was to sign six or eight agreements for exchange of germplasms to get down to the communities in response to..., and that happened" (interview, Bioversity International, project holder, 23 May 2019 (in English)). As a matter of fact, several Material Transfer

Agreements and STMAs were eventually concluded between the Fa.M.A. and Africa Rice (seven rice accessions), as well as with the National Agricultural Research Institute of Benin (INRAB) (four lima bean accessions from INRAB to Fa.M.A. and two bean accessions from the Fa.M.A. to IRAB). There was a further STMA between the Fa.M.A. and another community in Antavolobe (Madagascar).

32 We have used the "Group Analysis Method" widely used in Belgium, French-speaking Africa and Latin America (van Campenhoudt et al., 2005).

33 Group Analysis, 28 October 2019 (in Malagasy).

34 Mr R says, in Malagasy: "*sahala ny teny frantsay izany amiko*", i.e. literally, "it's all French to me"; the expression is not improper, but it is more common to say: "*sahala ny teny sinoa izany amiko*" – "it's all Chinese to me". French is the language (real or imagined) of the facilitator, the civil servants and the researchers.

35 Group Analysis, 28 October 2019 (in Malagasy).

References

Agrawal, A. (2005). Community, Intimate Government, and the Making of Environmental Subjects in Kumaon, India. *Current Anthropology*, *46*(2), 161–190.

Akrich, M. (1992). The De-Scription of Technical Objects. In W. E. Bijker & J. Law (Eds.), *Shaping Technology/Building Society: Studies in Sociotechnical Change* (pp. 205–224). The MIT Press.

Almekinders, C. J. M., Stone, G. D., Baranski, M., Carney, J. A., Hanspach, J., Krishna, V. V., Ramirez-Villegas, J., Etten, J. van, & Zimmerer, K. S. (2019). Socioecological Interactions Amid Global Change. In K. S. Zimmerer & S. de Haan (Eds.), *Agrobiodiversity Integrating Knowledge for a Sustainable Future* (pp. 117–143). MIT Press. https://research.wur.nl/en/publications/socioecological-interactions-amid-global-change

Altieri, M. A., Anderson, M. K., & Merrick, L. C. (1987). Peasant Agriculture and the Conservation of Crop and Wild Plant Resources. *Conservation Biology*, *1*(1), 49–58.

Badstue, L. B., Bellon, M. R., Berthaud, J., Juárez, X., Rosas, I. M., Solano, A. M., & Ramírez, A. (2006). Examining the Role of Collective Action in an Informal Seed System: A Case Study from the Central Valleys of Oaxaca, Mexico. *Human Ecology*, *34*(2), 249–273. https://doi.org/10.1007/s10745-006-9016-2

Banerjee, S. B. (2003). Who Sustains Whose Development? Sustainable Development and the Reinvention of Nature. *Organization Studies*, *24*(1), 143–180.

Bannister, K. (2009). Non-legal Instruments for Protection of Intangible Cultural Heritage: Key Roles for Ethical Codes and Community Protocols. In C. Bell & V. Napoleon (Eds.), *Protection of First Nations' Cultural Heritage: Laws, Policy and Reform* (pp. 278–308). UBC Press.

Bavikatte, K., Jonas, H., & von Braun, J. (2010). Traditional Knowledge and Economic Development: The Biocultural Dimension. In S. M. Subramanian & B. Pisupati (Eds.), *Traditional Knowledge in Policy and Practice: Approaches to Development and Human Well-Being* (pp. 294–326). United Nations University Press.

Bavikatte, K. S., Robinson, D. F., & Oliva, M. J. (2015). Biocultural Community Protocols: Dialogues on the Space Within. *IK: Other Ways of Knowing*, *1*(2), 1–31.

Beaujard, P. (1995). Les rituels en riziculture chez les tañala de l'ikongo (sud-est de Madagascar). Rituels, mythes et organisation sociale. In S. J. T. M. Evers & M. Spindler (Eds.), *Cultures of Madagascar: Ebb and Flow of Influences = Civilisations de Madagascar: Flux et reflux des influences* (pp. 249–273). IIAS.

Bierschenk, T., Chauveau, J.-P., Olivier de Sardan, J.-P., & Association euro-africaine pour l'anthropologie du changement social et du développement. (2000). *Courtiers*

en développement: Les villages africains en quête de projets. Karthala; Association euro-africaine pour l'anthropologie du changement social et du développement.
Bioversity International. (2014). *Darwin Initiative Application for Grant for Round 21: Stage 2, Mutually Supportive Implementation of the Nagoya Protocol and Plant Treaty*. Darwin Initiative. https://www.darwininitiative.org.uk/documents/22017/23405/22-017%20App%20St2%20-%20edited.pdf
Bioversity International. (2017). *Darwin Initiative Main Project Annual Report, Mutually supportive implementation of the Nagoya Protocol and Plant Treaty, 1 April 2016 to 31 March 2017, Annual Report 2*. Darwin Initiative. https://www.darwininitiative.org.uk/documents/22017/24002/22-017%20AR2%20-%20edited.pdf
Bioversity International. (2018). *Darwin Initiative, Mutually Supportive Implementation of the Nagoya Protocol and Plant Treaty, Final Report*. Darwin Initiative. https://www.darwininitiative.org.uk/documents/22017/24531/22-017%20FR%20edited.pdf
Blaser, M. (2009). Political Ontology: Cultural Studies without 'cultures'? *Cultural Studies, 23*(5–6), 873–896.
Bloch, M. (1994). *Placing the Dead: Tombs, Ancestral Villages and Kinship Organization in Madagascar*. Waveland Press.
Bonneuil, C. (2019). Seeing Nature as a 'Universal Store of Genes': How Biological Diversity Became 'Genetic Resources', 1890–1940. *Studies in History and Philosophy of Science Part C: Studies in History and Philosophy of Biological and Biomedical Sciences, 75*, 1–14.
Bridgewater, P., & Rotherham, I. D. (2019). A Critical Perspective on the Concept of Biocultural Diversity and Its Emerging Role in Nature and Heritage Conservation. *People and Nature, 1*(3), 291–304.
Cadena, M. de la, & Blaser, M. (2018). *A World of Many Worlds*. Duke University Press.
Condominas, G. (1961). *Fokon'olona et collectivités rurales en Imerina*. Berger-Levrault.
Cooper, H. D., Engels, J., & Frison, E. (1994). A Multilateral System for Plant Genetic Resources: Imperatives, Achievements and Challenges. *Issues in Genetic Resources No. 2*.
Delgado, N. A. (2016). Community Protocols as Tools for Resisting Exclusion in Global Environmental Governance. *Revista de Administração de Empresas, 56*(4), 395–410. https://doi.org/10.1590/S0034-759020160404
Deliége, A. (2012). *Pratiques économiques et transactions avec les ancêtres. Étude des cérémonies famadihana dans un village malgache*. Presses Universitaires de Louvain.
Delteil, P. (1931). *Le Fokon'olona et les Conventions de Fokon'olona*. F. Loviton et Cie.
Demeulenaere, É. (2014). A Political Ontology of Seeds: The Transformative Frictions of a Farmers' Movement in Europe. *Focaal, 2014*(69), 45–61.
Demeulenaere, É. (2016). Les terrains globalisés de l'inscription d'un Paysage culturel. Patrimoine mondial et bioculturalité en Éthiopie. *Terrains/Théories, 5*, Article 5.
Escobar, A. (1998). Whose Knowledge, Whose Nature? Biodiversity, Conservation, and the Political Ecology of Social Movements. *Journal of Political Ecology, 5*(1), 53.
Eyzaguirre, P. B., & Dennis, E. M. (2007). The Impacts of Collective Action and Property Rights on Plant Genetic Resources. *World Development, 35*(9), 1489–1498.
Frison, C. (2018). *Redesigning the Global Seed Commons: Law and Policy for Agrobiodiversity and Food Security*. Routledge.
Fujimura, J. H. (1992). Crafting Science: Standardized Packages, Boundary Objects, and 'Translation'. In A. Pickering (Ed.), *Science as Practice and Culture* (pp. 168–211). The University of Chicago Press.
Garine, É., Labeyrie, V., Violon, C., Wencélius, J., Leclerc, C., & Christine, R. (2018). Which Scale to Understand Seed Fluxes in Small-Scale Farming Societies? Snapshots of Sorghum from Africa. In F. Girard & C. Frison (Eds.), *The Commons, Plant*

Breeding and Agricultural Research: Challenges for Food Security and Agrobiodiversity (1st ed., pp. 159–172). Routledge. https://doi.org/10.4324/9781315110387

Genard, J.-L. (1999). *La grammaire de la responsabilité*. Cerf.

Girard, F., & Frison, C. (2018). *The Commons, Plant Breeding and Agricultural Research: Challenges for Food Security and Agrobiodiversity*. Routledge.

Girard, F., & Frison, C. (2021). From Farmers' Rights to the Rights of Peasants: Seeds and the Biocultural Turn. *Review of Agricultural, Food and Environmental Studies, 52*, 461–476.

Graddy, T. G. (2013). Regarding Biocultural Heritage: In situ Political Ecology of Agricultural Biodiversity in the Peruvian Andes. *Agriculture and Human Values, 30*(4), 587–604.

Grandidier, M. G. (1932). A Madagascar, anciennes croyances et coutumes. *Journal de la Société des Africanistes, 2*(2), 153–207. https://doi.org/10.3406/jafr.1932.1533

Gray, S. (2004). Imagination, Fraud and the Cultural Protocols Debate: A Question of Free Speech or Pornography? *Media and Arts Law Review, 9*(3), 23–37.

Gudeman, S. (2012). *Economy's Tension: The Dialectics of Community and Market*. Berghahn Books.

Halewood, M. (2013). What Kind of Goods Are Plant Genetic Resources for Food and Agriculture? Towards the Identification and Development of a New Global Commons. *International Journal of the Commons*, 7(2), 278–312.

Halewood, M., Bedmar Villanueva, A., Rasolojaona, J., Andriamahazo, M., Rakotoniaina, N., Bossou, B., Mikpon, T., Vodouhe, R., Fey, L., Drews, A., Kumar, P. L., Rasoanirina, B., Rasoazafindrabe, T., Aigbe, M., Agbahounzo, B., Otieno, G., Garforth, K., Kiene, T., & Nnadozie, K. (2021). Enhancing Farmers' Agency in the Global Crop Commons through Use of Biocultural Community Protocols. *Agriculture and Human Values, 39*, 579–594.

Harvey, D., & Haraway, D. (1995). Nature, Politics, and Possibilities: A Debate and Discussion with David Harvey and Donna Haraway. *Environment and Planning D: Society and Space, 13*(5), 507–527.

Hayden, C. P. (2003). *When Nature Goes Public: The Making and Unmaking of Bioprospecting in Mexico*. Princeton University Press.

Hughes, L., & Lamont, M. (2018). Cultural Rights and Constitutional Change. *African Studies*, 77(2), 159–170.

Jarvis, D. I., Hodgkin, T., Brown, A. H. D., Tuxill, J. D., López Noriega, I., Smale, M., Sthapit, B., & Samper, C. (2016). *Crop Genetic Diversity in the Field and on the Farm: Principles and Applications in Research Practices*. Yale University Press.

Kamau, E. C. (2013). Exploring Bases for Building Common Pools in Eastern Africa. In E. C. Kamau & G. Winter (Eds.), *Common Pools of Genetic Resources. Equity and Innovation in International Biodiversity Law* (pp. 373–398). Routledge.

Keller, E. (2008). The Banana Plant and the Moon: Conservation and the Malagasy Ethos of Life in Masoala, Madagascar. *American Ethnologist, 35*(4), 650–664.

Koehrsen, J. (2017). Boundary Bridging Arrangements: A Boundary Work Approach to Local Energy Transitions. *Sustainability, 9*(3), 424.

Latour, B. (2004). *Politics of Nature: How to Bring the Sciences into Democracy*. Harvard University Press.

Lefèvre, G. (2008). Lefèvre Gabriel, Médecines hybrides dans le sud et le sud-ouest de Madagascar. Les mots-plantes à Toliara. *Études Océan Indien, 40–41*, 287–293. https://doi.org/10.4000/oceanindien.1414

Lewis, D. J., & Mosse, D. (2006). *Development Brokers and Translators the Ethnography of Aid and Agencies*. Kumarian Press.

Löfmarck, E., & Lidskog, R. (2017). Bumping against the Boundary: IPBES and the Knowledge Divide. *Environmental Science & Policy*, *69*, 22–28.

MAEP, M., Secrétariat general, Direction générale de l'agriculture. (2008). *Document de stratégie nationale semencière (DSNS)*. MAEP. http://www.mpae.gov.mg/communication/wp-content/uploads/sites/2/2016/11/strategie_semence.pdf

MAEP, SAGE, & Ministère de l'environnement. (2017). *Document sur les plans d'investissement du site Antavolobe issu de l'atelier du 08 février 2017* (Doc SAGE /12). Darwin Initiative, CDB, ABS Initiative, Bioversity International, TIRPAA.

Ministère auprès de la Présidence en charge de l'agriculture et de l'élevage. (2016). *Rapport d'atelier sur le PBC 14–15 Octobre 2016 Analavory. 'Projet de mise en œuvre mutuellement soutenue du Protocole de Nagoya "PN" et du Traité International sur les Ressources Phytogénétiques pour l'Alimentation et l'Agriculture "TIRPAA"'* (N° 2016/ MPAE/SG/SECC). Repoblikan'i Madagasikara.

Ministère auprès de la Présidence en charge de l'agriculture et de l'élevage. (2017). *Stratégie nationale de développement rizicole (SNDR) 2016–2020*. MPAE.

Ministère de l'Agriculture et de l'Élevage. (2002). *État des lieux des Organisations Paysannes Professionnelles agricoles et des Organisations Paysannes de Madagascar* [Programme de Professionnalisation de l'Agriculture (P.P.d.A.)]. MAE.

Ministère de l'agriculture et de l'élevage. (2018). *Stratégie nationale et plan d'actions sur les ressources phytogénétiques pour l'alimentation et l'agriculture 2018–2025* [Projet FAO 'TCP/MAG/3605 - RPGAA aux bénéfices des populations locales']. MINAE, FAO.

Ministère de l'environnement, MPAE, Bioversity International, ABS Initiative, & Darwin Initiative. (2016). *Summary Field Report: Darwin Initiative's project on Mutually supportive implementation of ABS in Antavolobe and Analavory, Madagascar, 25th to 30th July, 2016*. Ministère de l'environnement, MPAE, Bioversity International, ABS Initiative, Darwin Initiative.

Mollinga, P. (2010). Boundary Work and the Complexity of Natural Resources Management. *Crop Science*, *50*(1), S1–S9.

Morgera, E., Tsioumani, E., & Buck, M. (2014). *Unraveling the Nagoya Protocol: A Commentary on the Nagoya Protocol on Access and Benefit-sharing to the Convention on Biological Diversity*. Brill; Nijhoff.

North, D. C. (1990). *Institutions, Institutional Change, and Economic Performance*. Cambridge University Press.

Palacios, X. F. (1998). *Contribution to the Estimation of Countries' Interdependence in the Area of Plant Genetic Resources*. FAO.

Parks, L. (2019). *Benefit-sharing in Environmental Governance: Local Experiences of a Global Concept* (1st ed.). Routledge.

Platteau, J.-P. (2004). Monitoring Elite Capture in Community-Driven Development. *Development and Change*, *35*(2), 223–246.

Rakoto Ramiarantsoa, H., & Lemoigne, N. (2014). 'La terre est chair, les roches grossissent': Gérer la vie là où elle se trouve. Pour une ethno-pédologie des savoirs paysans. *ELOHI. Peuples indigènes et environnement*, *5–6*, 89–116. https://doi.org/10.4000/elohi.747

Rakotomalala, M. (2006). La sorcellerie en Imerina. In S. Blanchy, J.-A. Rakotoarisoa, P. Beaujard & C. Randimilahy (Eds.), *Les dieux aux services du peuple. Itinéraires religieux, médiations, syncrétisme à Madagascar* (pp. 229–256). Karthala.

Robinson, C. J., & Wallington, T. J. (2012). Boundary work: engaging knowledge systems in co-management of feral animals on Indigenous lands. *Ecology and Society*, *17*(2): 16. http://dx.doi.org/10.5751/ES-04836-170216.

SAGE, & MAEP. (2016). *Rapport d'études sur les enquêtes de base du projet. 'Mise en œuvre mutuellement soutenue du Protocole de Nagoya et de Traité international sur les ressources Phytogénétiques pour l'alimentation et l'agriculture'*. MAEP, SAGE. http://www.bioversityinternational.org/fileadmin/user_upload/campaigns/Darwin/Baseline_study_year1_Madagascar.pdf

Schmietow, B. (2012). Conceptualizing the Commons: Genetic Resource Management, Property Claims, and Global Innovation Ethics. *St Antony's International Review, 8*(1), 70–87.

Scott, J. C. (2009). *Domination and the Arts of Resistance: Hidden Transcripts.* Yale University Press.

Scott, J., & Marshall, G. (Eds.). (2009). *A Dictionary of Sociology* (3rd ed. rev). Oxford University Press.

Sedjo, R. A. (1992). Property Rights, Genetic Resources, and Biotechnological Change. *The Journal of Law & Economics, 35*(1), 199–213.

Sharma, J. R. (2016). *Brokering in International Development. A Consideration of Analytical Issues* (New Norms and Forms of Development. Working Paper I). School of Social and Political Science, The University of Edinburgh.

Shrumm, H., & Jonas, H. (2012). *Biocultural Community Protocols: A Toolkit for Community Facilitators.* Natural Justice. https://naturaljustice.org/wp-content/uploads/2015/09/Biocultural-Community-Protocols-Toolkit.pdf

Swanson, T. M., Pearce, D. W., & Cervigni, R. (1994). *The Appropriation of the Benefits of Plant Genetic Resources for Agriculture: An Economic Analysis of the Alternative Mechanisms for Biodiversity Conservation* (Background Study Paper No. 1). Commission on Plant Genetic Resources, First Extraordinary Session Rome, 7–11 November 1994.

Tobin, B. M. (2013). Bridging the Nagoya Compliance Gap: The Fundamental Role of Customary Law in Protection of Indigenous Peoples' Resource and Knowledge Rights. *Law, Environment and Development Journal, 9*(2), 142–162.

Toledo, V. M. (1990). The Ecological Rationality of Peasant Production. In M. A. Altieri & S. B. Hecht (Eds.), *Agroecology and Small Farm Development* (pp. 53–60). CRC Press.

Toledo, V. M. (2001). Indigenous Peoples and Biodiversity. In S. A. Levin (Ed.), *Encyclopedia of Biodiversity* (pp. 451–463). Elsevier. https://doi.org/10.1016/B978-0-12-384719-5.00299-9

Tsing, A. (1999). Becoming a Tribal Elder, and Other Green Development Fantasies. In T. Li (Ed.), *Transforming the Indonesian Uplands Marginality, Power and Production* (pp. 157–200). Harwood.

Ulloa, A. (2005). *The Ecological Native: Indigenous Peoples' Movements and Eco-governmentality in Colombia.* Routledge.

van Campenhoudt, L., Chaumont, J.-M., & Franssen, A. (2005). *La méthode d'analyse en groupe: Applications aux phénomènes sociaux.* Dunod.

Watts, N., & Scales, I. R. (2015). Seeds, Agricultural Systems and Socio-natures: Towards an Actor–Network Theory Informed Political Ecology of Agriculture. *Geography Compass, 9*(5), 225–236.

Wynberg, R., & Laird, S. (2009). Bioprospecting, Access and Benefit Sharing: Revisiting the 'Grand Bargain'. In R. Wynberg, D. Schroeder, & R. Chennells (Eds.), *Indigenous Peoples, Consent and Benefit Sharing: Lessons from the San-Hoodia Case* (pp. 69–86). Springer Netherlands.

PART 3

Biocultural Jurisprudence, Sovereignty and Legal Subjectivity

11

BIOCULTURAL COMMUNITY PROTOCOLS AND THE ETHIC OF STEWARDSHIP

The Sovereign Stewards of Biodiversity

Reia Anquet and Fabien Girard[1]

Introduction

BCPs and the Western Ontological Matrix

Since the late 1980s onwards, countless declarations, reports, and scholarly works have underlined the crucial role of indigenous peoples and local communities (IPLCs) in conserving biodiversity. In parallel with the rise of the concept of biocultural diversity,[2] IPLCs are increasingly featured as *guardians* or *stewards* of rich landscapes and vital ecosystems. The Brundtland Report (Brundtland & World Commission on Environment and Development, 1987, para. 74) described these communities as "[...] the repositories of vast accumulations of traditional knowledge and experience that links humanity with its ancient origins". One year later, the Declaration of Belém (1988), under the lead of the anthropologist Darrell A. Posey, forcefully stated that "native peoples have been stewards of 99 percent of the world's genetic resources". Posey, and his team of lawyers, anthropologists, and indigenous activists (Graham Dutfield, Kristina Plenderleith, Addison Eugênio da Costa e Silva, and Alejandro Argumedo to name a few), played a considerable role in mainstreaming the "integrated rights approach" (Posey & Dutfield, 1996) as one of the progressive ways to synergistically advance the human rights of IPLCs and what they called "the right to development and environmental conservation" (Posey & Dutfield, 1996, p. 95). They emphasised biodiversity conservation with a strong and innovative focus on mutual adaptations and co-evolution between human cultures, languages, and the environment (Maffi & Woodley, 2010, p. 5). But the crux of their proposal was the maintenance of identity for IPLCs. More critical for Posey was the defence of indigenous peoples and the belief that better protection of their language and identity necessarily implied recognising the holistic nature of their community

DOI: 10.4324/9781003172642-14

life. This inevitably endowed IPLCs with a "bundle of rights" which comprised the "control over cultural, scientific, and intellectual property". Posey dubbed this "bundle", "traditional resource rights" and set as a fundamental prerequisite for its achievement the "rights to land and territory" (Posey, 2004, p. 163; Posey & Dutfield, 1996). Posey's inspirational work (Bavikatte, 2014, pp. 234–235) has paved the way for subsequent insightful "integrated" approaches to help IPLCs secure their ways of life.

International environmental and human rights lawyers Sanjay Bavikatte and Harry Jonas pushed further ahead with this concept by referring to IPLCs *as stewards* of nature. Through their theoretical work and grass-roots involvement, they succeeded in effectively harnessing the open-textured (see Posey, 2004, 163) language of Articles 8(j) and 10(c) of the Convention on Biological Diversity (CBD) (apparently intended by their main initiators – see Halewood, 1999), in light of the progressive provisions of the Agenda 21 and the Rio Declaration. The bundle of "biocultural rights"[3] – as they were to be conceptualised by Sanjay Bavikatte and colleagues – closely resembles the concept of "traditional resource rights". This bundle encompasses the rights to land, territory, and natural resources, the right to self-determination, i.e. self-government (Anaya, 2004), and the rights to culture and cultural heritage. Most importantly, biocultural rights are built upon two cornerstones: one relates to the direct interests of IPLCs (a group rights approach), while the other pertains to a more general interest of humankind (or the biotic community at large) in the conservation of the environment. Therefore, the recognition of biocultural rights does not take as its point of departure the right of a group or community to flourish, but rather the ethic of stewardship, i.e. the ethic entrenched in the role of IPLCs as conservationists or custodians of local ecosystems.

One of the core assumptions of biocultural jurisprudence is that much of this role of guardianship/stewardship of biodiversity is underwritten by an "ethic of stewardship", itself embodied in "a way of life" and rooted "within a moral universe" (Bavikatte, 2014, pp. 168–169). The last IPBES' report (IPBES, 2019, p. 42) leaves no doubt that it is through this idea of "stewardship of biodiversity" that IPLCs have recently gained a more substantial status in the international regime for the conservation of biodiversity. An important point to make is the recent reinforcement of the underlying theoretical canvas: both the Tkarihwaié:ri Code of Ethical Conduct[4] and the Atrato River case of the Constitutional Court of Colombia[5] (Macpherson et al., 2020) – one of the first decisions to formally recognise "biocultural rights" – make it clear that the ethic of stewardship touches upon the "holistic interconnectedness of humanity with ecosystems and obligations and responsibilities of indigenous and local communities".[6] Emphasis on possible different worldviews, ontologies, and epistemologies naturally produces deep philosophical tension when faced with "Euro-modernity" or "Western-modernity".

In the Access and Benefit-Sharing (ABS) regime, these theoretical changes have gone hand in hand with the advent and the consolidation of the concept of "biocultural diversity" (Maffi, 2001).[7] The ethic of stewardship has opened new avenues to "protect" IPLCs' "customary use of biological resources" and "knowledge, innovations and practices" within the meaning of Articles 8(j) and 10(c) of the CBD. Some have continued to explore the possibility of *sui generis* intellectual property rights (IPRs) vested in IPLCs (see Brush & Stabinsky, 1996; Greaves, 1996). The negotiation, adoption, and entry into force of the Nagoya Protocol, however, have given hope to all those who advocate holistic approaches and do not believe that IPRs and the attendant transformation of genetic resources and knowledge tradable on global markets would do any good to the communities concerned. Ongoing work on the interpretation of the relevant provisions of the CBD and Articles 5, 6, and 7 of the Nagoya Protocol is still encumbered by a significant level of uncertainty, but a great deal of hope has been pinned on community protocols – referred to here as Biocultural Community Protocols (BCPs) – as now enshrined in the Nagoya Protocol.[8] In particular, seen within a dynamic policy and advocacy context around "territories and areas conserved by indigenous peoples and local communities" (ICCAs),[9] Indigenous Biocultural Territories (Argumedo & Pimbert, 2008), rights of nature (Iorns Magallanes, 2019; Kotzé & Villavicencio Calzadilla, 2017; Tănăsescu, 2020), and biocultural rights (Bavikatte, 2014; Girard, 2019; Sajeva, 2018), BCPs have been heralded as able to solve the ABS conundrum for IPLCs. They ensure communities' rights to development, while supporting their role in biodiversity conservation and maintenance (Posey & Dutfield, 1996, p. 95), without letting their unique livelihoods and ways of life be shattered by the unfettered extension of disembedded markets and the language of trade.

There is no intention here to walk away from the "win-win" approach[10] at the core of ABS; the commodification of traditional and indigenous seeds and traditional knowledge (TK) is still believed to open new opportunities to development, thereby strengthening IPLCs' effectiveness in conservation. However, what is new is the procedural framework that goes along with BCPs, which is thought to give IPLCs more substantial control over their resources and TK. This control occurs through the right to say "no" and to set out the conditions of negotiations. It also enables IPLCs to uphold the market-inalienability (*extra commercium* quality) of certain aspects of their heritage – land, sacred sites, seeds, and language – on which communities depend for their survival, well-being, and to thrive (Bavikatte et al., 2010, p. 298).

It remains to be seen whether a more significant transformation can be expected from tools so far initiated and facilitated by NGOs within the constraints set by the States and international organisation funders (Parks, 2019, p. 82). They inevitably accept strong linkages between what are still mainly "moral economies" and disembedded global markets, without really pondering over the disruptive impact that the irruption of the market-oriented and instrumental rationality may have on many communities in the medium run (see Gudeman, 2001, pp. 27–29, 2012, p. 29).[11] At the very least, BCPs may help rebalance centuries-old asymmetrical relationships between IPLCs and (mainly) North-based bioprospectors, *extra commercium* things and tradable properties, and reinforce local or community prior and informed consent (PIC) procedures or support their recognition in domestic legislation where they are still lacking. As many BCPs included in our case study illustrate, BCPs may also be used as tools for the vindication of rights on land, territories, and resources. Finally, some believe that BCPs, insofar as they are undergirded by a new ethic of stewardship, can be used as "space opening" tools (Mulrennan & Bussières, 2020) to alternative, non-naturalistic ontologies. In this, BCPs would stand as standard-bearers in the struggle for making room for IPLCs' identities.

BCPs and Ecological Scripts: Eco-Governmentality, Counter-Narratives, and New Subjectivity

Is not such a reading excessively naïve? For such fundamental changes to take place, there would first have to be profound transformations in Western/European legal constructs and in their naturalistic underpinnings. The seven BCPs that this chapter analyses show that, even where a domestic legislation enshrines a local PIC procedure and further strengthens its application through the express recognition of BCPs (see Table 11.1, Annex), there is no apparent shift away from what Mario Blaser calls the Western or European "ontological matrix" (Blaser, 2009). The way this naturalistic "ontological matrix" unfolds in several of the BCPs under consideration, the interlacing of narratives and representations around biodiversity conservation, their apparent strategic reversals, together with references to the ethic of stewardship, are invitations to consider further what BCPs offer in terms of strategic use of hegemonic categories.

For this purpose, two central concepts will be used: scripts and eco-governmentality. The seven BCPs included in our case study were examined together with background documentation, and our analysis is informed by the extensive statutory and regulatory framework in which each was developed. These protocols were analysed as incorporating "scripts". Scripts partake in the "institutional and regulatory spaces in which the knowledges and practices are encoded, negotiated, and contested" (Peet & Watts, 1996, p. 11). They are involved in the construction of social practices and identities. We posit that scripts need to be grasped at the ontological level, which means that they reach into radical assumptions about what kinds of things do or exist, their conditions of existence, relationships, attachments, and connectivity (Blaser, 2009). To understand precisely what is at stake, let us follow the path of a script. Ontologies express something about the *real*, the "conditions of possibility we live with" (Mol, 1999); and for this reason, they are never given, "out there", ready to be picked up, but are the result of practices and "dynamic relations of hybrid assemblages" (Blaser, 2013, p. 552). They are always "in the making" (ibid.), "open and contested" (Mol, 1999, p. 75). They "manifest as 'stories'" or narratives and through non-discursive aspects (Blaser, 2009, p. 877). Discursive (e.g. texts and policies) and non-discursive elements (e.g. objects, conducts, and institutions) rest upon theoretical knowledge, a "body of doctrine" made of concepts, ideologies, and axioms. In European/Western ontologies, there is an ontological matrix

(Blaser, 2009) or meta-ontology (Blaser, 2013, p. 544) which is built along the "nature-culture divide". More importantly, storied practices (narratives) and theoretical knowledge stabilise into performative interpretations (Bonneuil, 2019, p. 9) of the world, which we define as "scripts" or "scenarios". As they are interpreting what is "out there", they necessarily take part in the enactment of what they narrate (Blaser, 2013, p. 552). This allows them to produce, in turn, "social practices and relationships of power" (Ulloa, 2005, p. 5). This is what is meant by eco-governmentality. Following Agrawal's Foucauldian-inspired analytic, eco-governmentality is defined as the variety of "knowledge-making apparatuses" (Brosius, 1999, n. 6) used to shape "[...] the conduct of specific persons and groups, including the mechanisms that such persons and groups use on themselves" (Agrawal, 2005a, n. 3). Eco-governmentality relies on the "techniques of the self" used (along with a State's regulatory strategies) to forge new practices and new links of political influence (e.g. the creation of a new local council, a division of space, or a provisioning of new crop varieties). The typology of these "techniques of the self" is not explored in the following paragraphs. What matters to our analysis are their outcomes: the effects of the regimes of power and the "intimate government" (Agrawal, 2005b). These techniques show what scripts

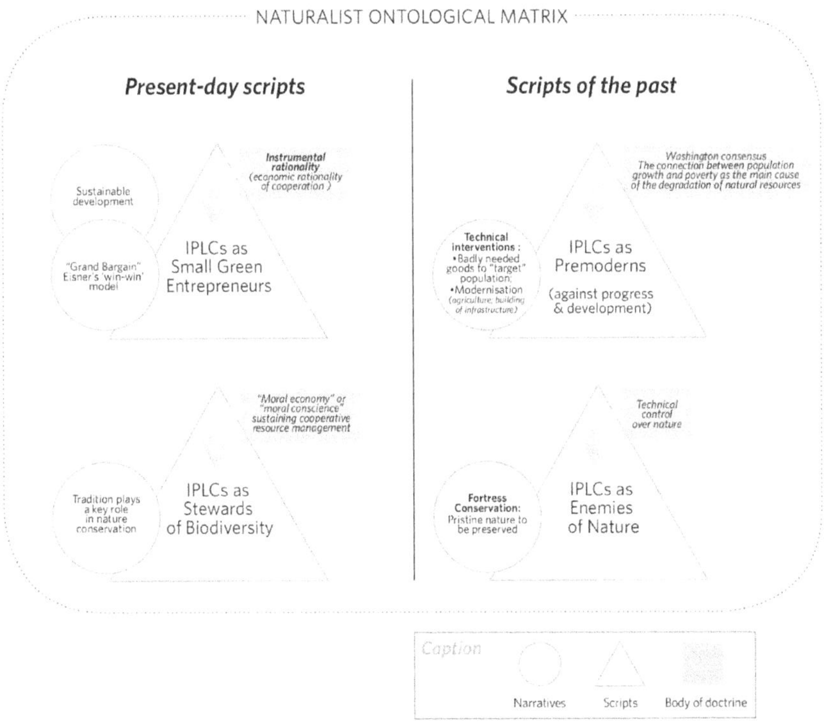

FIGURE 11.1 Naturalist scripts and narratives (Fabien Girard, original material for the book)

do, namely that they invade the political imagination and make local "subjects" pursue "goals that they imagine as their own" (Agrawal, 2005a, p. 179).

Four scripts were identified through the corpus of BCPs studied. We believe that they can all be related somehow to the "all-encompassing modernity" (or the naturalist ontological matrix) – and its series of foundational and oppositional pairs: nature/culture, non-humans/humans, non-moderns/moderns, objects/subjects, past/present (Blaser, 2013). Two of the scripts are present-day scripts, and the "steward of biodiversity" script is relied upon in all the BCPs under consideration. The other two scripts, those of the past, are mostly inherited from colonial occupation. These scenarios still have a bearing on development projects and environmental politics in several countries. Some of our case study BCPs illustrate this, though they are not used to defend IPLCs' claims.

In light of this, we first investigate what BCPs tell us of the kind of scripts, relations of power, and the specific eco-governmentality that IPLCs face in the international regime for biodiversity. Suppose scripts can "define actors" (Akrich, 1992, p. 207), allocate roles and tasks, and enrol and "enscript". In that case, this begs the question of the extent to which BCPs, despite all good intentions and promises, are deployed in domestic politics or international fora as a means to replace political negotiation with managerial efficiency (Brosius, 1999). Our second line of enquiry is that environmental policies and their scripts should not be seen as unrelenting machinery operating seamlessly on the undifferentiated mass of local actors. Supporting this argument is taking the risk of "ignoring the social relations through which technologies of control are formed, exercised, contested, and critiqued" (Cepek, 2011, pp. 544–545; also see Valdivia, 2015, p. 474; Peet & Watts, 1996, p. 16).

This is especially true of the "steward of biodiversity" script, a dominant script, which certainly has much to do with the "noble ecological savage" myth (Raymond, 2007; Redford, 1991), but which also distinctively hinges upon a new ethic capable of sustaining new representations and counter-discourses and thereby of creating cracks in the European-centred matrix. As we are reminded by Mario Blaser (2013, p. 558) and Sylvie Poirier (Poirier, 2008, p. 83),[12] indigenous peoples may at times outwardly indulge themselves in modern naturalist categories and use available "symbols of alterity" (e.g. the steward and caretaker of the environment), while making in fact great inroads into the nature-culture divide and opening the "door to the consideration of other ontologies as plausible and alternatives to the modern" ones (Blaser, 2013, p. 556). This opens our last line of thought, namely that as offshoots of the biocultural jurisprudence and the concept of "biocultural diversity" (in its political articulation: Brosius & Hitchner, 2010), and against the backdrop of ongoing legal innovations around "natural entities" and IPLCs' territories and heritage, BCPs have the potential to make room for "non-modern" worlds. In particular, through a strategic partnership with academics and civil society organisations and via the support of transnational networks, community protocols may contribute to unravelling the ontological threads of modernity and negotiating spaces for the advent of a new form of legal subjectivity. This is a subjectivity that would not be grounded in the aptitude of monadic agent to exert her dominion over inanimate nature. Instead, it would be based on the entanglements between plants, microorganisms, ecosystems, and a "person-in-community" (Gudeman, 1992) such as a carer, steward, custodian of a place where identities unfold. The result, it is argued, would be to open up space for IPLCs to sustain their identities by shrinking the reach of the naturalistic matrix.

These questions are addressed in the following sections of this chapter. We start with a critical appraisal of a string of BCPs developed in Africa and Latin America. We investigate the kind of scripts that are played out in their drafting and implementation. We analyse the type of counter-narratives and counter-hegemonic discursive practices that they may contain and deploy as an antidote. The chapter then moves on to examine the effects of entrenching BCPs in "stewardship", "custodian", "caretaker" representations and narratives on tradition, asking in particular whether the associated script is not *subjecting* peoples to new forms of eco-power (Brosius, 1999, p. 37). Finally, the chapter replaces BCPs and the "ethic stewardship" within the context of IPLCs' political struggle aimed at opening up space within Western ontologies. It is argued, in particular, that the use of the ethic of stewardship, together with bonds with transnational coalitions and networks, helps unveil and anchor a new form of legal subjectivity.

Unravelling the Imaginaries Bound Up with Environmental Scripts

The BCPs that we have examined bear testimony to the kind of ontological politics (Mol, 1999) that is now played out around the ABS regime and IPLCs. There are seven BCPs included in the scope of this study, covering both Africa (Benin, Kenya, Madagascar) and Latin America (Mexico, Panama). These BCPs offer a wide diversity of ABS regimes, legislative and regulatory frameworks on IPLCs' rights, experiences of colonialism, and socioeconomic and environmental conditions. The full list is provided in the Annex (see Table 11.1), which also contains a brief account of domestic legal regimes regarding local or community PIC and BCPs.

In what follows, we will examine BCPs through the lens of ontology and eco-power and pay attention to the discourses and narratives that unfold, understood as many performative scripts on IPLCs and the environment. The narratives and scripts that our chapter unveils are summarised in Figure 11.1. All these scripts are embedded in the Western or European "ontological matrix" (Blaser, 2009), but the two present-day scripts are much more important and therefore receive the greatest attention in the following pages for two reasons. First, the "stewardship of biodiversity" script is the dominant script, in the sense that it is pervasively present within the BCPs under study and beyond (see examples in Delgado, 2016). While our study illustrates how this script opens a new chapter in the history of the "[...] shift from the management of non-human nature to the management of people" (Roach et al., 2006, p. 60; also see Bavington, 2002), we endeavour to show that it also harbours new intellectual influences – first and foremost the "ethic of stewardship" – offering a promising potential to sustain IPLCs' identities. Second, the "small green entrepreneur" script holds a special position in biodiversity conservation, as it dates from the early days of the United Nations Conference on Environment and Development (UNCED) and the CBD, and is still the bedrock of community-based natural resource management (CBNRM) as an ample literature now illustrates.[13]

Two interrelated master narratives support the small green entrepreneur script: "sustainable development" and the "Grand Bargain"; they warrant a few comments. Sustainable development, which is enshrined in the 1987 Brundtland Report (Brundtland & World Commission on Environment and Development, 1987), means that conservation is based on utilisation, i.e. conservation cannot be separated from the idea of "economic and social development and poverty eradication" that are thought of as "the first and overriding priorities of developing countries" (CBD, Preamble). As could be read in an early draft version of the Rio Declaration outlining the headings to be included in the text,[14] the aim is to preserve biodiversity while reaching "optimum sustainable yield" or "optimum sustainable productivity" (Robinson, 1992, p. ci), what Ostrom will later coin as "long-term economic viability" (Ostrom, 1990, p. 31). The concept of the "Grand Bargain" (Boisvert & Vivien, 2012) is seen as the key to solve the conundrum of economic development within the confines of the carrying capacity of ecosystems. The ABS process is conceived of as a win/win situation (see Eisner, 1989)[15]: the "gene-poor" North retains its access to the remarkable bounty of genetic resources mainly located in the tropics. At the same time, the Global South captures part of the benefits arising out of IPRs on "biodiscovery". Or it can benefit directly as well from new ("environmentally sound") technologies through "technology-for-nature swaps".[16] In this, the Global South is deemed better equipped to tackle biodiversity erosion.[17] Of paramount importance in this model is the framing of ABS as a form of inducement-producing policy – together with a blend of measures known as institutional incentives (North, 1990) – to change how humans, understood as rational agents, "interact with their environment and how they use natural resources".[18]

In the next section of this chapter, we begin with an analysis of this script. We then move on to study the "enemies of nature" and "premoderns" scripts. These two scripts are never explicitly mentioned in the protocols but are still active as background influence in the development of certain BCPs and against which IPLCs are compelled to fight. Finally, we investigate the "stewardship of biodiversity" script. As this is the dominant script, we evaluate the BCPs according to its inclusion.

The "Small Green Entrepreneur" Script

In four out of seven BCPs studied, the main thrust of the community-based enterprise was to align farming communities and indigenous peoples with the "Grand Bargain" (Wynberg & Laird, 2009) and sustainable development narratives. Translated and stabilised in the "small green entrepreneurs" script, which owes much, as will be seen, to Ostrom's work and the Bloomington school of political economy, the two narratives inject certain assumptions about the local "agents" (tastes, competences, motives, aspirations – Akrich, 1992, pp. 207–208) and define their role within a set "framework of action" (ibid.) mainly geared towards incentivising biodiversity conservation.

The script is clearly discernible in the BCPs developed in Madagascar[19] and Benin within the framework of the Darwin Initiative-funded project on the "Mutually supportive implementation of the Nagoya Protocol and Plant Treaty".[20] This should not come as a surprise, as there is a blueprint laid out in the ABS framework of each country to connect BCPs with the intra-state benefit-sharing mechanism,[21] making clearly the instruments part of the strategy to foster both economic development and biodiversity conservation through bioprospection.

As much in the reports submitted by Bioversity International to the funding agency as in the contract holder's work (Halewood et al., 2021), several references are made to plant genetic resources as "new commons" (Halewood, 2013). The theoretical underpinnings have now a long pedigree in studies on plant genetic resources for food and agriculture (PGRFA) in particular. They can be summarised as follows: with their twofold economic attributes of low excludability and high rivalry, genetic resources embodied in seeds managed by farmers and IPLCs yield positive externalities, but which are hardly appropriated at the local level (Cooper et al., 1994; Swanson et al., 1994). Consequently, for farmers and local communities not to forgo their conservation practices for more lucrative activities (e.g. conversion to elite varieties and monoculture), incentives need to be developed (Bioversity International, 2018; Correa, 1999). BCPs, in setting up the conditions for ABS agreements with commercial partners (industry, researchers), stand as appropriate tools for influencing IPLCs' behaviour in biodiversity conservation. They are meant to show IPLCs that conserving genetic resources can pay off. BCPs entrenched in the script of "small green entrepreneurs" remain, of course, mostly unconcerned with the aspects of biocultural heritage and hardly touch upon the issue of the ethic of stewardship.

The BCP of the farming communities of Analavory (Madagascar), the BCP of Ampangalastary (Madagascar), and the BCP of the Municipality of Tori-Bossito (Benin) alike are prime examples of protocols instrumentally tailored to attract bioprospectors by giving them clear guidance about how to negotiate ABS agreements and the kind of benefits sought by the communities. They also set out the procedure whereby the communities can access genetic materials through the Multilateral System (MLS) of the International Treaty on

Plant Genetic Resources for Food and Agriculture (ITPGRFA). In the case of Analavory (pp. 7–9)[22] and Tori-Bossito (pp. 11–12), specific procedural rules are laid down to include landraces in the MLS, should the communities voluntarily decide to make materials that they hold available to the international community. To our knowledge, this linkage between a BCP and the ITPGRFA is unprecedented and warrants two observations. The first is that the communities of Analavory and Tori-Bossito are being located in low diversity areas[23]; the prospect of attracting bioprospectors is low.[24] The BCPs could not, therefore, entirely be articulated around the ABS framework, and much effort was invested in deploying an incentive framework for communities to engage in transnational germplasm transfers, test new cultivars through participatory plant breeding, and adopt new varieties (Halewood et al., 2021).

This brings us to our second observation: attempts at turning IPLCs into small entrepreneurs are always risky undertakings, especially where there are serious doubts about the kind of resources and knowledge that can attract the interest of researchers and bioprospectors. In this context, BCPs create expectations that they cannot deliver. Dashed hopes bring disappointment and frustration, a point already stressed by Pierre du Plessis some years ago, who stated that the "[t]he cost implications of pre-emptive ABS protocols are not worth it" (IIED et al., 2012, p. 4). Admittedly, the process of developing BCPs is an empowering process in itself. It contributes to raising the community's awareness about its values, resources, customs, and institutional organisations, thereby increasing its aptitude to manage its biocultural heritage and engage with third parties with greater bargaining power (Parks, 2019; Rutert, 2020). But these benefits can be wiped out by the state's posture of defiance vis-à-vis IPLCs. Significantly in the cases of Analavory and Ampangalastary, preliminary policy documents, field reports, and workshop reports frequently indulge in the colonial script of local communities as "enemies of nature". These documents stress in several sections how poverty and demography increase the anthropogenic pressure on watersheds. At times, they blame inappropriate agricultural practices (such as "[b]ush fires and irrational exploitation of the mountains [...]") for siltation, soil degradation, and water erosion (MAEP et al., 2017). Against this narrative, BCPs are necessarily less about capacity-building, empowerment, self-determination, and life plans, but rather about controlling how IPLCs interact with "nature" on the grounds they need to change and be forward-looking.

Of Two Scripts of the Past: IPLCs as "Enemies of Nature" or "Premoderns"

The previous point brings us to discuss age-old (but constantly repeated) scripts, among which that of IPLCs as "enemies of nature" is still pervasive in Africa.[25] This often coexists and merges with another script from the past, namely that IPLCs are "premodern", i.e. against progress and development or that they live at the mercy of nature.[26] The two scripts are generally deployed in postcolonial contexts to deny indigenous peoples rights to their ancestral lands or to justify forced displacements or encroachments upon indigenous territories and lands in the name of economic development.

Of particular interest here is the way some BCPs display alternative narratives and representations to refute any of those scripts. Kenya offers two such cases with the Ogiek BCP and the Lamu County BCP. The Ogiek people and the indigenous communities of Lamu County are facing different challenges. For decades, the Ogiek people have been seeking redress for the violent evictions from their ancestral homeland in the Mau Forest Complex. The BCP, whose second version was released in 2015, was part of a broader political campaign and a legal and judicial strategy that saw a landmark victory in the African Court on Human and Peoples' Rights in 2017. For their part, the five communities of the Lamu County – the Bajun, the Swahili, the Sanye, the Aweer (more commonly known as the Boni), and the Orma – are grappling with several development projects planned in the region (notably the Kenya Government's "LAPSSET project", including, among other things, a railway line, a 32-berth port, a motorway, a regional international airport). All these components of the project are a significant matter of concern to the indigenous communities living in the area. Their grievances range from crime and alcoholism, harm to the environment, dilution of the indigenous culture, to harm to national monuments, conflicts over scarce natural resources, and the marginalisation of indigenous communities; this is a context of still unaddressed historical injustice and endemic land insecurity.

These protocols, nevertheless, allow communities to cast themselves as champions of an alternative model of economic development, this time in tune with "nature", thereby dispelling competing images of backward-looking communities, allegedly hung on to retrogressive ideas about life. The Lamu County BCP expresses, in "sustainable development" terms, that the communities' vision is "[t]o build a culturally, socio-economically, and politically empowered community, striving to secure [their] natural resources and sustain a green environment" (p. 59). The protocol also insists on the "promotion of sustainable development" through "nature-based livelihoods", "small-scale industry", and "market for nature-based products" (p. 63). It also opposes the construction of the coal plant since "there are other means of generating electricity, some of which are clean and from renewable sources [...]" (p. 47). Similar rhetoric is deployed throughout the Ogiek BCP, where it refers to the allegedly "'more sustainable' economic livelihoods systems such as arable cultivation and livestock keeping" that the Ogiek have been forced to adopt, set against "sustainable development activities including beekeeping, commercial tree farming, grazing and tourism" (p. 22). The document links the ABS regime to the natural resources found on the Ogiek's ancestral lands, supporting strategies to target livelihood improvement, poverty alleviation, and sustainable development (p. 22). In the Ogiek case, a further challenge relates to the fact that the Kenyan government had grounded its decision to evict indigenous communities upon the alleged need to protect the Mau Forest Complex, an important water catchment. The African Court on Human and Peoples' Rights[27] explicitly ruled out this argument as unsubstantiated. The Ogiek's BCP strives to dispel the government's portrayal of Ogiek communities as "enemies of nature" by using a counter-narrative underlying their "vital role as guardians and conservators of biological diversity in Mau Forest Complex" (p. iv). The BCP also recalls that the "[...] word Ogiek means 'caretaker of all' plants and animals, or scientifically the flora and fauna", and that Ogiek communities "have always been among the most responsible stewards of forests owing to [their] historical links and attachment to it" (p. 3).

The interest of these two Kenyan cases is twofold. The first is to show that the IPLCs themselves can strategically use the scripts to defuse the most damaging representations, particularly those supported by narratives about pristine nature and the need to oust inhabitants from their biodiversity-rich territories (Doolittle, 2007). The second is that the modern scripts make up a repertoire of counter-narratives that can be drawn upon selectively depending on the script of the past to challenge. The choice is dictated by a principle of line symmetry that creates two different mirror images allowing narratives to move (i) either from "enemies of nature" to "stewards of biodiversity" (e.g. "'caretaker of all' plants and animals"); (ii) or from "premoderns" to "small green entrepreneurs" (e.g. "promotion of sustainable development"). This is due to the fact that each pair of "mirroring scripts" sits on the same

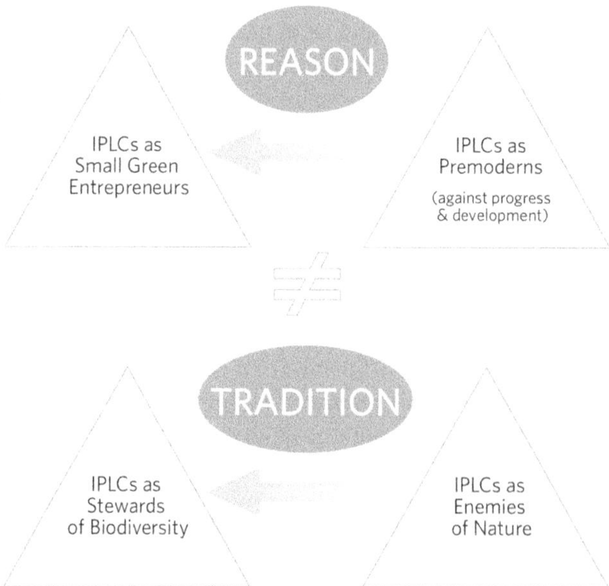

FIGURE 11.2 Mirroring scripts (Fabien Girard, original material for the book)

continuum of basic assumptions about IPLCs (inherently "traditional" in (i); and potentially "rational" in (ii) – on the pervasiveness of "reason" in the "premoderns" script – see Peet and Watts (1996, pp. 5–6)). Any other move is doomed to wield less discursive force and fell short of counteracting harmful scripts (see Figure 11.2). This can explain why the Ogiek's BCP, after several mentions to the ABS regime as a way to improve livelihood, moves on to articulate the people's relationship towards "nature" in terms of stewardship. It is to this script that we now turn our attention.

A Dominant Script: IPLCs as Stewards of Biodiversity

The scripts of IPLCs as "stewards" or "custodians" of ecosystems are found across almost all BCPs under study. This occurs where communities have to fight back antagonistic representations – such as "enemy of nature"– but also every time the communities are faced with assaults from outsiders on their lands or vindicate rights over their territories, genetic resources, and/or TK. What does this pervasive reference mean?

If it cannot be disputed that the "small green entrepreneur" script fits well within the international regime on ABS imbued with ideas about sustainable development, the same does not hold for the "stewardship of nature" script whose centrality in the BCPs is far more intriguing. In reality, this is precisely at this point where the intellectual ferment of the biocultural jurisprudence on BCPs appears most clearly (Bavikatte & Bennett, 2015; Bavikatte & Jonas, 2009; Bavikatte et al., 2015). And this is precisely where one may argue that BCPs hold out promises for IPLCs in terms of greater control over genetic resources/TK and land claims, and perhaps in terms of negotiating spaces for "non-modern" worlds to exist.

In light of the latter, it is little wonder that this script is heavily used in BCPs mainly concerned with land claims and tenure security. Stewardship practices are distinctively grounded in a sort of "land ethic" shaped by age-old links and "attachments" to the territory.[28] Admittedly, and particularly in the African context, a significant emphasis is fixed on cultural rights, given that "under international law, the connection between cultural rights and land rights for indigenous peoples forms a very important part of the international human rights legal framework" (Gilbert, 2016; Gilbert & Sena, 2018, p. 204). Thus, the Ogiek BCP forcefully stresses that: "We have a special relationship with our land and the natural resources in it. This relationship has special importance to our culture and spiritual values and ultimately for our continued existence as distinct peoples" (p. 3). The Lamu County BCP underlines in like manner: "We are the guardians of our environment. We have utilized and conserved our natural resources acknowledging their importance for future generations. Indeed, our cultural identity depends on it" (p. 11).

But in general, what singles out BCPs developed in a land claim context is the foregrounding of a holistic way of life, i.e. the substantial stress established on the inextricable links between the ecosystem, the land, the resources, and the knowledge, culture, and livelihoods of the communities. This is what the Lamu County BCP terms "bio-culture" (p. 45) and the BCP of the community of Degbe Aguinninnou (Benin) "bio cultural heritage <*héritage*>" (p. 18). In Benin, for example, despite it being mainly focused on the ABS process, the Degbe Aguinninnou BCP unveils a struggle for greater control over their sacred forests (Gbèvozoun and Gnanhouizoun) and the recognition of their traditional "social structures". In preliminary documents, "community rules and procedures for managing" their resources and forests (p. 18) were, indeed, the main concerns of this indigenous community.[29] The BCP registers a strong holistic approach revolving around what the community calls "areas of ancestral and cultural heritage". These are the "[...] the sacred forests 'Gbèvozoun' and 'Gnahouizou' and the sacred lake 'Houèdagba'", whose cultural, socioeconomic, and ecological roles and impact on the well-being of the community and the traditional healers are highlighted several times (pp. 10–11).[30]

The most glaring example of this holistic approach is the BCP of Capulálpam de Méndez, Oaxaca, developed by the Zapotec indigenous community, located in the Sierra Juárez (state of Oaxaca). Because of alleged past experiences of biopiracy, an important part of the BCP is about spelling out local ABS procedure (local PIC, mutually agreed terms – MAT – and benefit-sharing) based on customary law. It resembles the Ek Balam BCP, another Mexican protocol developed with the support of the same partners. There is, however, a substantial difference between the two. In the case of Capulálpam, the indigenous community is in dispute with the Mexican state and two neighbouring communities over the limits of their territory and communal land.[31] Thus, it emphasises the protection of ancestral territory as a precondition for their collective identity, sustenance, and survival; and the interconnection of the elements that populate it. As the community states,[32] we "have historically developed our life and our cosmovision around our territory and everything that coexists there" (p. 42).

This echoes the biocultural dimension ascribed to Mexican indigenous peoples' territories (Martínez Coria & Haro Encinas, 2015).[33] We should not, however, overlook the specific sociocultural situation of Capulálpam de Méndez. Capulálpam de Méndez is a town located in Oaxaca, a state accounting for the largest number of indigenous peoples in Mexico. Oaxaca's constitution recognised in 1995 the system of "*usos y costumbres*" (usages and customs) for local elections.[34] The constitutional amendment bears testimony to the local and subaltern struggles for greater territorial autonomy (akin to the nationwide Zapatista project that has grown into autonomous territories and practices of local governments) and more broadly for self-determination (Juan Martínez, 2013, pp. 136–137) within the state. The term "*usos y costumbres*" refers to an alternative system to the national electoral regime based on competition between political parties (ibid., p. 144). But this is also a system "in which civil life mingles with and is incorporated into religious practices. This system is rooted in a worldview according to which the individual forms a part of the community and has particular duties and rights", and covers community life in its civic, judicial, and territorial dimensions (Polo & Danielson, 2013, 170; also see Juan Martínez, 2013, 141).

The "*usos y costumbres*" system is used in Capulálpam de Méndez, but with a distinctive feature. The town comprises one single nucleus of human settlement,[35] i.e. one community only. This means that the town and the agrarian community[36] are governed, respectively, by a local assembly of citizens ("*asamblea general de ciudadanos*") and a community assembly governing the agrarian community ("*asamblea comunitaria*").[37] But both institutions are entrenched in the *cargos* system and *tequio* (unpaid work to benefit the community) as embodiments of the local "*usos y costumbres*".

Cargos are a system of rotating civic and religious responsibilities among community members (citizens for the municipality and *comuneros* for the agrarian community) based "on merit accumulated by service in a rising hierarchy of civic positions" (Antinori & Rausser, 2003, p. 5). Even if the cargo system originated in the colonial period, it is often considered a pre-Hispanic institution – mainly "indigenous" – that needs to be revived or reconstituted where it has disappeared (Blanco, 2012). This is of course part of political project geared towards territorial rights and political autonomy that goes along with attempts at giving an ethico-political foundation and tools for resistance to communities through such concepts as "communality" ("*comunalidad*")[38] that underlines a human's belonging to the land (Martinez Luna, 2010) and an individual's embeddedness in the community (Polo & Danielson, 2013).

In summary, most protocols under consideration in this chapter draw heavily on the idea that IPLCs are "stewards" of their lands, territories, and resources. This is less true for the Mexican protocols, where different philosophical foundations give greater weight to the land and acknowledge a foundational dimension for community life and cosmovision. All the other BCPs in this study state that the communities are "guardians" of their environment, "conservators of biodiversity", and at the very least that they have managed "traditionally" and "sustainably" their environment from generation to generation. In most instances, and it is where all the protocols can be said to coalesce, the "stewardship" role is described as stemming from historical, spiritual, and sociocultural attachments to the land, which translates into customary rules and institutional arrangements that regulate, and at times prohibit,[39] access to, and use of, resources and knowledge. Ultimately, these BCPs tell of a "natural conservationist" representation stressing that IPLCs have obligations and responsibilities towards their land, biocultural heritage, and future generations. The presence of this representation is no accident: it conveys that which makes IPLCs unique. It also eschews attempts to deprive IPLCs of history, culture, and agency – while at the same time opening spaces for discussing ontologically what is needed to support their distinctive identities and worlds.

BCPs and "Stewardship of Biodiversity": Eco-Power and Subject-Making

Studying BCPs through the lens of political ontology uncovers how IPLCs are torn between different environmental scripts. There is a prevalence of the present-day scripts ("stewardship of biodiversity" and "small green entrepreneur") which share the premise that IPLCs are key *actors* in biodiversity conservation and seek to distribute roles and responsibilities accordingly. This can be seen as the ideological attraction of most protocols. While this is mostly true, there are substantial differences between the BCPs that, in turn, inform their potential political and ontological reach.

The philosophical underpinnings of the "small green entrepreneur" representation hinge upon instrumental rationality. The rationality of "rational choice theory" in "commons" theory (Lara, 2015) is the building block of neo-classical economics. This theory posits that economic agents, though "fallible, norm-adopting individuals who pursue contingent strategies in complex and uncertain environments" (Ostrom, 1990, p. 185, Singleton, 2017), can *voluntarily* develop co-operative social norms, thereby successfully managing common-pool resources. As *adaptable*, *resilient*, and *robust* institutions depend on a complex "grammar of institutions" (Crawford & Ostrom, 1995; Ostrom, 2009), there needs to be a mix of institutional interventions and incentives. These incentives are vital to "achieve co-operative equilibrium outcomes" (Mosse, 1997, p. 469) among users of the common-pool resources.

Part of the same theoretical framing pervades the second modern script, that of the IPLCs as "stewards of biodiversity", which can at times coalesce into a mixed script. But references to "nature", "native", and "tradition", which feature prominently in international discourses and environmental laws, change the way communities are perceived and necessarily how they may perceive themselves. This is because they change the kind of expectations placed on IPLCs and the reasons for potential biodiversity local management failure (see Mosse, 1997, pp. 468–469). In this situation, IPLCs must not progress or develop or be connected to the global market; they are expected to "remain the same and not change" (Ulloa, 2005, p. 206). This may imply other types of managerial interventions, such as harnessing traditions and traditional ways of life, reviving them[40] when they are on the verge of decline or even, upon occasions, re-inventing and upholding them "by the force of law" (Mosse, 1997, p. 469).

These interventions are well-accounted for in the literature on CBNRM (Brosius et al., 1998; Mosse, 1997) and are likely to be replicated in BCP development since both rely on a people-centred approach of conservation. The involvement of local populations is primarily instrumental as they have a greater interest in the management of resources than the State or distant managers. IPLCs have profound knowledge of local ecological processes; use cost-effective methods, and have a stronger social acceptance of bottom-up conservation activities.

The international NGO and UN communities are aware of the intricate links between environmental degradation and social inequity, and therefore a concern for social justice for IPLCs exists (Brosius et al., 1998, p. 158). This is especially true of BCPs, which are also part of a process towards a greater recognition of IPLCs' rights over their genetic resources and TK, and are thus likely to meet several "justice challenges" associated with genetic resources and TK (see Deplazes-Zemp, 2019).

The above paragraph is proof of how politicised biodiversity conservation has become (Brosius & Hitchner, 2010, p. 146) and the consequences it has on people-centred conservation and BCPs ("who the people are" is a political question – see Peet & Watts, 1996, p. 27). Attention, therefore, needs to be paid to the way the "stewardship of biodiversity" script and its attendant narrative (the centrality of "tradition") may be used by multilateral lending agencies, donor institutions, and conservation organisations as a "disciplinary tool for national and regional planning" (Brosius et al., 1998, p. 163). In this section, we consider the concrete operation of the script and how actors are being recruited into it (Lewis & Mosse, 2006, p. 13). We also determine to what extent this script, through concepts such as tradition, stewardship, and nature, is involved in community-level changes that multilateral agencies, development institutions, NGOs, and facilitators may prompt (even inadvertently) through their dealing with a community. Framed in more radical terms, the problem that we wish to raise is whether BCPs, through the hypostasis of tradition, stewardship, and nature, are not a conduit for control dynamics and the reification of communities (Roach et al., 2006, p. 60).

At this point in our chapter, we could be accused of adopting a one-sided stance. We will, however, critically examine the political potential of the "ethic of stewardship" in the closing section of this chapter. From the maze of this complex script, we advance that there is a new form of legal subjectivity that BCPs have created; and that this legal subjectivity is more in tune with the worlds of the IPLCs.

Co-Opting Tradition, Managing Differences

At this juncture, it should come as no surprise that much of the cases making up our corpus show pervasive references to "tradition", together with appeals to territory, indigeneity (or autochthony as in Kenya),[41] and "communality" (see above in Mexico). These references are undoubtedly aimed at building "[...] images of coherent, long-standing, localized sources of authority tied to what are assumed to be intrinsically sustainable resources management regimes. They are also used to legitimize, and to render attractive [...]" (Brosius et al., 1998, pp. 164–165).

The BCPs of the community of Degbe Aguinninnou (p. 16) and of the Municipality of Tori-Bossito (p. 4) stress what they call "traditional management" of territories and natural resources. In the Biocultural Protocol of the community El Piro, this is referred to as the "traditional management of knowledge and the main species of fauna and plants" (p. 10).

The same general patterns are discernible across a large number of BCPs: traditional practices and TK of biodiversity conservation have been translated into traditional institutions (e.g. "traditional chief" in Ampangalatsary (p. 3) and El Piro (p. 9), "traditional authority" in the community of Degbe Aguinninnou (p. 16)), customs (also referred to as "traditions" – BCP of Ampangalatsary (p. 16)) and taboos and prohibitions (e.g. the prohibition to kill the boa, the taro, the turtledove, and the bat within the community of the Degbe Aguinninnou (p. 9), or to kill the jaguar and the *tepezcuincle* – paca – in Ek Balam (p. 35)). These institutions are meant to regulate the use of the resources and ensure their long-term survival.

This diversity of semantic beacons tends to be subsumed under the umbrella term (or genetic signifier) of a "traditional way of life". The "traditional" lifestyles, at times tied to specific worldviews ("*cosmovisión indígena*"/"indigenous worldview" in El Piro (p. 14), "*comunalidad*" in Capulálpam de Méndez)[42] or to distinctive cultural practices ("traditional dances" and "traditional dress" again in El Piro (p. 9)), constitute the ethical foundation in which conservation practices are rooted and from which they are projected onto real-life experiences and thus into the present. Traditional ways of life, however deeply they are embedded in specific relationships between humans and non-humans, are barely called upon in their own right, but for their instrumental ability to ethically sustain conservation practices that are deemed relevant to the *present* time.

This is unquestionably an effect of the "politics of difference" – which is also a "politics of the subjects" (Agrawal, 2005a, p. 164) – whereby "[…] the constitution of oneself as an 'authentic'[…]" self is "[…] conflated with specific ahistorical assumptions concerning the nature of indigeneity […]" or of a local community (Barcham, 2000, p. 138). It goes through a process wherein practices are progressively segregated, roles redistributed, and competencies rearranged (Agrawal, 1995, 2002). Unsurprisingly, the pivot of the process is the polarity between "tradition" and "modernity", allowing the shift of certain practices from the first category to the second. This process is often almost invisible: alleged unsustainable practices that need to be abandoned are simply passed off as "modern" agricultural practices.

The first step generally involves pitting "modernity" against "tradition", the former acting as the malign force driving the dissolution of the latter. In this respect, the siren voices of modern life ("Western academic training" or urban migration) are usually blamed for the youth's disinterest in TK and practices related to biodiversity (BCP, Comunidad El Piro, p. 10). For example, the increasing use of Western medicine is denounced as the reason for the decline in "traditional medicine" in El Piro (Ibid.). Some BCPs also state that "landraces" and "traditional knowledge" are put at risk by cash crops (pineapple, teak wood farming, firewood production, orchards) (BCP Tori-Bossito, p. 9), a hallmark of modern agriculture.

As a second step, authenticated tradition needs to be strengthened or revived. In the BCP for the community of El Piro, for example, a strong emphasis is anchored on "initiatives for the recovery of indigenous knowledge, biological resources associated with genetic resources", an inventory of the community's biological resources, genetic resources and TK, and the rolling out of an "environmental education programme based on traditional rules and practices" (p. 11). Along the same lines, four BCPs developed as part of the Darwin Project (Tori-Bossito, Degbe Aguinninnou, Analavory, and Ampangalatsary) include a section on Community Biodiversity Registers, set up as "repository of the community's past" (BCP of Degbe Aguinninnou, p. 25). They also have a section on community seed banks that preserve "traditional" varieties and farmers' landraces – especially those on the verge of extinction – and help reintroduce lost varieties.

The final step is to abandon all those practices that cannot qualify as "traditional" by accepted standards as they display too close a proximity with the Western world. In general, the process takes on a very subtle form. Targeted "traditional" practices are not pigeonholed – and therefore not disqualified – as "modern". Instead, they are left unlabelled and presented alongside impugned modern practices. This projects them instantaneously into a sphere of illegitimacy. In Tori-Bossito, for instance, shifting cultivation and bush fires are lumped together with industrial crops (p. 9) and placed under the heading of problematic practices.

There is certainly some unfairness in this broad-strokes account and one-sided criticism of protocols. While the relevance of some traditional practices to biodiversity conservation is indisputable, some others probably need to be challenged and, if needed, abandoned. There remains a problem of form, which reverberates further through every aspect of IPLCs' identities: creating the illusion that "tradition" is a vital lifeline for IPLCs, while at the same time doing away with practices (such as slash-and-burn or bush fires)[43] which may be deeply ingrained in cosmovisions and play a role in sustaining traditional institutions and livelihoods (Reyes-García et al., 2021; Whyte, 2013). This is at best a dangerous practice. Just as labelling communities "traditional" cannot be used to chain IPLCs to a romanticised past (Berkes, 2008, p. 239; Holt, 2005, p. 209),[44] so too it cannot be used for a sleight of hand to conceal managerial interventions and eco-power on subaltern groups.

Triggering Institutional Changes: Fixing Social Regularities, Rendering Legible

Through a sort of "cherry-picking" process, some BCPs tend to selectively retain as "traditional" only these practices that fit the funders' or facilitators' purposes or the vision of nature conservation embedded in global environmental agendas. This, it can be argued, is a tool for States and international institutions to conciliate the Western dream of *managing* biodiversity and the need to deal with differences.

Our corpus clearly confirms how this conciliation further relies on patterns already in use during colonial time to mediate relationship between coloniser and colonised, namely textualisation and codification (Pels & Salemink, 2000, p. 28). This pattern resurfaced with CBNRM, as to decentralise resource management which was felt to be an unacceptable loss in power by States that they could not bring themselves to accept without imposing "some degree of statutory uniformity for purposes of legal recognition" (Brosius et al., 1998, p. 166). In both cases, the ultimate goal is the same: fixing "regularities of social practices" and turning them "into the essence of local social order" (Pels & Salemink, 2000, p. 28).

The most sophisticated form of "statutory uniformization" is the imposition of new institutions or governance systems that replace customary authorities or community-based institutions. There are many well-researched accounts of this process in the context of CBNRM (Almeida, 2017, p. 9). Some BCPs included in our case study fit the same pattern, but what is worth highlighting is the way how strong references to tradition are used to conceal the magnitude of the institutional change or hold it out as small adjustments in the traditional governance structure. In Bonou (Benin), the BCP describes in detail the community's "internal decision-making system", which is a "college" made up of the king of Bonou, the "Gbévonon" (the chief of the community, also the guarantor of the divinity of the sacred forest "Gbévo"), the heads of families, and the guarantor of other deities. A consultative body, attached to the college, and made up of 16 young community's members, is tasked with monitoring and controlling the management of resources, especially those of the sacred forests (p. 17). Surprisingly enough, Section 8.3 of the BCP then goes on to describe the "management body" of the community, also called the "local committee", whose composition mostly departs from that of the "college" described above. It comprises the king of Bonou, three representatives of the community (also acting as the manager of the sacred forest), two landowners, and one representative of the village of Sotinkanmè.[45] The "local committee" is headed by a "bureau" made up of the king, a secretary, a treasurer, and two officers, respectively, tasked with planning and biodiversity management and cultural and religious activities. This "bureau" acts as the "community competent authority" on ABS (p. 28). Another striking feature emerges upon closer inspection: the "local committee" is modelled upon the committee referred to in Articles 41–44 of the Inter-ministerial Decree of 16 November 2012. It lays down the conditions for the sustainable management of the sacred forest in the Republic of Benin.[46] This text sets out the procedure for the transfer of sacred forest management to village communities, which implies a request for legal recognition of the sacred forest being submitted to the local authority by a village community, together with a local by-law passed by the commune establishing the "Sacred Forests Management Committee". Upon completing the complicated proceedings set up by the Inter-ministerial Order, the sacred forest is normally included in the communal forest estate by way of an Inter-ministerial Order of two ministers. To our knowledge, proceedings have never reached this last stage, but this account calls for some comments. First, a seemingly stable and diverse customary body, with a fairly broad social base, has been replaced by a small committee giving the king of Bonou broad powers over genetic resources – a typical "power over" situation (see Bannister, 2004, p. 3). Second, from the State's point of view, relationships with the community are significantly simplified, given that the governance structure is now clearly defined and regulated by a legal text. It also provides a template to be duplicated throughout the region. This example confirms that statutory uniformity is rarely sought through institutions created from scratch, but rather through the instrumental use of existing governance structures that can claim a "local pedigree" or be described as a "traditional community".

A more radical strategy consists of promoting institutions with all the trappings of "tradition", but which are traditional in name only. One of the two Malagasy BCPs was developed in the *fokontany* Ampangalatsary where, following Law No. 96-025 regarding the local management of renewable natural resources[47] (the so-called "GELOSE law"), management rights of the Iaroka Antavolobe forest were transferred by contract to a local natural resource management group, the VOI Firaisankina (in Malagasy *Vondron'Olona Ifotony* (VOI), or "Basic Community").[48] VOI are a perfect blend of instruments from modern law and traditionally inspired structures. For example, the "Basic Community" is endowed with legal personality,[49] and both the establishment and the operation of the body are stringently regulated by the GELOSE law of 1996 (see Pollini & Lassoie, 2011). This community is presented as an offshoot of an age-old traditional institution, the *fokon'olona*, which is a community whose members share a common kinship and a common territory.[50] This community is governed by a form of a "social contract" – *dina* – laying down rules for the main dimensions of economic and social lives. These customary rules are implemented by a traditional chief backed by an assembly of elders. The VOI tries to imitate the traditional *fokon'olona* and thus to maintain the illusion that there is a continuity between "restricted group of individuals willing to adopt management rules designed by the state and its partners" and communities (*fokon'olona*) that *de facto* manage the resources according to rules and institutions that have a long history (Pollini & Lassoie, 2011, p. 823).

A Final Note on the Ethic of Stewardship: Legal Subjectivation for Emancipation

The previous sections of this chapter have shown that BCPs can be taken as receptables revealing an array of scripts – all built on the modern naturalist matrix – that circulate within the international regime on biodiversity conservation and ABS. Furthermore, they show how these scripts, as "storied performativity", both recount and enact the relations between "agents/subjects" (humans and non-humans) and reach into practices of cultural belonging, i.e. identities. This script diffusion is unmistakable in the operation of the "steward of biodiversity" script which partakes in the construction and institutionalisation of what Astrid Ulloa calls the "ecological native" or "political-ecological agent" (Ulloa, 2005). This is exactly the dual phenomenon of institutionalising an ecological identity and this identity's anchoring in tradition that tends to crush the agency and historicity of IPLCs. It also conceals that these peoples are perfectly connected to, and influenced by, global socioecological and political changes. This is because it homogenises, naturalises, and reifies diverse cultural practices (Brosius & Hitchner, 2010, p. 146) that remain mediated by the evolving and adapting social sphere of values.

To counteract this opinion is Ulloa's observation: "Despite all the many negative connotations and implications of ecological native representations, indigenous peoples' movements are using them to transform non-indigenous peoples' ideas of their identities not only within the nation-state, but also in transnational arenas" (Ulloa, 2005, p. 215). It is this thought of Ulloa's that we would like to explore briefly in this last section. To do this, we place the question of the steward of nature or native ecological in the political ontology debate opened in the introduction. Is the mobilisation of "the ethic of stewardship" in particular likely to advance the cause of IPLCs, notably by severing the link with tradition? Is this ethic able to contribute forging a new ethical and political status as a crucial step towards the delineation of a new legal subjectivity for the holders of biocultural rights? What is this likely to yield in terms of space opening for discrete IPLCs' identities? Is a new process of legal subjectivation a potential antidote to the subject-making power of scripts?

The ethic of stewardship that can be traced back to Posey's (1999) work and environmental ethicists' breakthroughs, such as Callicott's (1994), has been mainstreamed in the scientific literature on property rights and biodiversity conservation and recently found its way into policy documents. The best example of this is undoubtedly the Tkarihwaié:ri Code of Ethical Conduct, which now includes the definition of "Traditional guardianship/custodianship", stressing the "holistic interconnectedness of humanity with ecosystems" and making the case that

> Indigenous and local communities may also view certain species of plants and animals as sacred and, as custodians of biological diversity, have responsibilities for their well-being and sustainability, and this should be respected and taken into account in all activities/interactions.[51]

While the risk that this ethic be construed in a way that would impose a "duty of stewardship" on IPLCs cannot be wiped out with a stroke of a pen,[52] we would like to bring forward a different interpretation. As both the Tkarihwaié:ri Code of Ethical Conduct and the Colombian Constitutional Court's decision in the

Atrato River case demonstrate,[53] references to duties remain too ambiguous, and there is no definitive conclusion that IPLCs are the bearers of a (legally enforceable) duty of stewardship. The Code of Ethical Conduct states that "[t]he traditional stewardship/custody **recognizes the obligations and responsibilities of indigenous and local communities** to protect and conserve their traditional role as stewards and guardians [...]".[54] Furthermore, the Constitutional Court of Colombia states that "[biocultural rights] imply that communities **must maintain their distinctive cultural heritage** [...]" ("*estos derechos implican que las comunidades deben* ***mantener su herencia cultural distintiva***").[55] We instead contend that both the Code of Ethical Conduct and the Atrato River case are to be understood as an obligation on States to take all steps necessary to enable IPLCs to sustain what is seen as their distinctive ethics, beyond the naturalistic infrastructure.

In reality, the reference to ethic of stewardship is a first way out of the game of "contiguous and oppositional concepts" (Grear, 2015, p. 83). The concepts subject/object, mind/body, and nature/culture have provided matrix for modern law and, above all, the basis for modern legal subjectivity. As we have seen, a pair of "oppositional concepts" that are particularly structuring for BCPs is the "modern/traditional" typology. The interest of the ethic of stewardship is to move beyond negativity in which the "modern/traditional" dyad maintains IPLCs.

The ethic of stewardship makes it possible to give shape and substance to those who can never be apprehended by the "modern", those who cannot be understood without being immediately referred to by its antithesis – namely the "traditional"; this "traditional" without which the "modern" cannot exist, but which quite ironically also directly threatens its existence and must be kept "at bay" (Blaser, 2009, p. 888). Let us put it this way: the "traditional" can only exist as an *indeterminate* and *reifying* category, inevitably destined to remain under the control of the "modern".

The whole point of using the "ethic of stewardship" and its power as a concept is to move away from the indeterminacy and reification of the "traditional" category. It is also more than that; it reinjects positivity and in so doing helps to clarify the "peoplehood" of "biocultural communities" (Bavikatte & Robinson, 2011). This additional step in the construction of biocultural jurisprudence is essential. It reemphasises the importance of placing BCPs within a broader theoretical and legal context. Focusing on the bundle of biocultural rights shifts the attention to legal subjectivation. It brings the debate back to its point of departure. In its wake comes the central question of what it is like to be, act, dwell, and dream as an indigenous people and a local community, i.e. on what it is like to be – positively – a subject of biocultural rights.

That the issue arises when the epistemic frameworks of modern law are being discussed with precedents such as the Atrato and the Whanganui Rivers cases (Tănăsescu, 2020) should come as no surprise. Just as in the 16th century, when on the Old Continent, profound socioeconomic and geopolitical upheavals had led, through a return to the debate on "man", the State, sovereignty, and property, to give rise to the rights of individuals ("*subjective rights*") and the *subject* of modern law; today, too, the entry into the Anthropocene ushers in a new "political ontology of the subject".

The link between the two periods cannot be overstated. Parallel to the installation of the modern subject, which forms its extremity, European/Western law has deployed an ontological matrix that has silenced nature and indigenous epistemologies and submerged "the agency and full ethical significance of all 'others' to the 'rational' master-subject" (Grear, 2015, pp. 86–87).

This lengthy process stretching from the 16th to the 18th centuries is worth remembering, if only briefly. During this period, humanity asserted its new dignity, or *dignitas hominis*, of which Pico Della Mirandola expressed the first requirement: that "man's own liberty to make of himself what he is" (Zarka, 1999, p. 245). Having emancipated himself from nature following a significant anthropological change, man (humankind) becomes more simply "naturally endowed with rights" (Zarka, 1999, p. 246). Furthermore, these new rights, which Grotius helped to define as "a moral quality of a person" (Grotius, 2012, bk. I, I, 4), form the embryo of what are now called subjective rights (Zarka: 247–8). The status of the person (*persona*) to whom the right as a moral quality relates remained to be defined. What was at stake then was the constitution of the *natural person* as a self (*ego*), the only one capable of rights and obligations, and who thus became the template of the *subject of law* (Zarka, 1999).

Here, we can see the crucial shift, whereby the legal person and the subject of law merge, as evidenced (but the examples are innumerable) in this formula written by Smith (1928, p. 283): "To be a legal person is to be the subject of rights and duties". There is a co-determination: the legal person is the subject of rights, i.e. the subject tailored to collect those specific rights defined as *moral qualities*. As such, the legal subject can hardly be anything other than the very person the law has constructed by recognising new rights (see Kurki, 2019, p. 121). Within the conceptual nexus that unfailingly ties new rights to a new way of seeing humans and culture – a new anthropology; the notion of legal personhood "acquires a sense of a sovereign, reflective subject, a being with his own self-determining personality" (Davies & Naffine, 2001, p. 57).

Nevertheless, two fault lines remain in modern law. The first is that there are still right-holders (like animals), i.e. legal subjects, which are not necessarily legal persons (Grear, 2010, p. 46; Kurki, 2019, pp. 122–124). The second fault line, and a much more important one, is that the "legal person" remains a construct (Grear, 2010, p. 51), even though it has been "naturalized" and "depoliticized" by a deep legal anthropomorphism that is so ingrained that it has ended up contaminating even the notion of legal subjectivity (Grear, 2010, p. 217, n. 35).

It is undoubtedly too difficult (or premature) to try to detach the legal personality from the matrix of the rational, sovereign "man" and, above all, from the attendant concept of private property (Davies, 2012). As Davis and Naffine forcefully note, modern legal personhood has been defined and constructed through property: "[t]o be a person is to be a proprietor and also to be property – the property of oneself" (Davies & Naffine, 2001, p. 5, n. 37; also see Grear, 2010, p. 51). The model of personhood remains that of Macpherson's "possessive individual" (Davies & Naffine, 2001, p. 56; Grear, 2010, p. 52; Naffine, 2003, p. 360) – one of the pillars of modern law which it seems premature to tackle head-on.

One can, however, break the link between "subject of law" and "legal person" (Kurki, 2019, p. 123; Pietrzykowski, 2017) and thus unleash the process of legal subjectivation. The precedent set by the constitution of the modern legal person in Europe between the 16th and 18th centuries shows that the pivotal moment is when, through the attribution of new rights, a new politico-legal identity can express itself. It is, therefore, necessary to follow the path that stretches from rights to identity to the new subject of law, according to a sequence that Yves Charles Zarka has already described: "The invention of the subject of the law does not precede the modern definition of natural law but follows it" (Zarka, 1999, p. 261). Of crucial interest is that while legal subjectivation is a "process triggered by legal norms", it "ultimately occurs outside the realm of law" (Urueña, 2012, p. 35). It therefore leaves open the debate, informed by insights from those concerned, the anthropology, the critical theory, on what is needed to sustain the pluralism of practices of belonging and the expression of polyphonic identities.

Returning to our thesis and conclusion, biocultural rights and the ethic of stewardship allow a strategic reversal from alienating subject-making practices to a new emancipatory process of legal subjectivation. At the very least, by prompting an investigation into the ethical-political status of the subject (the *persona communis*) to whom the new (biocultural) rights are granted, it fuels a process of critical reflection on how to find a way for non-naturalist ontologies in the international regime of biodiversity and international human rights laws.

As privileged inhabitants of the Global North, we feel a sense of unease when asked to specify what the ethic of stewardship and the ethical-political status should look like. But simple intuition won't prejudge any further avenues to be opened: the reversal mentioned above probably takes us back to an old tradition where the subject is a "sovereign" (*subjectum*) rather than a subservient subject or "*assujetti*" (*subjectus*) (Balibar, 2017). It therefore paves the way for IPLCs to be seen as "sovereign" over their territories, lands and resources, agents of their own past, present and future life, and able to forcefully express "claims for alternatives to modernity" (Blaser: 882–3). The *sovereign stewards of biodiversity* could be the missing link between environmental law and human rights law.

Annex

TABLE 11.1 BCPs covered by the study with insights into domestic legislation on local PIC procedures and BCPs

Country	*Biocultural community protocol*	*Local/community PIC (with domestic texts and materials)*	*Legal recognition of BCP*	*Source*
Benin	BPC of the Municipality of Tori-Bossito	Yes (but the scope of the local PIC is unclear) Relevant texts and materials: Decree No. 2018–405 of 7 September 2018 on national guidelines for Access and Benefit-Sharing arising from the use of genetic resources and associated traditional knowledge in the Republic of Benin (Articles 6 & 8)	Yes	https://www.bioversityinternational.org/fileadmin/user_upload/research/research_portfolio/policies_for_crop/BCP_ToriBossito_Benin_2018.pdf (French version).
	BCP of the community of Degbe Aguinninnou			https://www.bioversityinternational.org/fileadmin/user_upload/research/research_portfolio/policies_for_crop/BCP_Bonou_Benin_2018.pdf (French version).

Country	*Biocultural community protocol*	*Local/community PIC (with domestic texts and materials)*	*Legal recognition of BCP*	*Source*
Kenya	Ogiek Bio-Cultural Community Protocol The Lamu County Biocultural Community Protocol	Partially Relevant texts and materials: • Constitution of Kenya, 2010, ss. 11(1) & 11(2); s. 11(3)(a); s. 69(1)(c) • The Environmental Management and Coordination Act No. 8 (1999) (EMCA) • The Environmental Management and Co-ordination (Conservation of Biological Diversity and Resources, Access to Genetic Resources and Benefit Sharing) Regulations (2006) (s. 9(1)), s. 9(2), Part III 2.0(g)) *(Medaglia, Perron-Welch, & Phillips, 2014, 94–95) • Protection of Traditional Knowledge and Cultural Expressions Act, 2016 (No. 33 of 2016), s. 7, s. 10, ss. 18(1) & 18(2), s. 22(1), 25(3)(a), ss. 12 & 27 →*Modelled upon the Swakopmund Protocol for the Protection of Traditional Knowledge and Expressions of Folklore (Adopted by the Diplomatic Conference of ARIPO at Swakopmund (Namibia) on 9 August 2010 and amended on 6 December 2016.* *(Mwangi, 2019; Nwauche, 2017; Nzomo, 2015)	Partially	https://www.ogiekpeoples.org/images/downloads/Ogiek-Bio-Cultural-Protocol.pdf (English version). https://naturaljustice.org/wp-content/uploads/2019/04/Lamu-County-BCP-2018.pdf (English version).

(Continued)

Country	*Biocultural community protocol*	*Local/community PIC (with domestic texts and materials)*	*Legal recognition of BCP*	*Source*
Madagascar	BPC of Ampangalastary BPC of the farming communities of Analavory	Yes Relevant texts and materials: • Decree No. 2017–066 from 31 January 2017 on the regulation of Access and Benefit-Sharing arising out of the utilisation of genetic resources (Articles 9, 12 & 14) • Draft inter-ministerial order N°.../2019 laying down implementing rules for Decree N°2017–066 of 31 January 2017, Article 14, 22, 23 & Annex III	Yes	Not officially published Not officially published
Mexico	BCP of Capulálpam de Méndez, Oaxaca	No Relevant texts and materials: • The Political Constitution of the United Mexican States, Article 27, para. 1, 4 & 6, Art. 27.VII Article 2.A.VI, - to declare that the Nation has the direct dominion over natural resources is tantamount to saying that it has full ownership over them. • General Law on National Asset, Art. 6, para. I • Agrarian Law, DOF 25–06–2018 • National Water Law, DOF 06-01-2020, and also include the General Law of Sustainable Fishing and Aquaculture, DOF 24-04-2018	No	https://absch.cbd.int/api/v2013/documents/4D03DAC0-33C3-0F01-8370-A093EAABCE69/attachments/PCB%20Capula%CC%81lpam%20de%20Me%CC%81ndez%20pliego.pdf (Spanish version).

Country	*Biocultural community protocol*	*Local/community PIC (with domestic texts and materials)*	*Legal recognition of BCP*	*Source*
	Ek Balam BCP	• General Law of Ecological Balance and Environmental Protection, DOF 05–06–2018 • General Wildlife Law, DOF 19-01-2018 • General Law of Sustainable Forestry Development, DOF 5–06–2018, Article 2, XIII, Article 5, Art. 102 *(López Bárcenas, 2016, p. 42; Morineau & Morineau, 1997, p. 200 ; López Bárcenas & Espinoza Sauceda, 2006) Draft General Law on Biodiversity ("*Proyecto de Ley General de Biodiversidad*"), October 2016, introduced in Senate by the Senator Ninfa Salinas Sada and endorsed by the Ministry of Environment and Natural Resources (Semarnat).		https://absch.cbd.int/api/v2013/documents/4D03DAC0-33C3-0F01-8370-A093EAABCE69/attachments/PCB%20Capula%CC%81lpam%20de%20Me%CC%81ndez%20pliego.pdf (Spanish version).
Panama	Biocultural Protocol « Protection of the indigenous knowledge associated with genetic resources », El Piro Community, Ngäbe – Bugle Region, Panama	Yes Relevant texts and materials: • Constitution, Art. 127: "*comarca*". (Goodman, 2013; also see Herrera, 2015). • Ngäbe-Buglé: Law No. 10 of 7 March 1997, on the establishment of the Ngobe-Bugle comarca and other measures, *Gaceta Oficial* No. 23242, (11 March 1997), Art. 102, Art. 48, Art. 99 *(Wickstrom, 2003, p. 45). • Ngäbe-Buglé: Law No. 11 of 26 March 2012, which establishes a special regime for the protection of mineral resources, water and environmental resources in the Ngäbe-Buglé *comarca*, *Gaceta Oficial* N° 27001, (26 March 2012) • Law No. 41 of 1 July 1998, General Law of the Environment, Art. 100, Art. 104, art. 105	Yes	https://absch.cbd.int/api/v2013/documents/3B1A585E-DEE9-17F6-7840-14EFA8C10794/attachments/2017_Protocolo%20Biocultural%20Comunidad%20El%20Piro.pdf (Spanish version).

(*Continued*)

Country	*Biocultural community protocol*	*Local/community PIC (with domestic texts and materials)*	*Legal recognition of BCP*	*Source*
	Panama	• Law No. 20 of 26 June 2000, on the special intellectual property regime upon collective rights of indigenous communities, for the protection of their cultural identities and traditional knowledge, and whereby set forth other provisions, Art. 1, Art. 2, Art. 3, 4, 5 & 6) *(Romero, 2005, p. 321). Executive Decree No. 12 of 20 March 2001, Regulating Law No. 20 of 26 June 2000, Art. 3 *(Tonye, 2003, p. 171); *(International Institute for Environment and Development, n.d.). *(see De Obaldia, 2005 for a comprehensive study; also see Lixinski, 2013, p. 123). • Law No. 7 of 22 June 2016 Establishing the Protection of Knowledge of Indigenous Traditional Medicine, Art. 16; also see Art. 18; Art. 20; Art. 13; Art. 26 protects; Art. 23; Art. 22; Art. 20) • Executive Decree No. 19 of 24 March 2019 Regulating Access to and Control of the Use of Biological and Genetic Resources in the Republic of Panama		

Notes

1 The authors would warmly like to thank Fitiavana Ranaivoson and Manohisoa Rakotondrabe who aided in the gathering of data information in Madagascar, and Benjamin Coudurier who provided helpful research assistance. This chapter was greatly helped by the insightful comments of our colleagues Ingrid Hall, Christine Frison, and Mélanie Congretel. We would sincerely like to thank them. We are solely responsible for any remaining errors. *All URLs retrieved on 1 September 2021.
2 See Chapter 1 of this book.
3 This is examined at length in Giulia Sajeva's Chapter 6 of this book.
4 CDB, COP 10, Decision X/42. The Tkarihwaié:ri Code of Ethical Conduct to Ensure Respect for the Cultural and Intellectual Heritage of Indigenous and Local Communities, Nagoya, UNEP/CBD/COP/DEC/X/42, 29 October 2010, para. 20 (hereinafter: The Tkarihwaié:ri Code).
5 Corte Constitucional de Colombia, *Sentencia de revisión de tutela* T-622/16 (2016) (hereinafter, Atrato River case (2016)).
6 The Tkarihwaié:ri Code, para. 20.
7 Also see Chapter 1 of this book.
8 On the history of BCP negotiations, please refer to Chapter 1 of this book.
9 The term ICCA does not refer to a specific category of lands and territories. It is a generic term used by the ICCA consortium, but now also by the IUCN and CBD, to encompass a wide variety of lands, areas, and territories that share a number of common characteristics (Kothari et al., 2012, p. 16). According to Recommendation WPC Rec 5.26, these are "natural and modified ecosystems, including significant biodiversity, ecological services and cultural values, voluntarily conserved by indigenous and local communities through customary laws or other effective means" (IUCN & The World Conservation Union, 2005). Also see RES 3.049 Community Conserved Areas (IUCN & World Conservation Congress, 2005) and RES 4.049 & 4.050 (IUCN & World Conservation Congress, 2009). ICCAs are now supported by the COP of the CBD: COP, CBD, Decision X/31. Protected areas, UNEP/CBD/COP/DEC/X/31, 27 October 2010; Decision XI/24, UNEP/CBD/COP/DEC/XI/24, 5 December 2012, para. 1, e; Decision XIII/2. Progress towards the achievement of Aichi Biodiversity Targets 11 and 12, CBD/COP/DEC/XIII/2, 12 December 2016, para. 7; Decision 14/8, Protected areas and other effective area-based conservation measures, CBD/COP/DEC/14/8, Annex II. Also see the ongoing work on "other effective area-based conservation measures" (CBD/COP/DEC/14/, para. 2) (IUCN & World Commission on Protected Areas (WCPA), 2019).
10 This is elaborated below.
11 As the shadow of Manicheism looms large on these issues, it should be said that even (high) market economies hinge on what Gudeman calls the dialectic of trade and mutuality (Gudeman, 2012, p. 14). Admittedly, as Comaroff and Comarrof convincingly have shown, the irruption of commerce cannot straightforwardly be likened to "alienation-by-abstraction", "corrosion-by-commodification" (Comaroff & Comaroff, 2010, p. 25). From the perspective of the "agents" of moral economies, penetration of market economies, the emergence of "ethno-preneurialism" (Comaroff & Comaroff, 2010, p. 27; Rutert, 2020, p. 286) may also have a productive effect. In addition, Strathern made the important point that if market "disembeds what is usable", "the thrust of the indigenous IPR movement is to re-embed, re-contextualise, indigenous ownership in indigenous traditional culture" (Strathern, 1996, p. 22). That said, what is stressed here is the effect of "reification, cascading, and debasement" that may follow submersion of the "mutual realm" by the "market realm" (Gudeman, 2012, p. 57). In sum, if the conclusion of this chapter probably takes us closer to Comarrof and Comaroff's analysis than to Gudeman's, the disruptive effects of the "language of trade" on the "house economy" and the "base" (Gudeman, 2001, p. 5) must not be overlooked.

12 Sylvie Poirier (2008, p. 83) insists that this "[...] is a form of symbolic violence that is imposed on indigenous people"
13 See below.
14 UN Doc. A/CONF.151/PC/78 of 26 July 1991.
15 UNEP/Bio. Div. 3/12, 13 August 1990, para. 7.
16 UNEP/Bio.Div.3/6 20 June 1990, para. 9 – drawn from UNEP/Bio.Div.3/Inf.4, para. 40.
17 UNEP, Governing Council, Decision 15/34 of 25 May 1989, (A/44/25), p. 161.
18 UNEP/CBD/COP/3/7, para. 7.
19 Refer to Chapter 10 in this book.
20 The programme was funded by the Darwin Initiative, the UK government grants scheme focusing on biodiversity protection. Bioversity International was the contract holder, with funding covering a period of three years (1/04/2015-31/03/2018). It is further described in Chapter 10 of this book.
21 For these two countries, see Chapter 10 (this book) and Annex of this chapter. In Kenya, see the Protection of Traditional Knowledge and Cultural Expressions Act, 2016 (No. 33 of 2016) (also see the Annex of this chapter).
22 For the sake of brevity, only the page referred to in each BCP is kept between brackets. Further information about each BCP is provided in the Annex of this chapter.
23 See Chapter 10 (this book).
24 Semi-directed interview, Bioversity International, Project Leader, 23/05/2019.
25 And more broadly when it comes to mitigation projects funded in the context of climate finance (Special Rapporteur on the rights of indigenous peoples & Human Rights Council, 2017).
26 See, for instance, *Simion Swakey Ole Kaapei & Others v Commissioner of Lands & Others* [2014] eKLR (Ole Kaapei), para. 33, where the High Court of Kenya at Nakuru alludes to the need to protect a community of pastoralists from "the vagaries of nature by ensuring pastoralist venture or way of life does not condemn the pastoralist to a life entirely subjected to nature".
27 *African Commission on Human and Peoples' Rights* v Kenya, App 006/2012, (006/2012) [2017] AfCHPR 28; (26 May 2017), para 130.
28 See the direct reference to "attachment to the land" under the UN Declaration on the Rights of Peasants and Other People Working in Rural Areas (2018), Art. 1.1.
29 See below.
30 The notion of sacred spaces can be strategically used by IPLCs and their advocates (Borman, 2017), as natural sacred sites enjoy a certain level of protection under international human rights law, notably through the right to freedom of religion. This is especially true in the African human rights system (Gilbert, 2018, pp. 138–141).
31 The community claims an area of about 8,000 Ha following the "primordial titles" (pre-existent or historic rights) granted in 1599, and recognising their *bienes comunales*, i.e. their communal property system. The community therefore disputes those land rights recognised through agrarian reform in 1952, and ratified through the so-called "partial titling" of 1995 which transferred only 3,843 Ha of land (*Tribunal unitario agrario* no. 21, 1995 - Consejo Municipal de Desarrollo Rural Sustentable & Morales Santiago, 2009, p. 38). The 1952 reform was revoked in August 1952 on appeal filed by the neighbouring communities of Yotao and Tepanzacoalco. The same judgment ordered the application for titling to be filed again and the case reconsidered for the purpose of solving the conflict over boundaries. The title deed issued in 1995 is said to be "partial" as the communities are still in dispute over the border area (López Bárcenas, 2016, pp. 105–106). The community has also been fighting since 2015 with private companies over mining operations on its lands, and seeks the termination of all mining concessions.
32 There are two forms of communal land tenure in Mexico, also referred to as "social" forms of property (as against private or public property). The agrarian reform of 1917 established the *Ejidos* (as in the community of Ek Balam) and the

"agrarian communities" ("*comunidad agraria*" as in Capulálpam de Méndez). There is an important difference between the two: agrarian communities

> are based on the recognition, restitution, confirmation, or a combination of these factors of the property rights of population centers – *núcleos de población or pueblos* – that have communally possessed land, water, and forest since precolonial times; their members – *comuneros* – have presumably had possession since immemorial times, and have followed customs and communal practices often recognized by the Spanish crown in colonial times [...]. Land recognition, confirmation, or restitution was often made upon submission of colonial land titles so called *títulos primordiales*
>
> *(Gutiérrez-Zamora & Hernández Estrada, 2020, n. 1)*

Landless peasants or *núcleos de población* (nuclei of settlements) unable to prove their possession were granted land through the *ejido* system. Both *ejidos* and agrarian communities are legal entities with legal personality, types of land endowments (in the former, it is a *dotación*, in the latter a *restitución*), land-tenure arrangements, and institutional organisations. On the institutional side, *ejidos* and agrarian communities alike have a general assembly as its highest level of governance, and the executive bodies are, respectively, the commissary *ejidal* – designated by the *ejidatarios* – and the commissary of communal goods – designed by the *comuneros* (each executive organ comprises a president, a secretary, and a treasurer). Their work is supervised by a vigilance committee. Until 1992, the property regime of both *ejidos* and agrarian communities was a sort of *usufructus* (collectively or individually held property depending on whether it applied to areas for collective use or to surfaces under cultivation that were parcelled out). As a consequence, lands could not be sold, rented, used as collaterals for loans, or subjected to any market transactions. Following the new Agrarian Law of 1992 (The Political Constitution of the United Mexican States, Art. 27, § VII; *Ley Agraria* D.O.F., 26-02-1992), a liberal-inspired counter-reform, land redistribution to landless communities was discontinued. Furthermore, the rights to *ejido* members were extended. Now, *ejidatarios* can rent or even sell their land to other persons in the *ejido*. Plots assigned for housing in the *ejido* are allotted as a private property. Finally, the individual *ejidatarios* may grant as security the usufruct of the lands for common use and of their parcels of land for cultivation (Kelly, 1993, p. 563). Currently, the only practical difference that exists between *ejidos* and agrarian communities is that in the latter, farming plots are never granted personally (whether or not they are farmed individually) and *comuneros* cannot sell their lands. Nevertheless, following a vote of the assembly, *comuneros* can choose to shift to the *ejido* system and "thus gain access to individual plots and, even, to their later sale if it is decided by a qualified assembly" (Morett-Sánchez & Cosío-Ruiz, 2017).

33 In Mexico, indigenous peoples are not subjects of public law, but "entities of public interest" ("*entidades de interés público*") (Constitution, para. A(VIII) of article 2). A limited consultation process is now included in Art. 2, para. B(IX) of the constitution (amended DOF 29-01-2016), which obliges authorities to "[c]onsult indigenous peoples in the preparation of the National Development Plan and the plans of the federative entities, municipalities and, where appropriate, the boroughs of Mexico City; and, where appropriate, to incorporate the recommendations and proposals they make". Significant progress was nevertheless achieved recently. For instance, the *Ley General de Desarrollo Forestal Sustentable* (General Law for Sustainable Forest Development), 5 June 2018 (*Nueva Ley DOF 05-06-2018*), recently amended by Decree 26 April 2021 (*Decreto por el que se reforman diversas disposiciones de la Ley General de Desarrollo Forestal Sustentable*, D.O.F. 25-04-21, Art. 93), which states: "In the case of land located in indigenous territories, the authorisation for change of land use must be accompanied by measures of prior, free, informed, culturally-appropriate and *bona fide* consultation, subject to the terms of the

applicable legislation". The weakness of the text lies in the absence of said "applicable legislation" – either in the constitution, federal law, or state law – in respect of free, prior and informed consent (FPIC). Some federative entities already offer a more robust protection – e.g. San Luis Potosí, Durango, and Oaxaca, or Chihuahua, Hidalgo, and Morelos: https://infosen.senado.gob.mx/sgsp/gaceta/64/3/2020-12-03-1/assets/documentos/Inic_PAN_Sen_Xochitl_art_5_expide_ley_federal_de_consulta_indigena.pdf). In recognition of these fundamental weaknesses, Senator Xóchitl Gálvez Ruiz (Partido Acción Nacional - PAN) has presented a draft *Ley Federal de Consulta a los Pueblos y Comunidades Indigenas y Afromexicanas* (General Law of Consultation of Indigenous and Afro-Mexican Peoples and Communities) on 26 November 2020, adopted by the House of Deputies (*cámara de diputados*) on 20 April 2021. The bill is currently before the Senate for review and voting.

34 In 1998, following the Law on the Rights of Indigenous Peoples and Communities, it came to be known as the "*normas de derecho consuetudinario*", i.e. norms of customary law.

35 As a matter of practice, municipalities generally consist of various population centres (Juan Martínez, 2013, p. 150).

36 On which see n. (32) above.

37 Called the "*Assemblea General de comuneros*".

38 The concept was coined by two indigenous Oaxaca intellectuals, Floriberto Díaz, a member of the Mixe indigenous community, and Jaime Martínez Luna, a Zapotec. It is said to be built on four foundational principles: communal territory, governance through communally appointed leadership roles (*cargos*), communal labour, and enjoyment (*fiesta*). They are undergirded by respect and reciprocity (Esteva, 2012; Martinez Luna, 2010). See n. (42) below on the importance of *comunalidad* for the community of Capulálpam de Méndez.

39 See below.

40 Interestingly enough, the Ek Balam BCP stresses that

> [d]uring the last decades the community of Ek Balam has taken up again traditional Mayan ritual practices such as the *Ch'a Cháak* or rainmaking ritual, the *Hanal Pixán* which is the festivity dedicated to the dead and the *Looj Ka Ta* or rite of protection against bad winds.
>
> (*p. 27*)

41 See Lynch (2012).

42 On the importance of this concept in the Oaxaca State, see n. (38). In Capulálpam de Méndez, *comunalidad* refers to three fundamental components: a *mode of social organisation* that orders and develops in a *residential structure* (the community itself) and that stems from a *collective mindset*. *Comualidad* is pitted against "individualism" as a counteracting force and is struggling against forms of "internal colonialism, characterized as a totalitarian system that refuses dialogue with the diverse" (p. 18). Elsewhere, the emphasis is put on festivities that are linked to agricultural and hydrological cycles and that aim at "asking and thanking God for food, rain, crops, to avoid natural disasters, droughts" (p. 29).

43 Fires are a multifaceted phenomenon in Madagascar (Kull, 2002) and provide an excellent example of the point in hand. Swidden agriculture (*tavy*), for instance, is still commonly referred to by state agents to explain deforestation. In tune with the "premodern" script, it is believed to be driven by population growth and poverty. Yet, it has repeatedly been shown that *tavy*, whose impact on the deforestation on the Great Red Island won't be discussed here, ensures basic food provision and is deeply engrained in the "Malagasy ethos of growth" (Keller, 2008, p. 652), and therefore cannot boil down to overly simplistic factors. *Tavy* certainly ought not to be outlawed and discarded without accounting for its social, economic, and cultural role and without a deep understanding of the drivers behind conversion of forests into other forms of land cover (Scales, 2014).

44 An example of one such distortion is provided by *Delgamuukw v. British Columbia*, 1997 CanLII 302 (SCC), [1997] 3 SCR 1010:

> Accordingly, in my view, lands subject to aboriginal title cannot be put to such uses as may be irreconcilable with the nature of the occupation of that land and the relationship that the particular group has had with the land which together have given rise to aboriginal title in the first place.

Inversely, there are other cases acknowledging that "traditional" activities are not frozen; see *Ilmari Länsman et al. v Finland*, Comm No. 511/1992, UN Doc. CCPR/C/52/D/511/1992 (1994), para. 9.3; *Apirana Mahuika v New Zealand*, Comm No 547/1993, UN Doc CCPR/C/70/D/547/1993 (27 October 2000), para. 9.4. *African Commission on Human and Peoples' Rights v Kenya*, App 006/2012, (006/2012) [2017] AfCHPR 28; (26 May 2017), para. 185.

45 Sotinkanmè is one of the 34 villages that make up the commune of Bonou (the commune is itself divided into five districts (Gouvernement de la République du Bénin & PNUD, 2015).

46 Arrêté interministériel N° 0121/MEHU/MDGLAAT/DC/SGM/DGFRN/ SA du 16/11/2012 *fixant les conditions de gestion durable de la forêt sacrée en République du Bénin*.

47 Loi n° 96-025 *relative à la gestion locale des ressources naturelles renouvelables*.

48 GELOSE contracts are concluded between the community, the rural municipality it belongs to, and a decentralised state service (for instance, water and forest administration) (Pollini & Lassoie, 2011, p. 817).

49 GELOSE Law no. 96-025, Article 3.

50 On which see Chapter 10 of this book.

51 The Tkarihwaié:ri Code, para. 20.

52 See also Chapter 6 of this book, by Giulia Sajeva.

53 Atrato River case, see above n. (5).

54 The Tkarihwaié:ri Code, para. 20 (our emphasis).

55 Atrato River case, para. 5.14 (with the Court's emphasis).

References

Agrawal, A. (1995). Dismantling the Divide Between Indigenous and Scientific Knowledge. *Development and Change*, *26*(3), 413–439.

Agrawal, A. (2002). Classification des savoirs autochtones: La dimension politique. *Revue internationale des sciences sociales*, *173*(3), 325–336. https://doi.org/10.3917/riss.173.0325

Agrawal, A. (2005a). Community, Intimate Government, and the Making of Environmental Subjects in Kumaon, India. *Current Anthropology*, *46*(2), 161–190.

Agrawal, A. (2005b). *Environmentality: Technologies of Government and the Making of Subjects*. Duke University Press. https://doi.org/10.1515/9780822386421-008

Akrich, M. (1992). The De-Scription of Technical Objects. In W. E. Bijker & J. Law (Eds.), *Shaping Technology/Building Society: Studies in Sociotechnical Change* (pp. 205–224). The MIT Press.

Almeida, F. (2017). *Legislative Pathways for Securing Community-Based Property Rights*. Rights and Resources Initiative.

Anaya, S. J. (2004). *Indigenous Peoples in International Law* (2nd ed.). Oxford University Press.

Antinori, C. M., & Rausser, G. C. (2003). *Does Community Involvement Matter? How Collective Choice Affects Forests in Mexico* (qt83j385n0; Department of Agricultural & Resource Economics, UC Berkeley, Working Paper Series). Department of Agricultural & Resource Economics, UC Berkeley.

Argumedo, A., & Pimbert, M. (2008). *Protecting Farmers' Rights with Indigenous Biocultural Heritage Territories: The Experience of the Potato Park* (p. 12). International Institute for Environment and Development (IIED), Asociacion ANDES.

Balibar, É. (2017). *Citizen subject: Foundations for Philosophical Anthropology* (S. Miller, Trans.). Fordham University Press.

Bannister, K. (2004). *Lessons for ABS: Academic Policies, Community Protocols and Community-level PIC.* International Expert Workshop on Access to Genetic Resources and Benefit Sharing, Cuernavaca, Mexico. https://polisproject.org/PDFs/Bannister%202004_ABS%20Lessons.pdf

Barcham, M. (2000). (De)Constructing the Politics of Indigeneity. In D. Ivison, P. Patton, & W. Sanders (Eds.), *Political Theory and the Rights of Indigenous Peoples* (pp. 137–151). Cambridge University Press.

Bavikatte, K., Jonas, H., & von Braun, J. (2010). Traditional Knowledge and Economic Development: The Biocultural Dimension. In S. M. Subramanian & B. Pisupati (Eds.), *Traditional Knowledge in Policy and Practice: Approaches to Development and Human Well-Being* (pp. 294–326). United Nations University Press.

Bavikatte, K. S. (2014). *Stewarding the Earth: Rethinking Property and the Emergence of Biocultural Rights*. Oxford University Press.

Bavikatte, K. S., & Bennett, T. (2015). Community Stewardship: The Foundation of Biocultural Rights. *Journal of Human Rights and the Environment*, *6*(1), 7–29. https://doi.org/10.4337/jhre.2015.01.01

Bavikatte, K. S., & Jonas, H. (2009). A Bio-cultural Critique of the CBD and ABS. In *Bio-Cultural Community Protocols. A Community Approach to Ensuring the Integrity of Environmental Law and Policy.* United Nations Environment Programme, Natural Justice.

Bavikatte, K. S., & Robinson, D. F. (2011). Towards a Peoples History of the Law: Biocultural Jurisprudence and the Nagoya Protocol on Access and Benefit Sharing. *Law, Environment and Development Journal (LEAD Journal)*, 7(1), 35–51.

Bavikatte, K. S., Robinson, D. F., & Oliva, M. J. (2015). Biocultural Community Protocols: Dialogues on the Space Within. *IK: Other Ways of Knowing*, *1*(2), 1–31.

Bavington, D. (2002). Managerial Ecology and Its Discontents: Exploring the Complexities of Control, Careful Use and Coping in Resource and Environmental Management. *Environments*, *30*(3), 3–22.

Berkes, F. (2008). *Sacred Ecology* (2nd ed.). Taylor & Francis.

Bioversity International. (2018). *Darwin Initiative, Mutually supportive implementation of the Nagoya Protocol and Plant Treaty, Final Report.* Darwin Initiative. https://www.darwininitiative.org.uk/documents/22017/24531/22-017%20FR%20edited.pdf

Blanco, I. I. (2012). *Autonomy, Community and Caciquismo: Ethnicity, Social Control and Violence in a Mixe Village in Oaxaca, Mexico* (ICIP Working Papers:2012/2, p. 62). Institut Català Internacional per la Pau.

Blaser, M. (2009). Political Ontology: Cultural Studies without 'cultures'? *Cultural Studies*, *23*(5–6), 873–896.

Blaser, M. (2013). Ontological Conflicts and the Stories of Peoples in Spite of Europe: Toward a Conversation on Political Ontology. *Current Anthropology*, *54*(5), 547–568. https://doi.org/10.1086/672270

Boisvert, V., & Vivien, F.-D. (2012). Towards a Political Economy Approach to the Convention on Biological Diversity. *Cambridge Journal of Economics*, *36*(5), 1163–1179. JSTOR.

Bonneuil, C. (2019). Seeing Nature as a 'Universal Store of Genes': How Biological Diversity Became 'Genetic Resources', 1890–1940. *Studies in History and Philosophy of Science Part C: Studies in History and Philosophy of Biological and Biomedical Sciences, 75*, 1–14.

Borman, R. (2017). Whose Sacred Sites? Indigenous Political Use of Sacred Sites, Mythology, and Religion. In F. Sarmiento & S. Hitchner (Eds.), *Indigeneity and the Sacred. Indigenous Revival and the Conservation of Sacred Sites in the Americas* (pp. 3–10). Berghahn.

Brosius, J. P. (1999). Green Dots, Pink Hearts: Displacing Politics from the Malaysian Rain Forest. *American Anthropologist, 101*(1), 36–57.

Brosius, J. P., & Hitchner, S. L. (2010). Cultural Diversity and Conservation. *International Social Science Journal, 61*(199), 141–168. https://doi.org/10.1111/j.1468-2451.2010.01753.x

Brosius, J. P., Tsing, A. L., & Zerner, C. (1998). Representing Communities: Histories and Politics of Community-Based Natural Resource Management. *Society & Natural Resources, 11*(2), 157–168.

Brundtland, G. H. & World Commission on Environment and Development. (1987). *Report of the World Commission on Environment and Development: 'Our common future'*. United Nations.

Brush, S. B., & Stabinsky, D. (Eds.). (1996). *Valuing Local Knowledge: Indigenous People and Intellectual Property Rights*. Island Press.

Callicott, J. B. (1994). *Earth's Insights: A Survey of Ecological Ethics from the Mediterranean Basin to Australian Outback*. University of California Press.

Cepek, M. L. (2011). Foucault in the Forest: Questioning Environmentality in Amazonia. *American Ethnologist, 38*(3), 501–515. https://doi.org/10.1111/j.1548-1425.2011.01319.x

Comaroff, J. L., & Comaroff, J. (2010). *Ethnicity, Inc.* University of Chicago Press.

Consejo Municipal de Desarrollo Rural Sustentable, & Morales Santiago, E. (2009). *Plan de Desarrollo Municipal Capulálpam de Méndez*. SAGARPA.

Cooper, H. D., Engels, J., & Frison, E. (1994). A Multilateral System for Plant Genetic Resources: Imperatives, Achievements and Challenges. *Issues in Genetic Resources No. 2*, 1–40.

Correa, C. (1999). In situ Conservation and Intellectual Property Rights. In S. Brush (Ed.), *Genes in the Field* (pp. 239–260). Lewis Publisher.

Crawford, S. E. S., & Ostrom, E. (1995). A Grammar of Institutions. *The American Political Science Review, 89*(3), 582–600.

Davies, M. (2012). Persons, Property, and Community. *Feminists@law, 2*(2). https://doi.org/10.22024/UniKent/03/fal.37

Davies, M., & Naffine, N. (2001). *Are Persons Property?: Legal Debates about Property and Personality*. Ashgate.

De Obaldia, I. (2005). Western Intellectual Property and the Indigenous Cultures: The Case of the Panamanian Indigenous Intellectual Property Law. *Boston University International Law Journal, 23*(2), 337–394.

Delgado, N. A. (2016). Community Protocols as Tools for Resisting Exclusion in Global Environmental Governance. *Revista de Administração de Empresas, 56*(4), 395–410. https://doi.org/10.1590/S0034-759020160404

Deplazes-Zemp, A. (2019). Challenges of Justice in the Context of Plant Genetic Resources. *Frontiers in Plant Science, 10*. https://doi.org/10.3389/fpls.2019.01266

Doolittle, A. A. (2007). Fortress Conservation. In P. Robbins (Ed.), *Encyclopedia of Environment and Society* (pp. 704–705). Sage.

Eisner, T. (1989). Prospecting for Nature's Chemical Riches. *Issues in Science and Technology, 6*(2), 31–34.

Esteva, G. (2012). Hope from the Margins. In D. Bollier & S. Helfrich (Eds.), *The Wealth of the Commons: A World beyond Market & State*. Levellers Press.

Gilbert, J. (2016). *Indigenous Peoples' Land Rights Under International Law: From Victims to Actors*. Brill Nijhoff.

Gilbert, J. (2018). *Natural Resources and Human Rights: An Appraisal*. Oxford University Press.

Gilbert, J., & Sena, K. (2018). Litigating Indigenous Peoples' Cultural Rights: Comparative Analysis of Kenya and Uganda. *African Studies*, 77(2), 204–222.

Girard, F. (2019). Communs et droits fondamentaux: La catégorie naissante des droits bioculturels. *Revue Des Droits et Libertés Fondamentaux, 28*. http://www.revuedlf.com/droit-fondamentaux/communs-et-droits-fondamentaux-la-categorie-naissante-des-droits-bioculturels/

Goodman, B. (2013). Panama, the Comarca, and the Kuna. *Village Rights International*. https://www.villagerights.com/Panama-The-Comarca-And-The-Kuna

Gouvernement de la République du Bénin, & PNUD. (2015). *Projet commune du millénaire de Bonou pour un développement durable (PCM-Bonou)*. Gouvernement de la République du Bénin, PNUD.

Grear, A. (2010). *Redirecting Human Rights: Facing the Challenge of Corporate Legal Humanity*. Palgrave Macmillan.

Grear, A. (2015). The Closures of Legal Subjectivity: Why Examining 'Law's Person' is critical to an Understanding of Injustice in an Age of Climate Crisis. In A. Grear & L. J. Kotzé (Eds.), *Research Handbook on Human Rights and the Environment* (pp. 79–101). Edward Elgar.

Greaves, T. (1996). Tribal Rights. In S. B. Brush & B. Stabinsky (Eds.), *Valuing Local Knowledge: Indigenous Peoples and Intellectual Property Rights* (pp. 25–40). Island Press.

Grotius, H. (2012). *Hugo Grotius on the Law of War and Peace* (S. C. Neff, Ed.). Cambridge University Press.

Gudeman, S. (1992). Remodeling the House of Economics: Culture and Innovation. *American Ethnologist, 19*(1), 141–154.

Gudeman, S. F. (2001). *The Anthropology of the Economy: Community, Market, and Culture*. Blackwell.

Gudeman, S. F. (2012). *Economy's Tension: Dialectics of Community and Market*. Berghahn.

Gutiérrez-Zamora, V., & Hernández Estrada, M. (2020). Responsibilization and State Territorialization: Governing Socio-Territorial Conflicts in Community Forestry in Mexico. *Forest Policy and Economics, 116*, 102188.

Halewood, M. (1999). Indigenous and Local Knowledge in International Law: A Preface to Sui Generis Intellectual Property Protection. *McGill Law Journal, 44*(2), 953–996.

Halewood, M. (2013). What Kind of Goods Are Plant Genetic Resources for Food and Agriculture? Towards the Identification and Development of a New Global Commons. *International Journal of the Commons*, 7(2), 278–312.

Halewood, M., Bedmar Villanueva, A., Rasolojaona, J., Andriamahazo, M., Rakotoniaina, N., Bossou, B., Mikpon, T., Vodouhe, R., Fey, L., Drews, A., Kumar, P. L., Rasoanirina, B., Rasoazafindrabe, T., Aigbe, M., Agbahounzo, B., Otieno, G., Garforth, K., Kiene, T., & Nnadozie, K. (2021). Enhancing Farmers' Agency in the Global Crop Commons through Use of Biocultural Community Protocols. *Agriculture and Human Values, 39*, 579–594.

Herrera, F. (2015). La evolución de las demandas indígenas sobre la tierra y las respuestas del Estado en Panamá. *Tareas, 150*, 39–62.

Holt, F. L. (2005). The Catch-22 of Conservation: Indigenous Peoples, Biologists, and Cultural Change. *Human Ecology, 33*(2), 199–215.

IIED, Asociacion ANDES, & Natural Justice. (2012). *Using Biocultural Community Protocols to Implement MEAs and UNDRIP at the Local Level for Sustainable Development* (Workshop of the ISE Global Coalition for Biocultural Diversity). 13th ISE Congress. https://pubs.iied.org/pdfs/G03388.pdf

International Institute for Environment and Development. (n.d.). *National and Local Policy and Law for Protecting Biocultural Heritage | Biocultural Heritage.* Retrieved 23 July 2020, from https://biocultural.iied.org/national-and-local-policy-and-law-protecting-biocultural-heritage

Iorns Magallanes, C. (2019). From Rights to Responsibilities using Legal Personhood and Guardianship for Rivers. In B. Martin, L. Te Aho, & M. Humphries-Kil (Eds.), *ResponsAbility: Law and Governance for Living Well with the Earth* (pp. 216–239). Routledge.

IPBES. (2019). *Summary for Policymakers of the Global Assessment Report on Biodiversity and Ecosystem Services* (Summary for Policy Makers). Zenodo. https://doi.org/10.5281/ZENODO.3553579

IUCN, & The World Conservation Union. (2005). *Benefits Beyond Boundaries Proceedings of the Vth IUCN World Parks Congress, Durban, South Africa, 8–17 September 2003.* IUCN.

IUCN, & World Commission on Protected Areas (WCPA). (2019). *Recognising and Reporting Other Effective Area-Based Conservation Measures.* IUCN. https://portals.iucn.org/library/node/48773

IUCN, & World Conservation Congress. (2005). *Resolutions and Recommendations: World Conservation Congress, Bangkok, Thailand, 17–25 November 2004.* IUCN.

IUCN, & World Conservation Congress. (2009). *Resolutions and Recommendations: World Conservation Congress, Barcelona, 5–14 October 2008.* IUCN.

Juan Martínez, V. L. (2013). What We Need Are New Customs. In T. A. Eisenstadt, M. S. Danielson, M. J. Bailon Corres, & C. Sorroza Polo (Eds.), *Latin America's Multicultural Movements* (pp. 135–168). Oxford University Press.

Keller, E. (2008). The Banana Plant and the Moon: Conservation and the Malagasy Ethos of life in Masoala, Madagascar. *American Ethnologist, 35*(4), 650–664.

Kelly, J. J. (1993). Article 27 and Mexican Land Reform: The Legacy of Zapata's Dream. *Columbia Human Rights Law Review, 25*, 541–570.

Kothari, A., Corrigan, C., Jonas, H., Neumann, A., Shrumm, H., & Secretariat of the Convention on Biological Diversity. (2012). *Recognising and Supporting Territories and Areas Conserved by Indigenous Peoples and Local Communities: Global Overview and National Case Studies.* Secretariat of the Convention on Biological Diversity, ICCA Consortium, Kalpavriksh, and Natural Justice. http://www.deslibris.ca/ID/242837

Kotzé, L. J., & Villavicencio Calzadilla, P. (2017). Somewhere between Rhetoric and Reality: Environmental Constitutionalism and the Rights of Nature in Ecuador. *Transnational Environmental Law, 6*(3), 401–433. https://doi.org/10.1017/S2047102517000061

Kull, C. A. (2002). Madagascar Aflame: Landscape Burning as Peasant Protest, Resistance, or a Resource Management Tool? *Political Geography, 21*(7), 927–953.

Kurki, V. A. J. (2019). *A Theory of Legal Personhood.* Oxford University Press.

Lara, A. (2015). Rationality and Complexity in the Work of Elinor Ostrom. *International Journal of the Commons, 9*(2), 573–594.

Lewis, D. J., & Mosse, D. (2006). *Development Brokers and Translators the Ethnography of Aid and Agencies*. Kumarian Press.
Lixinski, L. (2013). *Intangible Cultural Heritage in International Law*. Oxford University Press.
López Bárcenas, F. (2016). *Análisis de los instrumentos y vacíos jurídicos para reconocer y aplicar normas, protocolos y procedimientos comunitarios*. Conabio-GIZ.
López Bárcenas, F., & Espinoza Sauceda, G. (2006). Recursos genéticos y conocimiento tradicional indígena la regulación internacional y su impacto en la legislación mexicana. In L. Concheiro Bórquez & F. López Bárcenas (Eds.), *Biodiversidad y conocimiento tradicional en la sociedad rural: Entre el bien común y la propiedad privada* (pp. 105–131). Centro de Estudios para el Desarrollo Rural Sustentable y la Soberanía Alimentaria.
Lynch, G. (2012). Becoming Indigenous in the Pursuit of Justice: The African Commission on Human and Peoples' Rights and the Endorois. *African Affairs, 111*(442), 24–45.
Macpherson, E., Torres Ventura, J., & Clavijo Ospina, F. (2020). Constitutional Law, Ecosystems, and Indigenous Peoples in Colombia: Biocultural Rights and Legal Subjects. *Transnational Environmental Law, 9*(3), 1–20.
MAEP, SAGE, & Ministère de l'environnement. (2017). *Document sur les plans d'investissement du site Antavolobe issu de l'atelier du 08 février 2017* (Doc SAGE /12). Darwin Initiative, CDB, ABS Initiative, Bioversity International, TIRPAA.
Maffi, L. (Ed.). (2001). *On Biocultural Diversity: Linking Language, Knowledge, and the Environment*. Smithsonian Institution Press.
Maffi, L., & Woodley, E. (2010). *Biocultural Diversity Conservation: A Global Sourcebook*. Earthscan.
Martínez Coria, R., & Haro Encinas, J. A. (2015). Derechos territoriales y pueblos indígenas en México: Una lucha por la soberanía y la nación. *Revista Pueblos y Fronteras Digital, 10*(19), 228.
Martinez Luna, J. (2010). The Fourth Principle. In L. Meyer & B. Maldonada Alvarado (Eds.), *New World of Indigenous Resistance. Noam Chomsky and Voices from North, South, and Central America* (pp. 85–100). City Lights.
Medaglia, J. C., Perron-Welch, F., & Phillips. (2014). *Overview of National and Regional Measures on Access and Benefit Sharing: Challenges and Opportunities in Implementing the Nagoya Protocol, Third Edition* (CISDL Biodiversity & Biosafety Law Research Programme). Centre for International Sustainable Development Law (CISDL).
Mol, A. (1999). Ontological Politics. A Word and Some Questions. *The Sociological Review, 47*(S1), 74–89. https://doi.org/10.1111/j.1467-954X.1999.tb03483.x
Morett-Sánchez, J. C., & Cosío-Ruiz, C. (2017). Panorama de los ejidos y comunidades agrarias en México. *Agricultura, Sociedad y Desarrollo, 14*, 125–152.
Morineau, O., & Morineau, M. (1997). *Los derechos reales y el subsuelo en México* (2nd ed.). Inst. de Investigaciones Jurídicas.
Mosse, D. (1997). The Symbolic Making of a Common Property Resource: History, Ecology and Locality in a Tank-irrigated Landscape in South India. *Development and Change, 28*(3), 467–504.
Mulrennan, M. E., & Bussières, V. (2020). Indigenous Environmental Stewardship: Do Mechanisms of Biodiversity Conservation Align with or Undermine It? In N. J. Turner (Ed.), *Plants, People, and Places: The Roles of Ethnobotany and Ethnoecology in Indigenous Peoples' Land Rights in Canada and Beyond* (pp. 282–312). McGill-Queen's University Press.
Mwangi, C. K. (2019). *Protecting Traditional Knowledge in Kenya through African Customary Law: An Analysis of Inclusive Subordination*. PhD Dissertation Strathmore University.

Naffine, N. (2003). Who are Law's Persons? From Cheshire Cats to Responsible Subjects. *The Modern Law Review, 66*(3), 346–367.

North, D. C. (1990). *Institutions, Institutional Change, and Economic Performance.* Cambridge University Press.

Nwauche, E. (2017). The Positive Protection of Traditional Cultural Expressions by Intellectual Property Rights in Africa. In E. Nwauche (Ed.), *The Protection of Traditional Cultural Expressions in Africa* (pp. 97–150). Springer International Publishing.

Nzomo, V. (2015, December 16). Comments on the Protection of Traditional Knowledge and Traditional Cultural Expressions Bill, 2015. *IP Kenya.* https://ipkenya.wordpress.com/2015/12/16/comments-on-the-protection-of-traditional-knowledge-and-traditional-cultural-expressions-bill-2015/

Ostrom, E. (1990). *Governing the Commons: The Evolution of Institutions for Collective Action.* Cambridge University Press.

Ostrom, E. (2009) *Understanding Institutional Diversity.* Princeton University Press.

Parks, L. (2019). *Benefit-sharing in Environmental Governance: Local Experiences of a Global Concept* (1st ed.). Routledge.

Peet, R., & Watts, M. (Eds.). (1996). Liberation Ecology. Development, Sustainability, and Environment in an Age of Market Triumphalism. In *Liberation Ecologies. Environment, Development and Social Movements* (0 ed., pp. 1–45). Routledge. https://doi.org/10.4324/9780203235096

Pels, P., & Salemink, O. (2000). *Colonial Subjects: Essays on the Practical History of Anthropology.* University of Michigan Press.

Pietrzykowski, T. (2017). The Idea of Non-personal Subjects of Law. In V. A. J. Kurki & T. Pietrzykowski (Eds.), *Legal Personhood: Animals, Artificial Intelligence and the Unborn* (Vol. 119, pp. 49–67). Springer International Publishing.

Poirier, S. (2008). Reflections on Indigenous Cosmopolitics—Poetics. *Anthropologica, 50*(1), 75–85.

Pollini, J., & Lassoie, J. P. (2011). Trapping Farmer Communities within Global Environmental Regimes: The Case of the GELOSE Legislation in Madagascar. *Society & Natural Resources, 24*(8), 814–830.

Polo, C. S., & Danielson, M. S. (2013). Political Subsystems in Oaxaca's Usos y Costumbres Municipalities. In T. A. Eisenstadt, M. S. Danielson, M. J. Bailon Corres, & C. Sorroza Polo (Eds.), *Latin America's Multicultural Movements* (pp. 169–191). Oxford University Press.

Posey, D. A. (1999). Introduction: Culture and Nature – The Inextricable Link. In D. A. Posey (Ed.), *Cultural and Spiritual Values of Biodiversity* (pp. 1–19). UNEP.

Posey, D. A. (2004). Traditional Resources Rights – De facto self-determination for indigenous peoples. In D. A. Posey & K. Plenderleith (Eds.), *Indigenous Knowledge and Ethics: A Darrell Posey Reader* (pp. 155–168). Routledge.

Posey, D. A., & Dutfield, G. (1996). *Beyond Intellectual Property: Toward Traditional Resource Rights for Indigenous Peoples and Local Communities.* International Development Research Centre.

Raymond, H. (2007). The Ecologically Noble Savage Debate. *Annual Review of Anthropology, 36*(1), 177–190. https://doi.org/10.1146/annurev.anthro.35.081705.123321

Redford, K. H. (1991). The Ecologically Noble Savage. *Cultural Survival Quarterly, 15*(1), 46–48.

Reyes-García, V., Fernández-Llamazares, Á., Aumeeruddy-Thomas, Y., Benyei, P., Bussmann, R. W., Diamond, S. K., García-del-Amo, D., Guadilla-Sáez, S., Hanazaki, N., Kosoy, N., Lavides, M., Luz, A. C., McElwee, P., Meretsky, V. J., Newberry,

T., Molnár, Z., Ruiz-Mallén, I., Salpeteur, M., Wyndham, F. S., … Brondizio, E. S. (2021). Recognizing Indigenous Peoples' and Local Communities' Rights and Agency in the Post-2020 Biodiversity Agenda. *Ambio*. https://doi.org/10.1007/s13280-021-01561-7

Roach, C. M., Hollis, T. I., McLaren, B. E., & Bavington, D. L. Y. (2006). Ducks, Bogs, and Guns: A Case Study of Stewardship Ethics in Newfoundland. *Ethics & the Environment*, *11*(1), 43–70.

Robinson, N. A. (Ed.). (1992). *Agenda 21 & the UNCED Proceedings, Volume I*. Oceana Publ.

Romero, T. L. (2005). Sui generis Systems for the Protection of Traditional Knowledge. *International Law: Revista Colombiana de Derecho Internacional*, *3*(6), Article 6. https://revistas.javeriana.edu.co/index.php/internationallaw/article/view/14043

Rutert, B. (2020). *Contested Properties. Peoples, Plants and Politics in Post-Apartheid South Africa*. transcript Verlag.

Sajeva, G. (2018). *When Rights Embrace Responsibilities: Biocultural Rights and the Conservation of Environment* (1st ed.). Oxford University Press.

Scales, I. R. (2014). The Drivers of Deforestation and the Complexity of Land Use in Madagascar. In I. R. Scales (Ed.), *Conservation and Environmental Management in Madagascar* (pp. 105–125). Routledge.

Singleton, B.E. (2017). What's missing from Ostrom? Combining design principles with the theory of sociocultural viability. *Environmental Politics*, *26*(6), 994–1014.

Smith, B. (1928). Legal Personality. *Yale Law Journal*, *37*(3), 283–299.

Special Rapporteur on the Rights of Indigenous Peoples, & Human Rights Council. (2017). *Report of the Special Rapporteur on the Rights of Indigenous Peoples, Human Rights Council, Thirty Sixth Session, 11–29 September 2017, Agenda Item3* (A/HRC/36/46; Promotion and Protection of All Human Rights, Civil, Political, Economic, Social and Cultural Rights, Including the Right to Development). Human Rights Council, United Nations, General Assembly.

Strathern, M. (1996). Potential Property: Intellectual Rights and Property in Persons. *Social Anthropology: The Journal of the European Association of Social Anthropologists*, *4*, 17–32.

Swanson, T. M., Pearce, D. W., & Cervigni, R. (1994). *The Appropriation of the Benefits of Plant Genetic Resources for Agriculture: An Economic Analysis of the Alternative Mechanisms for Biodiversity Conservation* (Background Study Paper No. 1). Commission on Plant Genetic Resources, First Extraordinary Session Rome, 7–11 November 1994.

Tănăsescu, M. (2020). Rights of Nature, Legal Personality, and Indigenous Philosophies. *Transnational Environmental Law*, *9*(3), 429–453.

Tonye, M. M. (2003). Sui Generis Systems for the Legal Protection of Traditional Knowledge and Biogenetic Resources in Cameroon and South Africa. *The Journal of World Intellectual Property*, *6*(5), 763–774.

Ulloa, A. (2005). *The Ecological Native: Indigenous Peoples' Movements and Eco-governmentality in Colombia*. Routledge.

Urueña, R. (2012). *No Citizens Here: Global Subjects and Participation in International Law*. Nijhoff.

Valdivia, G. (2015). Eco-Governmentality. In T. Perreault, G. Bridge, & J. McCarthy (Eds.), *The Routledge Handbook of Political Ecology* (pp. 467–480). Routledge.

Whyte, K. P. (2013). Justice Forward: Tribes, Climate Adaptation and Responsibility. *Climatic Change*, *120*(3), 517–530. https://doi.org/10.1007/s10584-013-0743-2

Wickstrom, S. (2003). The Politics of Development in Indigenous Panama. *Latin American Perspectives*, *30*(4), 43–68.

Wynberg, R., & Laird, S. (2009). Bioprospecting, Access and Benefit Sharing: Revisiting the 'Grand Bargain'. In R. Wynberg, D. Schroeder, & R. Chennells (Eds.), *Indigenous Peoples, Consent and Benefit Sharing: Lessons from the San-Hoodia Case* (pp. 69–86). Springer Netherlands.

Zarka, Y. C. (1999). The Invention of the Subject of the Law. *British Journal for the History of Philosophy*, 7(2), 245–262.

12

CONCLUDING THOUGHTS

Biocultural Jurisprudence in Hindsight: Lessons for the Way Forward

Fabien Girard, Christine Frison and Ingrid Hall[1]

This concluding chapter resituates each contribution within the broader intellectual framework canvassed in the introduction. In so doing, it follows the three thematic focuses that make up the three parts of the book.

Conceptual Insights: Biocultural Diversity, Biocultural Rights, and Space Making

The first part of the book aimed to provide the reader with some historical background on BCPs/CPs, their philosophical and political underpinnings, as well as the expected aims and functions within and beyond ABS laws.

Scholarly literature on "cultural protocols" and "community research protocols" insists that they stand as "community-level strategies for protecting intangible cultural heritage" (Bannister, 2009, pp. 278–279). However, they fall short of merely codifying customary laws and traditional governance systems. Besides, diversity is the rule rather than the exception. As Kelly Bannister states in a previous study based on a substantial body of protocols that she reiterates in her chapter (Chapter 2, "A Biocultural Ethics Approach to Biocultural Rights: Exploring Rights, Responsibilities and Relationships through Ethics Initiatives in Canada"):

> [s]ome [protocols] are general and overarching, others are specific. Some are wholly or partly derived from traditional beliefs and practices, others are negotiated between contracting parties as a strategic tool for governance of a specific project or economic advantage in the market economy.
>
> *(ibid., 288)*

This echoes what Miri (Margaret) Raven and Daniel Robinson write in their chapter (Chapter 8, "Biocultural Rights and Protocols in the

DOI: 10.4324/9781003172642-15

Pacific"): BCPs "can be derived from existing cultural norms and customs or be newly constructed to serve a particular purpose".

There is little doubt that BCPs/CPs are never a faithful transposition or codification of what Bavikatte et al. once called the community "space within", i.e. the "ethical grammar" of the relationships communities have with their ecosystems, "a grammar that is coded in culture, values, practices, and customary laws" (Bavikatte et al., 2015). From early on in discussions on the inclusion of BCPs/CPs in a strengthened ABS framework, the notion was to prevent misappropriation of TK and genetic resources held by IPLCs through the laying down of local PIC procedures informed by customary law and traditional decision-making processes. The Mo'otz Kuxtal Voluntary Guidelines[2] highlighted this focus on local PIC, stressing the importance of community values, worldviews, and praxis. Nevertheless, BCPs/CPs do not just reflect local values or worldviews. They are embedded in different levels and scales of normativities (local, domestic, international), they are tools for negotiating "with a variety of actors",[3] and finally, they are outward-looking. Thus, BCPs/CPs morph into "hybrid" objects for cross-cultural engagement and translation processes. In no way can BCPs/CPs eschew the conceptual language of access law (or natural resource legislation beyond ABS contexts). This is particularly true of domestic legislation entrenching BCPs as part of the community PIC and MAT process, as seen below. It equally holds true whenever a community wishes to add a layer of customary rules to the domestic procedure for community PIC and benefit-sharing; or struggles to have a local PIC formally enshrined in law. An example of the former is the Bushbuckbridge Community Protocol (The Kukula Traditional Health Practitioners Association & Natural Justice, 2017),[4] one of the two emblematic South African BCPs with the much-celebrated Rooibos Biocultural protocol studied by Leslé Jansen and Rayna Sutherland (Chapter 9, "The Khoikhoi Community's Biocultural Rights Journey with Rooibos"). The Bushbuckbridge Community Protocol, intending to unite customary law and state law, refers to the National Environmental Management: Biodiversity Act (2004) and the Bioprospecting, Access, and Benefit Sharing (BABS) Regulations (2008).[5] The latter is exemplified by a string of Mexican BCPs developed with the support of the *Secretaría de Medio Ambiente y Recursos Naturales*, the civil organisation *Red Indígena de Turismo de México*, and with funding from the UNDP.[6] Some of these BCPs strive to put Mexican law in line with ILO's Convention No. 169, the jurisprudence of the Inter-American Court of Human Rights, and Articles 5 (benefit-sharing), 6 (access to genetic resources), and 7 (access to TK) of the Nagoya Protocol (López Bárcenas, 2016; López Bárcenas & Espinoza Sauceda, 2006).

The outcome is that extant procedures or protocols never quite emerge unscathed from these encounters with ABS norms and the language of trade, an undeniable fact when one considers truly ad hoc protocols.

At this point in reasoning, the agency of IPLCs emerges. Interlinkages between customs and state laws – legally recognised or sought for – between subsistence economies and global markets point towards the local as a *locus* of political agency, legal innovation, and creative normative strategies driven by subaltern actors (de Sousa Santos & Rodríguez-Garavito, 2005). Subaltern actors, social movements, and indigenous activists have been able to sustain a "cosmopolitanism from below" (Appadurai, 2013), or at times an "insurgent cosmopolitanism" (de Sousa Santos, 2006) advocated locally for more recognition of communities' customary laws – an overarching aim of the African Group throughout the Nagoya negotiations – and sovereignty over natural resources. At the same time, they have shaped a counter-narrative with a global reach allowing them to break free from Western hegemony and making room for non-Western epistemologies and ontologies in international fora. As Kelly Bannister puts it, "[i]ntentional and thoughtful unearthing and uprooting of entrenched western ethical traditions and assumptions, and how these are institutionalized in scientific, legal and educational systems, can make space and create more hospitable ground for co-creating a shared biocultural ethic that respects the different values, worldviews and systems involved" (Chapter 2, this book). In her chapter, exploring consent from a bioethics perspective, she illustrates the depth of these locally driven initiatives within the specific context of the process of reconciliation in Canada (see already, Bannister, 2018). She shows how the relational emphasis of biocultural ethics can complement biocultural rights-based approaches by encouraging a "justice of rights" and a "relational justice" through right relationships. Looking at cultural protocols, ethical codes, and codes of practice across different cases, Bannister brings the different aspects of "consent" to the forefront. She shows that it can become more than a transactional process with an outcome (an exchange of goods and services) – to being a truly "relational" one, i.e. reflecting a "community's broader principles and values", its worldview and its understanding of relationships among humans, and between humans and the natural world.

The terminology of "collaborative consent" and "meaningful consenting" proposes to reflect and reform the relationship between Indigenous and non-Indigenous actors on ABS and beyond, "representing more of an interactive verb [i.e. consenting] rather than a transactional noun". As Phare et al. (2018, p. 1) state, "collaborative consent" implies a "committed engagement between Indigenous and non-Indigenous governments— acting as equal partners, each with their asserted authority—to secure mutual consent on proposed paths forward related to matters of common concern and all aspects of governance". This form of consent enables "Indigenous values and principles to be living and embodied (rather than merely aspirational) through co-creation of intentional and meaningful relationships and fostering concomitant relational accountability" (Bannister, this volume).

Ingrid Hall's Chapter 4 ("Unmaking the Nature/Culture Divide: The Ontological Diplomacy of Indigenous Peoples and Local Communities at the CBD") highlights the interconnectedness of local creativity, locally driven and indigenous-inspired concepts, tools and counter-narratives on the one hand,

and ontological diplomacy at work in different ABS fora, on the other. With greater available space for local voices at the global level, IPLCs' representatives have challenged naturalistic Western ontology and its attendant nature regimes (Escobar, 1999). As Hall illustrates, these institutional spaces are used as "hybrid fora" and ways to engage in "socio-technical controversies" (Callon et al., 2011) over ABS-related issues, which use a diversity of strategies. Hall points to two mutually reinforced forms of "ontological diplomacy", creating fault lines widening the negotiation space available to IPLCs.

This process of space making enabled the African Group to succeed in getting community protocols adopted within the Nagoya Protocol and for social movements and Indigenous activists to promote and implement BCPs/CPs around the world. This process has also unleashed creativity at the local level where initiatives now abound: biocultural heritage territories (Argumedo & Pimbert, 2008), biocultural approaches to conservation (Gavin et al., 2015), and the indigenous, Black agrarian, and grassroots seed sovereignty initiatives that Garrett Graddy-Lovelace chronicles in her chapter (Chapter 5, "From Obstruction to Decolonization? Contested Sovereignty, the Seed Treaty, and Biocultural Rights in the U.S./Turtle Island and Beyond"). She redirects our attention to place-based and plant-based mobilisations within the US/Turtle Island, a nation-state that she sees as a paragon of "colonial settler racial capitalism and (neo)liberalist orientation". These significant biases, as she contends, are visible in the recent US ratification of the International Treaty on Plant Genetic Resources for Food and Agriculture (ITPGRFA), which might well mean the stalemate on such critical issues as DSI and farmers' rights will go on. Nevertheless, seizing on the 2019 Atateken North American Regional Declaration on Biocultural Diversity and Recommended Actions,[7] she sees all the potentialities of the CBD as a "key forum" wherein IPLCs have been able to "voice their claims and even actualize their worldviews". In turn, this has percolated through to the grassroots level in the US and fostered several biocultural-based initiatives.

At this juncture, there is no doubt that the process described above is not one-directional – from global to local – but perhaps best understood as a feedback loop. The spaces opening within hybrid fora give IPLCs more leeway to accommodate their worldviews and needs and to negotiate new rights at the national and local levels. What these new spaces trigger or sustain at the local level, in turn, provides living proof of alternatives, perhaps marginal ones but inestimable ways of life for the world to see. This is what the Constitutional Court of Colombia alluded to in its seminal *Atrato River* case: the "exemplarity" that Indigenous peoples represent for humanity, in the "singularity and at the same time the universality" of their existence.[8] These "exemplary examples" and experiences subsequently feed back into discussions and negotiations within hybrid fora. The Potato Park Biocultural Heritage Territory in Peru and its inter-community agreement developed by six communities is one of these inspirational precedents (local, place-based "exemplary" examples) used to promote BCPs/CPs by NGOs such as Natural Justice and the African Group.[9] BCPs/CPs are now showcased in the ongoing and laborious work of the Ad Hoc Technical Expert Group on Farmers' Rights as suitable mechanisms to implement farmers' rights.[10]

In Chapter 3 ("Sumaq kawsay (Good living) and indigenous potatoes: On the delicate exercise of ontological diplomacy"), Ingrid Hall reveals that different philosophical underpinnings and fundamental assumptions about IPLCs and the conservation of nature underwrite current discussions about BCPs/CPs. She indicates that, in the Potato Park case, different biocultural concepts, methods, and tools have been deployed and experimented with in order to establish the emblematic Potato Park biocultural heritage territory as a project of in situ conservation; the biocultural protocol is only one of these tools.

The notion of "territory" remains the magnet, very much in line with the belief held by Darrell Posey, Graham Dutfield, and Alejandro Argumedo that the right to land and territory is an absolute prerequisite for the continuation of IPLCs' identities (Posey, 1995). As Posey asserts, "[c]ontrol over cultural, scientific, and intellectual property is de facto self-determination, although only after rights to land and territory are secured by law and practice [...]" (Posey, 2004b, p. 156). To achieve this goal, Posey et al. championed a "bundle of rights" approach – dubbed "traditional resources rights" (TRRs) – understood as an "integrated rights concept that recognises the inextricable link between cultural and biological diversity" (Posey & Dutfield, 1996, p. 95). However, they firmly believed that TRRs aimed first to secure the human rights of Indigenous peoples and local communities, and then only by extension to sustain the right to development and environmental conservation (ibid.).

In contrast, Giulia Sajeva carries out a critical appraisal of the jurisprudential underpinnings of BCPs in Chapter 6, "The Legal Framework behind Biocultural Rights: An Analysis of Their Pros and Cons for Indigenous Peoples and for Local Communities". She exposes how much of the recent literature on protocols draws on Kabir Bavikatte's breakthrough book on biocultural rights, *Stewarding the Earth* (Bavikatte, 2014), and his work within the South African NGO Natural Justice, co-founded with Harry Jonas. Their work has been highly influential in the process leading up to the adoption of the Nagoya Protocol. Unlike TRRs, biocultural rights are overtly built upon two foundations: one relates to the interests of IPLCs – as a group rights approach – and the other pertains to a more general interest of humankind (or the biotic community at large) in the conservation of the environment – a universal rights approach (also see Sajeva, 2018, p. 99 and seq.). In other words, their international recognition is premised on their instrumental role in the conservation and sustainable use of the environment, taking the ethic of stewardship as its central node. This new intellectual protection stresses the primordial role of the ethic of stewardship and the continuation of sustainable practices and lifestyles. For Giulia Sajeva, this is not without risks as

> [b]iocultural rights may be claimed for only by sustainable Indigenous peoples and local communities and bind them to exercise the rights in their

> basket in ways that do not harm the environment, but actually promote its protection. They are environmentally-conditioned rights.
>
> *(Saveja, this volume, also see, Sajeva, 2021)*

In Chapter 8, "Biocultural Rights and Protocols in the Pacific", Miri (Margaret) Raven and Daniel Robinson make the point that, apart from reactivating the "noble savage" myth (Alcorn, 1994; Raymond, 2007; Redford, 1991), "overly prescriptive uses of the term [stewardship] risks simplifying the everyday lived realities of Indigenous peoples". Importantly, existing uses of the "stewardship" concept may prove at odds with specific ontological schemes (Descola & Sahlins, 2014), such as totemism, wherein humans entertain a kinship network and a relational ethic with plants and animals (Robinson & Raven, 2020).

Biocultural Community Protocols, Access and Benefit-Sharing, and Beyond

The book's second section embarks upon a thorough examination of recent BCPs/CPs informed by practical experience, community-based action research, and ethnographic work. The chapters included under this heading mainly focus on the ABS regime, although it strives to paint a broader picture through an incursion into a community protocol developed outside the Nagoya Protocol framework. This part addresses the content and functions of the BCPs/CPs, as well as the process of negotiation.

Mapping BCPs/CPs: Content and Functions

The chapter written by Pía Marchegiani and Louisa Parks (Chapter 7, "Community Protocols as Tools for Collective Action beyond Legal Pluralism – the Case of Tracks in the Salt") studies the protocol produced by the Atacama communities in the Salinas Grandes and Laguna de Guayatayoc area of northwest Argentina as an enlightening example beyond ABS laws. It confirms a deep-rooted trend towards the increasing use of CPs as a tool for protecting a territory against the extractive industry or massive state-led development programmes. Several CPs have been developed in Africa[11] and Latin America[12] by IPLCs along these lines to avoid or halt any encroachment upon, or breach of, their customary tenure and resource rights, as well as innovation, practices, sacred sites, rituals, and customary laws. In doing so, these CPs share a common feature with the BCPs/CPs developed in an ABS context. In all but a tiny number of cases, land tenure and resource rights are mentioned as a matter of concern or a contentious issue. They are the backbone of most BCPs/CPs, operating as a living matrix connecting the communities' spiritual, cultural, and reciprocal relationships to nature.

Leslé Jansen and Rayna Sutherland's sensitive account (Chapter 9, "The Khoikhoi Community's Biocultural Rights Journey with Rooibos") of the San and Khoikhoi's struggle in post-apartheid South Africa for recognition as an Indigenous people also bears testimony to the centrality of historical land connections and land claims in the community protocol process. This BCP points towards the "biocultural heritage" approach that, in the wake of the studies by Posey and the IIED (especially in the case of the Potato Park, see Argumedo, 2008; Graddy, 2013; Swiderska et al., 2020), places great emphasis on securing rights to land and territory as a precondition to any further consideration, notably IPLCs' engagement with external actors.[13]

Together, the chapters uncover significant commonalities cutting across the divide between access law and non-ABS contexts. For instance, most of the cases studied stress that BCPs/CPs rely on a detailed account of communities' beliefs, lifestyles, and worldviews while emphasising their spiritual, cultural, and reciprocal relationships with the environment. Most BCPs/CPs address this question by circumscribing the collective under consideration, which might include the non-humans with whom the people interact. For instance, as stressed above, Miri (Margaret) Raven and Daniel Robinson insist in their chapter on Aboriginal and Torres Strait Island peoples on the need to align BCPs with the Aboriginal originary matrix, sustaining a continuity between both interiority and physicality across a very large array of beings, including plants and animals (as totemic species) (also see Robinson & Raven, 2020). In the Malagasy setting, Manohisoa Rakotondrabe and Fabien Girard (Chapter 10, "Biocultural Community Protocols and Boundary Work in Madagascar: Enrolling Actors in the Messy World(s) of Global Biodiversity Conservation") unveil the embeddedness of traditional crops in a symbolic system revolving around an ancestor (Deliège, 2012, pp. 187–188; Keller, 2008). It explains the Analavory farmers' constant obliviousness to the scope of their protocol, particularly the fact that it also covers "genetic resources other than the Plant Genetic Resources for Food and Agriculture (PGRFA) used for food and agriculture". For the peasants, indeed, genetic resources which are not PGRFA – i.e. are not cultivated – do not belong to them but to "those with healing gifts". The piece on the Khoikhoi Community's Rooibos Biocultural Community Protocol, by Leslé Jansen and Rayna Sutherland, also illustrates the foundational role of the relationship between the communities and the rooibos plant and how it was reinstated or re-actualised through the development and drafting process.

Finally, BCPs are very often inscribed into a political agenda or read as political tools. BCPs/CPs give impetus to communities "to use international and national law to support the local manifestations of their right to self-determination" (Morgera, Tsioumani, & Buck, 2014, 223). More often than not, BCPs justify their existence and mandates based on existing international and domestic instruments. One way to explain this is to re-emphasise the fact that BCPs/CPs are, for their advocates, tools helping to "[…] uphold the rights of indigenous peoples and other communities by filling a space at the nexus of international, national and customary law and policy" (Parks, 2018, p. 88). International instruments and domestic legislation are drawn upon as a political means to signal all of the necessary rights and entitlements to protect IPLCs' distinct ways of life. Even within the framework of the Nagoya Protocol, it is assumed that IPLCs are not in a position to maintain their practices relating to the conservation of biodiversity. Nor is it easy for them to negotiate fairly and equitably the terms and conditions under which their genetic resources and TK are accessed or to preserve the part of their heritage that must remain *extra commercium* (insusceptible of being traded). Before these processes, IPLCs must first acknowledge and then counter the piecemeal nature of law. The value of BCPs/CPs, as Kabir Bavikatte contends, lie precisely

> [...] in their ability to act as the glue that holds together the total mosaic of a community life that is fragmented under different laws and policies, with the understanding that the conservation of Nature is a result of a holistic way of life.
>
> *(Bavikatte, 2014, p. 233)*

However crucial the preceding clauses may be in the context of ABS, IPLCs' primary efforts are often expended in articulating, based on customary law and traditional institutions, clear terms and conditions to regulate access to their knowledge and resources as well as benefit-sharing. In the BCP covering the farming communities of Analavory, in Madagascar, the protocol first sets out conditions and procedures for accessing PGRFA (for a detailed account of this prong of the protocol, see Halewood et al., 2021). It then moves on to set out the conditions and procedures for accessing genetic resources other than PGRFA, distinguishing in the process two separate regimes depending on whether the access is intended for commercial use or not. A similar pattern is found in most BCPs, notably in the Khoikhoi Community's Rooibos Biocultural Community Protocol, while at times, less sophisticated documents can be used (i.e. different media, posters, and signs).

Most significantly, from the standpoint of legal scholars, these terms, conditions, and processes form the core of BCPs/CPs in an ABS context. They first address relations between the community and putative or would-be bioprospectors by laying down a comprehensive framework to seek the community's PIC and negotiate the MAT and sharing of benefits. To a certain extent, this allows a "community to prepare in advance for negotiations of an ABS arrangement, rather than enter into such negotiations in an ad hoc manner, contributing thus to a more level playing field among the parties" (Morgera & Tsioumani, 2010, p. 157). Second, the core part of BCPs/CPs is also intended as a governance tool whereby the community can tackle intra-community equity and prevent conflicts (Ibid., 158). This is especially important, as Miri (Margaret) Raven and Daniel Robinson demonstrate in Vanuatu, where communities are heavily involved in "biotrade" or are inclined towards developing or improving an existing ABS value chain (Oliva et al., 2012).

At this juncture, several caveats must be borne in mind: compliance with the provisions of BCPs usually remains voluntary, unless domestic legislation makes it mandatory for bioprospectors to abide by the terms and conditions outlined in the protocol (e.g. in the Madagascar case, even though the Malagasy government is seemingly unwilling to recognise BCPs developed by communities that are not constituted as legal persons).

Except in those few instances where BCPs/CPs can be part of the PIC process set up by domestic ABS regulations, what is then the value of these protocols? The most straightforward answer is to consider these protocols as mainly ethical in nature (Grey, 2004), i.e. they communicate ethical norms and depend on moral adherence (Bannister, 2009, p. 297). Nevertheless, even in this legally inchoate form, somehow lingering in the anteroom of law, their importance cannot be overstated. They stand as "art of speaking" tools (Bavikatte et al., 2015) or "tools of conviviality" (Illich, 1973) – i.e. emancipatory or empowering tools collectively developed by a community and through which the group communicates its values, concerns, and interests. At the end of this analytical overview of the book's content, we offer the reader a comprehensive typology of BCPs/CPs by referring to their functions.

Negotiating BCPs/CPs: Brokers and the Making of Communities

Part 2 of this book also endeavours to address two thorny issues commonly associated with BCPs/CPs.

First, the realisation that the process of developing a protocol counts as much – if not more – than the content itself (for a critical view, see Rutert, 2020, pp. 255–257). The question is significant because the process of developing the protocol is featured as the building block of a more extensive process aimed at community empowerment. Even further, it is vital as, however important the community's role may have been, the process is almost always initiated and supported by external actors (this seems less true in the Canadian context; see Kelly Bannister's chapter in this book). This phenomenon begs the question of how much the development of BCPs/CPs is shaped, even unintentionally, by international aid donors, international or local NGOs (operating as "brokers"), professional facilitators, and scholars involved in the preparation, drafting, and implementation process. Also raised is the question of "communities becoming dependent on non-state actors' support" (Parks, 2019, p. 172). As Pía Marchegiani and Louisa Parks' chapter tends to show, even though in this case the CP fell short of advancing communities' rights, the fact that the development process was firmly in communities' hands allowed them to react quickly and use the protocol as a platform to circumvent some dead-ends and realign their collective action in the face of institutional bottlenecks. At the same time, we can also sense that the level of external engagement is likely to be inversely proportional to the degree of disenfranchisement and cultural disruption the community faces, a crucial issue that Leslé Jansen and Rayna Sutherland address regarding the San and Khoikhoi in South Africa.

Nevertheless, it remains to be seen whether the way most of these projects are funded through and entangled in vast bureaucratic machinery is not a fundamental structural defect constantly threatening to thwart the earnest efforts and benevolent endeavours of those committed to advancing the communities' positions. The ethnographic work conducted in Madagascar by Manohisoa Rakotondrabe and Fabien Girard confirms what Louisa Parks had already observed in her previous work, namely the "NGOization" of "many non-state actors that build relationships with local communities" (Parks, 2019, p. 195). These "brokers", albeit acting as *go-betweens* or *mediators* (ibid.), are increasingly

> part of transnational networks that are not only geographically extensive and tie global institutions, ideas, discourses and professionals to national and local places but [are] also institutional in that they link various international, national, local, governmental, non-governmental, for-profit and non-profit institutions, disciplines, technologies frameworks, knowledge systems and professionals.
>
> *(Sharma, 2016, p. 5)*

Furthermore, as funds are allocated through competitive bids issued by international aid agencies, global NGOs appropriate the lion's share of international funding. Furthermore, new aid frameworks are increasingly evidence-based or results-based and subjected to far-reaching goals such as the "Sustainable Development Goals" (SDGs) assessed against targets and global development indicators. Meanwhile, the project's "outcomes" and "outputs" are monitored through log-frames, value for money indicators, theory of change diagrams, and so on (ibid., 7). This new quantitative paradigm creates biases. The hegemony of global brokers may exacerbate the mismatch between local realities and representations of "the local" by those actors "whose competence is not place-based" (Demeulenaere, 2016). To be sure, these global brokers can still contract projects and programmes out – as Bioversity International did with Natural Justice in Madagascar and Benin – to NGOs far more experienced in operating on a local footing and deploying solutions tailored to on-the-ground realities. However, at the end of the aid chain, service providers (such as local NGOs acting as facilitators or sub-contractors) are "shaped" and "constrained" by the rules imposed on international brokers, and then they "[...] pass these constraints on to local communities [...]" (Parks, 2019, p. 82). As epitomised in the Malagasy case, it may also be the case that disagreements around the goals and representations of communities' role between the "principal" (broker) and the sub-contractor cause continued tensions and result in the community being caught between competing and largely irreconcilable aims.

The last point worth considering brings us back to one of the most disputed and contentious issues that arose amid community-based natural resource management programmes: "what makes community?". Two decades ago, Arun Agrawal and Clark Gibson (Agrawal & Gibson, 1999, p. 633) warned "commons" scholars that:

> [t]he vision of small, integrated communities using locally-evolved norms and rules to manage resources sustainably and equitably is powerful. But

> because it views community as a unified, organic whole, this vision fails to attend to differences within communities, and ignores how these differences affect resource management outcomes, local politics, and strategic interactions within communities, as well as the possibility of layered alliances that can span multiple levels of politics.

First and foremost, this observation has relevance for communities that can be understood as *Gemeinschaft*, i.e. those "defined on an ethnic basis" (Parks, 2019, p. 80). Admittedly, "internal recognition" appears less problematic when a community can self-define as a cohesive indigenous or ethnic group. Pía Marchegiani and Louisa Parks report that there were almost no discussions about the definition of the community's contours in the case of the "tracks in the salt" community protocol, although a record-number 33 indigenous communities located around the salt planes of Salinas Grandes and Laguna de Guallatayoc in the Andes were involved. The communities were arguably able to coalesce around a shared worldview and the ethical grammar of *Buen Vivir*.

At the same time, we must be wary of falling back into the romance of community – as "small, harmonious, integrated, isolated groups" (Coombe, 2011, p. 84), whose members are indefectibly tied by common interests and shared norms to manage resources sustainably. It has already been highlighted that BCPs/CPs are rarely a mere codification of traditional governance systems. First of all, it may be, for reasons most infamously illustrated by the San-Hoodia case (Munyi & Jonas, 2013), that some communities were historically left unable to maintain their traditional institutions. In these instances, developing a BCP/CP is mainly about (re)building something that does not quite exist outside an elusive collective memory and that in fact can only take shape through aspirations and claims for the future (and this may sometimes prove insufficient – Bannister, 2009, p. 288). Second, it is hardly the case that having traditional governance systems is enough for a community to automatically navigate the complex landscape of community-level PIC successfully. In particular, it is doubtful whether benefit-sharing can be foreseen by traditional internal rules, given that, for the community in most cases, access to genetic resources and TK constitute an unprecedented form of engagement with external actors (Bannister, 2009, p. 288). Therefore, more often than not, BCPs/CPs are "hybrid" tools (ibid.). The hybridity is not objectionable in and of itself. However, it implies a sort of reconfiguration or rebuilding of what has existed (sometimes for a long time). Furthermore, whenever the BCP/CP lays down a new process for obtaining PIC, which then translates into new institutions operating as gatekeepers, this necessarily raises the issue of power relations and potential appropriation by the local elites. Sometimes, the emerging institution is a socially legitimate and "respected political body" (such as a group of elders). This being said, sometimes it "is a political body that is divorced from traditional practices and/or not supported by the traditional knowledge holders of the community", leading to the intractable questions of "who has the right to speak for who" and "who can speak about what?" (Bannister, 2004, p. 3). This is not to say that communities, as bounded groups with a clear sense of territorial possession and traditional institutions, are solely things of fiction – or social construction at best. Rosemary J. Coombe has warned against two possible shortcomings: "simple allegations of essentialism (strategic or otherwise), sitings of social construction, and accusations of romanticism" – as both "[…] reveal a profound lack of political sensitivity to the fields of power and leverage in which peoples struggle for recognition, resources, and opportunity" (Coombe, 2011, p. 85).

To prevail over these shortcomings, the concept of "community of interest" ("*Gesellschaft*" in Tönnies' (2001) analytic) is of some help. The point is not to signal the existence of different types of communities as in Tönnies' writing, but to take into account the fact that a BCP/CP arises out of certain types of interests that the community's members might not equally share. As the BCP drafted by the Kukula Traditional Health Practitioners Association (Parks, 2019, p. 80) or the Lamu County Biocultural Community Protocol (Chapter 11, this book) illustrate, a tension permeates the collective. Here, a community shows its Janus-faced nature: it stands as a "site of *regulation* and management" as much as "a source of *identity* and a repository of *tradition*; the embodiment of various *institutions* […]" (Watts, 2000, p. 37 italics in original). The truth is that a "community" is never "out there", but continually negotiated and actualised in its flexible borders, membership, and institutional structures. "Political dilemmas", "livelihood struggles", "environmental needs", and "local aspirations" (Coombe, 2011, p. 85) may be pulling inwards towards the ethnic core, but constant assembling and continuous reassembling, at times fuelled by engagement with external actors (Almeida, 2017), speak to the fragility of a community. In the Malagasy case presented in Manohisoa Rakotondrabe and Fabien Girard's chapter, this process of tension and emergence manifests through elite appropriation, institutional reconfiguring, and social exclusion. This highlights the fact that the identification of different interests and their repartition inside the community involved should be considered when working towards the success of a BCP/CP.

Biocultural Jurisprudence, Sovereignty and Legal Subjectivity

This last and shorter part of this book is intended as an avenue to new issues and ideas that could emerge with the wide-ranging development of BCPs/CPs and to provide a platform for a more critical appraisal of the philosophy underlying these protocols. We also seek to examine the main question implicitly addressed throughout the book: how effective are BCPs/CPs as political tools in the context of strengthening IPLCs' rights and making more room for self-policy, self-governance, and bottom-up decision-making? As BCPs/CPs are always initiated by external actors and developed with their help, what community-level interventions and changes have multilateral agencies, developments institutions, NGOs, and facilitators (even inadvertently) prompted? Does the emphasis on stewardship and tradition conceal a new kind of politics of control, managerial intervention, or eco-governmentality that is only distinguishable from past politics in the fact that there is "[...] a shift from the management of non-human nature to the management of people" (Roach et al., 2006)? In short, are BCPs, through the hypostasis of "nature" and "tradition", a conduit for control dynamics and community reification?

In their chapter, Fabien Girard and Reia Anquet (Chapter 11, "Biocultural Community Protocols and the Ethic of Stewardship: The Sovereign Stewards of Biodiversity") explore these different avenues.

They begin by resituating BCPs in a context, shaped by the nature-culture mould, wherein we have seen the proliferation of community-based initiatives aimed at protecting biocultural heritage (e.g. ICCAs, Biocultural Territories, and biocultural rights). The study examines seven BCPs developed in Africa (Benin, Kenya, and Madagascar) and Latin America (Mexico and Panama), all of which strive to a greater or lesser extent to open space (Mulrennan & Bussières, 2020; Nemogá et al., 2018) to non-naturalistic ontologies.

Grounding its analysis in political ontology, the chapter questions this optimistic vision, arguing that the ABS framework is embedded in Western/European legal constructs and their naturalistic underpinnings. For example, if BCPs certainly endeavour to protect IPLCs' biocultural heritage, they remain within the economist model ("win-win" approach) underwriting the ABS regime. There is much proof of this in the grey and scholarly literatures alike, making the point that BCPs are conciliatory tools allowing economic development, biodiversity maintenance, and the preservation of IPLCs' identities. This model also reflects Bavikatte's deepest convictions (Bavikatte et al., 2010, p. 298), when he saw in BCPs a powerful device to navigate the two extreme poles of commodification and inalienability in an ABS context (building on Radin's "incomplete modification", 1987).

Drawing further on the sociology of sciences and Science and Technology Studies (STS), the authors identify different "scripts", defined as stories and non-discursive elements (e.g. objects, conducts, and institutions), that are able to "define actors" (Akrich, 1992, p. 207), allocate roles and tasks, and finally enrol and "enscript". Scripts are powerful, for they partake in the "institutional and regulatory spaces in which the knowledge and practices are encoded, negotiated, and contested" (Peet & Watts, 1996, p. 11) – further, they can mould "social practices and relationships of power" (Ulloa, 2005, p. 5).

Fabien Girard and Reia Anquet unveil four scripts in the corpus of BCPs under review. The authors demonstrate that they can all be traced back to the naturalist ontological matrix (Blaser, 2013), even though some are present-day scripts (IPLCs as stewards of nature and IPLCs as small green entrepreneurs). Others, though, belong to the past but regularly resurface in post- or neo-colonial contexts. Furthermore, the authors show that protocols may result in reification processes whereby the only practices allowed to subsist are those conforming to the developer's or conservationist's vision of what development and conservation should be.

The authors pay specific attention to the "steward of nature" script in its multifaceted nature. This script is undergirded by an "ethics of stewardship" that the introduction likens to a "situated ethics" (Puig de la Bellacasa, 2017, p. 150) stemming from daily engagements with, and care of, all those which – visible, invisible, humans, and non-humans – populates the land (the "land community": Leopold, 1991, p. 310). Within a new legal context marked by the Te Awa Tupua (Whanganui River Claims Settlement), Act 2017 (Byrnes, 2004; Iorns Magallanes, 2019; Magallanes, 2008; O'Bryan, 2017; Rodgers, 2017)[14] or the Atrato River case (Calzadilla, 2019; Cano Pecharroman, 2018), BCPs may be read as part of a broader political strategy. This context has enabled IPLCs to construct their ecological identities and to reopen discussions, this time informed by ontological investigations, about the political and legal status of IPLCs.

Concluding this chapter, the authors argue that the mobilisation of the ethic of stewardship allows for specifying the ethico-political status of the new subjects, i.e. the being of the *persona* (in this case a "group") to whom these new rights relate (Zarka, 1999). Furthermore, it makes room for more-than-humans and new assemblages cutting across the subject-object divide.

Conclusion

Based on the reflections presented in this book, we propose now to distinguish between different protocols according to the main functions they can perform and their political significance. Indeed, some of them are primarily technical tools and have a very limited scope, while at the other end of the spectrum, others carry political claims of the utmost importance and go so far as to put forward alternative ontologies to the dominant naturalism. The following reading grid allows us to explore how BCPs/CPs can become powerful tools and go beyond the status of purely technical instruments to constitute political and ontological claims.

We have distinguished the following types of protocols:

- **Purely technical instruments.** In this case, BCPs/CPs aim to connect local PIC and MAT with customary rules and procedures to ensure that community values and inalienable parts of their heritage are respected throughout the negotiations. Some legislation now clearly envisions these interlinkages, as Fabien Girard, Ingrid Hall, and Christine Frison recall for Madagascar and Benin in Chapter 1 ("Community Protocols and Biocultural Rights: Unravelling the Biocultural Nexus in ABS").
- **Call for the recognition of local procedures in ABS.** Next, one may well imagine a law foreseeing a community PIC and benefit-sharing process – as will be the case very soon in the Cook Islands (see Miri (Margaret) Raven's and Daniel Robinson's chapter) – while falling short of recognising BCPs.

- **Political claim**. A notch above, references to community PIC and benefit-sharing and their reliance on customary law may signal a political claim to community's rights over genetic resources, TK, and/or their territory. BCPs/CPs, therefore, must be read as a direct challenge to the State-centric approach to sovereignty over land and natural resources. BCPs/CPs' real value is thus political, activating claims for self-determination over natural resources (see Gilbert, 2016, p. 26; Jansen and Sutherland, this book).
- **Political-ontological claims.** Finally, BCPs/CPs touch on fundamental issues of sovereignty, but this time shifting the debate to the ontological terrain. Claims to local PIC and benefit-sharing are associated with customary law and community's cultural, spiritual, ecological, and economic values and are attached to a specific land and territory. Thus, they necessarily push for more encompassing recognition of any rights deemed necessary for IPLCs' survival, flourishing, and the maintenance of their alleged role in biodiversity maintenance. Biocultural-based initiatives – such as the biocultural labels studied by Garrett Graddy-Lovelace (Anderson & Hudson, 2020), the "territories and areas conserved by indigenous peoples and local communities" (Kothari et al., 2012) (ICCAs), or food sovereignty (Edelman et al., 2014; Sarrazin & Scott, 2020) – are exemplars of these fresh attempts at circumnavigating Westphalian sovereignty and giving a firmer foothold to IPLCs' claims to land, territory, culture, and self-government. Importantly, underwriting these claims is the consideration of non-naturalistic ontologies via the ethic of stewardship.

  This last type of protocol calls for a comment on the notion of Stewardship. The idea that connexions between the ethic of stewardship and biodiversity conservation could support political-ontological (i.e. sovereignty entrenched in non-naturalistic ontologies) claims was envisioned by Bavikatte and was not entirely foreign to Posey's efforts to promote a new bundle of rights (see, in particular, Posey, 2001, pp. 385–386, 2004a, p. 135). However, it also carries risks: that of seeing the ethic of stewardship converting into a contentious "duty of stewardship"[15] (Grear, 2015; Sajeva, 2018, 2021); and that of IPLCs being trapped in an essentialising fiction, entrenching their subaltern status.

Overall, this classification highlights how BCPs/CPs can range from technical instruments to political claims and even become enriched by ontological perspectives. This way, in addition to the recognition of their worldviews, IPLCs are able to bolster their political claims by bringing to the fore their cosmological specificities. This growing complexity and richness makes BCPs and CPs important and decisive tools for the recognition of the communities in their uniqueness. As the debates on the post-2020 Biodiversity Agenda get underway (Reyes-García et al., 2021), we can only hope that these discussions will fuel the creativity and inform the work of those in charge of taking the next steps forward.

Notes

1 The chapter is part of research project n° ANR-18-CE03-0003-01 funded by the French National Research Agency (ANR). *All URLs retrieved on 1 September 2021.
2 COP CBD, Decision XIII/18, CBD/COP/13/25 (17 December 2016).
3 Mo'otz Kuxtal Voluntary Guidelines; see footnote **2**.
4 Another example is the BCP for territories of the Community Council of Alto San Juan – ASOCASAN (Colombia) (2012) (Protocolo Comunitario Biocultural para el Territorio del Consejo Comunitario Mayor del Alto San Juan Asocasan, Tado, Chocó, Colombia, 2012) (English version: http://www.pnuma.org/publicaciones/PCB%20 ASOCASAN_espanol_2012.pdf), which sets out in detail, in the Appendices, the legal standards (including international ones) on territorial rights of Colombian Black communities, the right of ownership over natural resources, the rights related to valuing, protecting, and controlling TK, and the right to prior consultation and FPIC.
5 Along the same lines, the Raika biocultural protocol (India) calls on the National Biodiversity Authority of India and the CBD Secretariat to recognise substantial elements of the instrument, notably their "local breeds and associated traditional knowledge as set out in the Raika Biodiversity Register and to include it in the Peoples Biodiversity Register (under Rule 22(6) of the Biological Diversity Rules)" (http://www.abs-initiative.info/uploads/media/Nairobi-0909-Raika_02.pdf).
6 Protocolo Comunitario de Capulálpam de Méndez, Oaxaca, para la gestión de los recursos genéticos y su conocimiento tradicional en el ámbito del Protocolo de Nagoya. México (2018) (ABSCH-CPP-SCBD-240739-3); Protocolo Comunitario de Ek Balam, X-Kumil, Yucatán, para la gestión de los recursos genéticos y su conocimiento tradicional en el ámbito del Protocolo de Nagoya. México (2018) (ABSCH-CPP-SCBD-240737-4); Protocolo Comunitario Biocultural de Isla Yunuén, Pátzcuaro, Michoacán, para la gestión de los recursos genéticos y su conocimiento tradicional en el ámbito del Protocolo de Nagoya. México (2018) (ABSCH-CPP-SCBD-240740-4); Protocolo Comunitario Biocultural del Territorio Comcáac, Sonora, México (2018) (ABSCH-CPP-SCBD-240738-3); Protocolo Comunitario Biocultural de Vicente Guerrero, Españita, Tlaxcala. México (2020) (https://absch.cbd.int/api/v2013/documents/D8188070-3FD1-EA8B-360B-0EEF80D38ECD/attachments/PCB%20Vicente%20Guerrero%20M%C3%A9xico.pdf); (Hernández Ordoñez, 2019).
7 https://www.cbd.int/portals/culturaldiversity/docs/north-american-regional-declaration-on-biocultural-diversity-en.pdf
8 Corte Constitucional República de Colombia (2016) *Centro de Estudios para la Justicia Social "Tierra Digna" contra la Presidencia de la República y otros*, Sentencia T-622/16 (hereinafter *Atrato River* case). Here, the Court cites Chen & Gilmore (2015). English version available at http://files.harmonywithnatureun.org/uploads/upload838.pdf
9 Ad Hoc Open-ended Working Group on Access and Benefit-Sharing, Seventh meeting, Provisional Agenda item 3, Study on Compliance in Relation to the Customary Law of Indigenous and Local Communities, National Law, Across Jurisdictions, and International Law, UNEP/CBD/WG-ABS/7/INF/5, 6 March 2009, para. 26–27.
10 FAO, Ad Hoc Technical Expert Group on Farmers' Rights, IT/GB-9/AHTEG-FR-3/20/2 (June 2020), pp. 13, 27-28, referring to two of the four BCPs developed in Madagascar and Benin as part of the Darwin Initiative project evoked in Chapter 11 and Chapter 12 (this book).
11 See, for instance, in Kenya, the Ogiek Bio-Cultural Community Protocol (Ogiek Peoples' Development Program (OPDP), 2005) and the Lamu County Biocultural Community Protocol (Save Lamu, 2018) (also see, this book, Chapter 11).
12 See the BCP for territories of the Community Council of Alto San Juan – ASOCASAN (Colombia) (Consejo Comunitario Mayor del Alto San Juan, 2012).
13 See below and, this book, Chapter 1.

14 Te Awa Tupua (Whanganui River Claims Settlement) Act 2017, http://www.legislation.govt.nz/act/public/2017/0007/latest/whole.html.
15 On this issue, Bavikatte's *Stewardship the Earth* maintains a great deal of ambiguity as can be seen from a comparison of the following pages (Bavikatte, 2014, pp. 141, 143, 149, 164, 168–169).

References

Agrawal, A., & Gibson, C. C. (1999). Enchantment and Disenchantment: The Role of Community in Natural Resource Conservation. *World Development*, *27*(4), 629–649. https://doi.org/10.1016/S0305-750X(98)00161-2

Akrich, M. (1992). The De-Scription of Technical Objects. In W. E. Bijker & J. Law (Eds.), *Shaping Technology/Building Society: Studies in Sociotechnical Change* (pp. 205–224). The MIT Press.

Alcorn, J. B. (1994). Noble Savage or Noble State?: Northern Myths and Southern Realities in Biodiversity Conservation. *Etnoecológica : Revista Internacional, Trilingüe, Dedicada al Estudio Del Manejo Tradicional y Concervación de La Naturaleza*, *2*(3), 6–19.

Almeida, F. (2017). *Legislative Pathways for Securing Community-Based Property Rights*. Rights and Resources Initiative.

Anderson, J., & Hudson, M. (2020). The Biocultural Labels Initiative: Supporting Indigenous Rights in Data Derived From Genetic Resources. *Biodiversity Information Science and Standards*, *4*, e59230. https://doi.org/10.3897/biss.4.59230

Appadurai, A. (Ed.). (2013). *The Future as Cultural Fact: Essays on the Global Condition*. New York: Verso Books.

Argumedo, A. (2008). The Potato Park, Peru: Conserving Agrobiodiversity in an Andean Indigenous Biocultural Heritage Area. In T. Amend, J. Brown, A. Kothari, A. Phillips, & S. Stolton (Eds.), *Protected Landscapes and Agrobiodiversity Values*. Kaspareg Verlag, IUCN&GTZ, Vol. 1, pp. 45–58.

Argumedo, A., & Pimbert, M. (2008). *Protecting Farmers' Rights with Indigenous Biocultural Heritage Territories: The Experience of the Potato Park* (p. 12). International Institute for Environment and Development (IIED), Asociacion ANDES.

Bannister, K. (2004). *Lessons for ABS: Academic Policies, Community Protocols and Community-level PIC*. International Expert Workshop on Access to Genetic Resources and Benefit Sharing, Cuernavaca, Mexico. https://polisproject.org/PDFs/Bannister%202004_ABS%20Lessons.pdf

Bannister, K. (2009). Non-legal Instruments for Protection of Intangible Cultural Heritage: Key Roles for Ethical Codes and Community Protocols. In C. Bell & V. Napoleon (Eds.), *Protection of First Nations' Cultural Heritage: Laws, Policy and Reform* (pp. 278–308). UBC Press.

Bannister, K. (2018). Ethical Guidance for Access and Benefit-Sharing: Implications for Reconciliation. In C. Oguamanam (Ed.), *Genetic Resources, Justice and Reconciliation* (1st ed., pp. 212–230). Cambridge University Press. https://doi.org/10.1017/9781108557122.013

Bavikatte, K. S. (2014). *Stewarding the Earth: Rethinking Property and the Emergence of Biocultural Rights*. Oxford University Press.

Bavikatte, K., Jonas, H., & von Braun, J. (2010). Traditional Knowledge and Economic Development: The Biocultural Dimension. In S. M. Subramanian & B. Pisupati (Eds.), *Traditional Knowledge in Policy and Practice: Approaches to Development and Human Well-Being* (pp. 294–326). United Nations University Press.

Bavikatte, K. S., Robinson, D. F., & Oliva, M. J. (2015). Biocultural Community Protocols: Dialogues on the Space Within. *IK: Other Ways of Knowing, 1*(2), 1–31.

Blaser, M. (2013). Ontological Conflicts and the Stories of Peoples in Spite of Europe: Toward a Conversation on Political Ontology. *Current Anthropology, 54*(5), 547–568. https://doi.org/10.1086/672270

Byrnes, G. (2004). *The Waitangi Tribunal and New Zealand History.* Oxford University Press.

Callon, M., Barthe, Y., & Lascoumes, P. (2011). *Agir dans un monde incertain: Essai sur la démocratie technique.* Ed. du Seuil.

Calzadilla, P. V. (2019). A Paradigm Shift in Courts' View on Nature: The Atrato River and Amazon Basin Cases in Colombia. *Law, Environment and Development Journal, 15*(1), 51–59.

Cano Pecharroman, L. (2018). Rights of Nature: Rivers That Can Stand in Court. *Resources,* 7(1), 13. https://doi.org/10.3390/resources7010013

Chen, C. W., & Gilmore, M. (2015). Biocultural Rights: A New Paradigm for Protecting Natural and Cultural Resources of Indigenous Communities. *International Indigenous Policy Journal, 6*(3). https://doi.org/10.18584/iipj.2015.6.3.3

Consejo Comunitario Mayor del Alto San Juan. (2012). *Protocolo Comunitario Biocultural para el Territorio del Consejo Comunitario Mayor del Alto San Juan Asocasan. Tado, Chocó.* ASOCOSAN, PNUMA, IIAP, Natural Justice.

Coombe, R. J. (2011). Cultural Agencies: 'Constructing' Community Subjects and their Rights. In M. Biagioli, P. Jaszi, & M. Woodmansee (Eds.), *Making and Unmaking Intellectual Property* (pp. 79–98). University of Chicago Press.

de Sousa Santos, B. (2006). Globalizations. *Theory, Culture & Society, 23*(2–3), 393–399. https://doi.org/10.1177/026327640602300268

de Sousa Santos, B., & Rodríguez-Garavito, C. A. (2005). Law, politics, and the subaltern in counter-hegemonic globalization. In B. de Sousa Santos & C. A. Rodríguez-Garavito (Eds.), *Law and Globalization from Below* (1st ed., pp. 1–26). Cambridge University Press. https://doi.org/10.1017/CBO9780511494093.001

Deliège, A. (2012). *Pratique économique et transactions avec les ancêtres. Études des cérémonies dans un village malgache.* Presses universitaires de Louvain.

Demeulenaere, É. (2016). Les terrains globalisés de l'inscription d'un Paysage culturel. Patrimoine mondial et bioculturalité en Éthiopie. *Terrains/Théories, 5,* Article 5.

Descola, P., & Sahlins, M. D. (2014). *Beyond Nature and Culture* (J. Lloyd, Trans.; Paperback edition). The University of Chicago Press.

Edelman, M., Weis, T., Baviskar, A., Jr, S. M. B., Holt-Giménez, E., Kandiyoti, D., & Wolford, W. (2014). Introduction: Critical Perspectives on Food Sovereignty. *The Journal of Peasant Studies, 41*(6), 911–931. https://doi.org/10.1080/03066150.2014.963568

Escobar, A. (1999). After Nature: Steps to an Antiessentialist Political Ecology. *Current Anthropology, 40*(1), 1–30. https://doi.org/10.1086/515799

Gavin, M. C., McCarter, J., Mead, A., Berkes, F., Stepp, J. R., Peterson, D., & Tang, R. (2015). Defining Biocultural Approaches to Conservation. *Trends in Ecology & Evolution, 30*(3), 140–145. https://doi.org/10.1016/j.tree.2014.12.005

Gilbert, J. (2016). *Indigenous Peoples' Land Rights under International Law: From Victims to Actors.* Brill Nijhoff.

Graddy, T. G. (2013). Regarding Biocultural Heritage: In situ Political Ecology of Agricultural Biodiversity in the Peruvian Andes. *Agriculture and Human Values, 30*(4), 587–604.

Grear, A. (2015). Editorial: The Discourse of 'biocultural' Rights and the Search for New Epistemic Parameters: Moving Beyond Essentialisms and Old Certainties in an Age of Anthropocene Complexity? *Journal of Human Rights and the Environment, 6*(1), 1–6. https://doi.org/10.4337/jhre.2015.01.00

Grey, S. (2004). Imagination, Fraud and the Cultural Protocols Debate: A Question of Free Speech or Pornography? *Media & Arts Law Review, 9*(23), 23–37.

Halewood, M., Bedmar Villanueva, A., Rasolojaona, J., Andriamahazo, M., Rakotoniaina, N., Bossou, B., Mikpon, T., Vodouhe, R., Fey, L., Drews, A., Kumar, P. L., Rasoanirina, B., Rasoazafindrabe, T., Aigbe, M., Agbahounzo, B., Otieno, G., Garforth, K., Kiene, T., & Nnadozie, K. (2021). Enhancing Farmers' Agency in the Global Crop Commons through Use of Biocultural Community Protocols. *Agriculture and Human Values, 39*, 579–594.

Hernández Ordoñez, S. R. (2019). El protocolo de Nagoya en México: Un análisis legal del cumplimiento y el papel de los protocolos comunitarios bioculturales. *Revista de La Facultad de Derecho de México, 69*(275–2), 611. https://doi.org/10.22201/fder.24488933e.2019.275-2.69422

Illich, I. (1973). *Tools for Conviviality*. Calder & Boyars.

Iorns Magallanes, C. (2019). From Rights to Responsibilities Using Legal Personhood and Guardianship for Rivers. In B. Martin, L. Te Aho, & M. Humphries-Kil (Eds.), *ResponsAbility: Law and Governance for Living Well with the Earth* (pp. 216–239). Routledge.

Keller, E. (2008). The Banana Plant and the Moon: Conservation and the Malagasy Ethos of Life in Masoala, Madagascar. *American Ethnologist, 35*(4), 650–664.

Kothari, A., Corrigan, C., Jonas, H., Neumann, A., Shrumm, H., & Secretariat of the Convention on Biological Diversity. (2012). *Recognising and Supporting Territories and Areas Conserved by Indigenous Peoples and Local Communities: Global Overview and National Case Studies*. Secretariat of the Convention on Biological Diversity, ICCA Consortium, Kalpavriksh, and Natural Justice. http://www.deslibris.ca/ID/242837

Leopold, A. (1991). Conservation: In Whole or in Part? In S. L. Flader & J. B. Callicott (Eds.), *The River of the Mother of God and Other Essays by Aldo Leopold [1944]* (pp. 310–319). The University of Wisconsin Press.

López Bárcenas, F. (2016). *Análisis de los instrumentos y vacíos jurídicos para reconocer y aplicar normas, protocolos y procedimientos comunitarios*. Conabio-GIZ.

López Bárcenas, F., & Espinoza Sauceda, G. (2006). Recursos genéticos y conocimiento tradicional indígena la regulación internacional y su impacto en la legislación mexicana. In L. Concheiro Bórquez & F. López Bárcenas (Eds.), *Biodiversidad y conocimiento tradicional en la sociedad rural: Entre el bien común y la propiedad privada* (pp. 105–131). Centro de Estudios para el Desarrollo Rural Sustentable y la Soberanía Alimentaria.

Magallanes, C. J. I. (2008). Reparations for Maori Grievances in Aotearoa New Zealand. In F. Lenzerini (Ed.), *Reparations for Indigenous Peoples: International and Comparative Perspectives* (pp. 523–564). Oxford University Press.

Morgera, E., & Tsioumani, E. (2010). The Evolution of Benefit Sharing: Linking Biodiversity and Community Livelihoods. *Review of European Community & International Environmental Law, 19*(2), 150–173.

Morgera, E., Tsioumani, E., & Buck, M. (2014). *Unraveling the Nagoya Protocol: A Commentary on the Nagoya Protocol on Access and Benefit-Sharing to the Convention on Biological Diversity*. Brill; Nijhoff.

Mulrennan, M. E., & Bussières, V. (2020). Indigenous Environmental Stewardship: Do Mechanisms of Biodiversity Conservation Align with or Undermine It? In N. J. Turner (Ed.), *Plants, People, and Places: The Roles of Ethnobotany and Ethnoecology in Indigenous Peoples' Land Rights in Canada and Beyond* (pp. 282–312). McGill-Queen's University Press.

Munyi, P., & Jonas, H. (2013). Implementing the Nagoya Protocol in Africa: Opportunities and Challenges for African Indigenous Peoples and Local Communities. In E. Morgera, M. Buck, & E. Tsioumani (Eds.), *The 2010 Nagoya Protocol on Access and Benefit-sharing in Perspective* (pp. 215–245). Brill; Nijhoff.

Nemogá, G., Domicó, J., & Molina, A. (2018). Designing Biocultural Protocols with the Embera People of Colombia. *Terralingua*, 7(1), 20–24.
O'Bryan, K. (2017). Giving a Voice to the River and the Role of Indigenous People: The Whanganui River Settlement and River Management in Victoria. *Australian Indigenous Law Review, 20*, 48–77.
Ogiek Peoples' Development Program (OPDP). (2005). *Ogiek Bio-Cultural Community Protocol (BCP). Safeguarding Rights and Managing Resources to Improve Livelihoods.* Ogiek Peoples' Development Program (OPDP). https://www.ogiekpeoples.org/images/downloads/Ogiek-Bio-Cultural-Protocol.pdf
Oliva, M. J., von Braun, J., & Salinas Lanao, G. (2012). Biocultural Community Protocols and Ethical Biotrade: Exploring Participatory Approaches in Peru. In K. Swiderska (Ed.), *PLA 65—Biodiversity and Culture: Exploring Community Protocols, Rights and Consent* (pp. 166–172). IIED.
Parks, L. (2018). Challenging Power from the Bottom Up? Community Protocols, Benefit-Sharing, and the Challenge of Dominant Discourses. *Geoforum, 88*, 87–95.
Parks, L. (2019). *Benefit-sharing in Environmental Governance: Local Experiences of a Global Concept* (1st ed.). Routledge.
Peet, R., & Watts, M. (1996). Liberation Ecology. Development, Sustainability, and Environment in an Age of Market Triumphalism. In R. Peets & M. Watts (Eds.), *Liberation Ecologies. Environment, Development and Social Movements* (pp. 1–45). Routledge. https://doi.org/10.4324/9780203235096
Phare, M. A., Simms, R., Brandes, O. M., & Miltenberger, M. (2018). *Collaborative Consent and Water in British Columbia: Towards Watershed Co-Governance.* POLIS Project on Ecological Governance and Centre for Indigenous Environmental. www.poliswaterproject.org/polis-research-publication/ collaborative-consent-water-british-columbia-towards-watershed-co-governance/
Posey, D. A. (1995). *Indigenous Peoples and Traditional Resource Rights: A Basis for Equitable Relationships?* Green College Centre for Environmental Policy and Understanding, University of Oxford. https://www.ecolex.org/fr/details/literature/indigenous-peoples-and-traditional-resource-rights-a-basis-for-equitable-relationships-mon-057188/
Posey, D. A. (2001). Posey Biological and Cultural Diversity. The Inextricable, Linked by Language and Politics. In L. Maffi (Ed.), *On Biocultural Diversity: Linking Language, Knowledge, and the Environment* (pp. 379–396). Smithsonian Institution Press.
Posey, D. A. (2004a). Indigenous Knowledge, Biodiversity, and International Rights. Learning about Forests from the Kayapó Indians of the Brazilian Amazon. In K. Plenderleith (Ed.), *Indigenous Knowledge and Ethics: A Darrell Posey Reader* (pp. 133–140).Routledge.
Posey, D. A. (2004b). Traditional Resources Rights – De facto Self-determination for Indigenous Peoples. In D. A. Posey & K. Plenderleith (Eds.), *Indigenous Knowledge and Ethics: A Darrell Posey Reader* (pp. 155–168). Routledge.
Posey, D. A., & Dutfield, G. (1996). *Beyond Intellectual Property: Toward Traditional Resource Rights for Indigenous Peoples and Local Communities.* International Development Research Centre.
Puig de la Bellacasa, M. (2017). *Matters of Care: Speculative Ethics in More than Human Worlds.* University of Minnesota Press.
Radin, M. J. (1987). Market-Inalienability. *Harvard Law Review, 100*(8), 1849–1937.
Raymond, H. (2007). The Ecologically Noble Savage Debate. *Annual Review of Anthropology, 36*(1), 177–190. https://doi.org/10.1146/annurev.anthro.35.081705.123321
Redford, K. H. (1991). The Ecologically Noble Savage. *Cultural Survival Quarterly, 15*(1), 46–48.
Reyes-García, V., Fernández-Llamazares, Á., Aumeeruddy-Thomas, Y., Benyei, P., Bussmann, R. W., Diamond, S. K., García-del-Amo, D., Guadilla-Sáez, S., Hanazaki,

N., Kosoy, N., Lavides, M., Luz, A. C., McElwee, P., Meretsky, V. J., Newberry, T., Molnár, Z., Ruiz-Mallén, I., Salpeteur, M., Wyndham, F. S., … Brondizio, E. S. (2021). Recognizing Indigenous Peoples' and Local Communities' Rights and Agency in the Post-2020 Biodiversity Agenda. *Ambio*. https://doi.org/10.1007/s13280-021-01561-7

Roach, C. M., Hollis, T. I., McLaren, B. E., & Bavington, D. L. Y. (2006). Ducks, Bogs, and Guns: A Case Study of Stewardship Ethics in Newfoundland. *Ethics & the Environment*, *11*(1), 43–70.

Robinson, D. F., & Raven, M. (2020). Recognising Indigenous Customary Law of Totemic Plant Species: Challenges and Pathways. *The Geographical Journal*, *186*(1), 31–44. https://doi.org/10.1111/geoj.12320

Rodgers, C. (2017). A New Approach to Protecting Ecosystems: The Te Awa Tupua (Whanganui River Claims Settlement) Act 2017. *Environmental Law Review*, *19*(4), 266–279. https://doi.org/10.1177/1461452917744909

Rutert, B. (2020). *Contested Properties. Peoples, Plants and Politics in Post-Apartheid South Africa*. transcript Verlag.

Sajeva, G. (2018). *When Rights Embrace Responsibilities: Biocultural Rights and the Conservation Of Environment* (1st ed.). Oxford University Press.

Sajeva, G. (2021). Environmentally Conditioned Human Rights. A Good Idea? In D. P. Corrigan & M. Oksanen (Eds.), *Rights of Nature. A Re-examination* (pp. 87–100). Routledge.

Sarrazin, M.-L., & Scott, C. (2020). Livelihoods, Food Sovereignty, and Well-Being. *Policy Brief 2: Nurturing Biocultural Diversity*. https://cicada.world/files/pb/CICADA_PolicyBrief2_Livelihoods.pdf

Save Lamu. (2018). *The Lamu County Biocultural Community Protocol. The Lamu Indigenous Community and their Rights to the Preservation of their Natural Resources, Cultures, Heritage and Management of Biodiversity*. Save Lamu. https://naturaljustice.org/wp-content/uploads/2019/04/Lamu-County-BCP-2018.pdf

Sharma, J. R. (2016). *Brokering in International Development. A Consideration of Analytical Issues* (New Norms and Forms of Development. Working Paper I). School of Social and Political Science, The University of Edinburgh.

Swiderska, K., Argumedo, A., & Pimbert, M. (2020). *Biocultural Heritage Territories: Key to Halting Biodiversity Loss, IIED Briefing*. iied, Asociación ANDES, INMIP, Coventry University.

The Kukula Traditional Health Practitioners Association, & Natural Justice. (2017). *Biocultural Protocol of the Kukula Traditional Health Practitioners Association* (2nd ed.). Kukula Traditional Health Practitioners Association; Natural Justice; Kruger to Canyons Biosphere Region. https://za.boell.org/sites/default/files/bcp-kukula-traditional-health-practitioners-association-english.pdf

Tönnies, F. (2001). *Community and Civil Society* (J. Harris, Ed.; M. Hollis, Trans.). Cambridge University Press.

Ulloa, A. (2005). *The Ecological Native: Indigenous Peoples' Movements and eco-Governmentality in Colombia*. Routledge.

Watts, M. J. (2000). Contested Communities, Malignant Markets, and Gilded Governance: Justice, Resource Extraction, and Conservation in the Tropics. In C. Zerner (Ed.), *People, Plants, and Justice: The Politics of Nature Conservation* (pp. 21–51). Columbia University Press.

Zarka, Y. C. (1999). The Invention of the Subject of the Law. *British Journal for the History of Philosophy*, 7(2), 245–262.

INDEX

Note: Bold page numbers refer to tables; *italic* page numbers refer to figures and page numbers followed by "n" denote endnotes.

For Product Safety Concerns and Information please contact our EU representative GPSR@taylorandfrancis.com
Taylor & Francis Verlag GmbH, Kaufingerstraße 24, 80331 München, Germany

www.ingramcontent.com/pod-product-compliance
Ingram Content Group UK Ltd.
Pitfield, Milton Keynes, MK11 3LW, UK
UKHW020155200926
475702UK00016B/2719
9781032000817